EARTH SCIENCE

Frontispiece photo courtesy of NASA

EARTH SCIENCE

JOHN GABRIEL NAVARRA

JOSEPH S. WEISBERG

FRANK MICHAEL MELE

Jersey City State College

John Wiley & Sons, Inc.　　New York · London · Sydney · Toronto

Preface

OUR EARTH IS A VERY, VERY SMALL OBJECT SUSPENDED IN the vastness of space. There are much larger planets journeying around the sun, and there are much larger stars than our sun dotting the black emptiness of space. But it is on this planet, Earth, circling around an insignificant star, that life was spawned and that we find ourselves. Man, originally a part of the Earth's succession of life, has gradually set himself apart from it. The brotherhood of man has prospered, has spread across the planet and, through its activities, has affected weather, climate, topography, balances in nature, and the condition of the physical environment.

The most important attributes that distinguish man from other animals are his ability to think, to communicate, and to plan. Throughout the ages, some of man's best thought has attempted to unravel the story of the origin of the Earth and of man himself. Man's natural inquisitiveness has prompted him to search for evidence that clarifies the relationship among the various forms of life and the physical environment, as well as the relationship of the Earth to the other planets, to the sun, and to the universe. Man's study of this evidence is called *Earth Science*.

This text is a nontechnical introduction to the study of earth science. It is designed to give an account of man's progress in his quest to understand the planet on which he lives.

The story is written for the liberal arts student who recognizes the importance of this information to his development as an articulate, literate citizen who has a responsibility to understand his planet home and to plan for ways of preserving and saving it from the destructive activities that are polluting it and destroying its natural balances. The theme of genuine interest in our environment and in man's effect on it imbued our thinking as we wrote the story. There is frequent evidence of this theme throughout the book.

The study of earth science is like the study of a play—but a play in which all of the characters and scenery are real. For convenience, we have divided the story into five acts—each one is an important area of concern for the earth scientist. Act I is called the *lithosphere.* It deals with the study of the solid portion of the Earth. Act II, the *atmosphere,* examines the gaseous envelope that surrounds the Earth. Act III, the *hydrosphere,* is concerned with the liquid envelope of the Earth. Act IV, the *biosphere,* discusses the theories and concepts dealing with the origin and evolution of the Earth and the life on it. The final act is called *space.* This act deals with the motions of the Earth and other planets in our solar system.

John Gabriel Navarra
Joseph S. Weisberg
Frank Michael Mele

August, 1970

Contents

The Lithosphere

THE EARTH IS MAN'S HOME. AS FAR AS WE CAN DETER-mine, it is the only inhabited planet in our entire solar system. Yet man has not always treated the planet in as careful a manner as it deserves. Throughout his history, man has exploited the Earth without really understanding the effects brought about by his exorbitant use of its resources. But with the increasing demands being placed upon the planet by expanding populations and the con-current depletion of the Earth's natural resources, man has finally begun to examine the Earth on a scale never before seen or imagined. What are the origins of our planet? How has it changed during the eons of its existence? What will be its future? These questions and many others of a similar nature are presently being asked. The answers are com-ing from data gathered through the efforts of hun-dreds of thousands of geoscientists who are pres-ently engaged in the greatest scientific explorations our planet has ever seen. From these studies and the theories and practical applications they suggest, man will be able to husband the Earth's future re-sources much more efficiently and carefully than in the past. As we increase our numbers, so must we also increase our knowledge and understanding of this home of ours and the future it promises to our unborn generations.

CHAPTER 1

Our Planet in Space: The Earth

WHEN WE CONSIDER THE VASTNESS of space and the number of stars in our own galaxy, the Earth truly seems an insignificant object. But the Earth is our home. Civilization upon civilization has arisen and passed into oblivion while man has raised himself from the depths of ignorance to new heights of understanding.

From the very beginnings, man has wondered about the Earth on which he lives. He has sought to understand it and to fathom how the Earth relates to other objects in the universe.

The Earth took form about five billion years ago. We do not know whether its birth was an isolated event or was one small part of an immense universal creation. For a long time, man considered his home as a rather special place in the universe. We are no longer quite so sure.

Although the Earth formed these five billion years ago, man has been present—curious, investigating, and seeking answers—for less than two million years. And, furthermore, modern experimental science can only be traced back about 400 years. This is despite the fact that science as a human endeavor is several thousand years old.

Investigating the Planet Earth

Man's curiosity has led him to investigate every part of the Earth's structure that is available to him, and to seek ways of examining those parts that are beyond his immediate grasp.

As a result of his investigations, man has separated the Earth into three principal divisions: the lithosphere, the hydrosphere, and the atmosphere. The *lithosphere* is the outer skin of the Earth; the *hydrosphere* is the liquid envelope of the Earth; and the *atmosphere* is the gaseous envelope surrounding the lithosphere and the hydrosphere.

We call the modern study of the Earth the Earth Sciences. In the broadest terms, the Earth Sciences include geology, meteorology, oceanography, paleontology, and space science. Each of these disciplines attempts to study the interacting forces that have resulted in present conditions on the Earth. The Earth Sciences also attempt to relate the Earth to the larger universe.

One of the oldest of the Earth Sciences is the science concerned with the lithosphere. This is the science of geology. The word "geology" is derived from the Greek—"ge" from the Greek word that means Earth, "-logy" from "logos," which implies a written record. The science of geology is the systematic study of the dynamic forces of nature and the changes that develop in the landforms of the Earth.

Geologists use the principles of many sciences in their study of the Earth; as a result, the field has been subdivided into many specialties. Some geologists study the formation

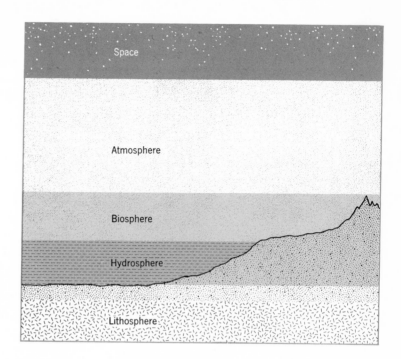

FIGURE 1.1 Because the earth is related to all the other objects in the universe, the information collected by space scientists is useful in probing the origins of our own environment.

of rocks and minerals as well as the manner in which rock formations are bent, broken, raised, and lowered. Other geologists devote themselves to deciphering the history of the Earth. Still others use physics and chemistry to examine the interior of the Earth and the physical and chemical changes that bring about changes within rocks. They seek clues to valuable mineral, oil, and natural gas deposits that have many industrial applications.

The study of the Earth's gaseous atmosphere is referred to as meteorology. Meteorologists are most concerned with the physics of the moving atmosphere and its associated phenomena. Their investigations include studies of winds, cloud formations, precipitation, and storms which affect the Earth's surface.

The hydrosphere is an important part of the Earth since nearly three-fourths of the globe is covered with water. Water flows over the ground as streams and accumulates as pools, lakes, and the large bodies of water known as the oceans. Water also moves below

the surface, filtering through the soil and rocks in a variety of ways.

All the divisions of the Earth are interrelated by the various living things which inhabit the land, water, and air. The paleontologist interprets the history of the Earth largely through the study of fossils and through reconstructing the conditions in which ancient life forms must have lived.

Throughout the ages it has become increasingly apparent that the Earth is not an isolated object in space. Scientists have recognized that in order to understand the Earth they must investigate space. The new field of space science has become a reality with the advent of rockets. Space vehicles are now able to carry instruments and people into space for the purpose of collecting data (see Figure 1.3). Space probes have already looked at the moon, Mars, and Venus. In the years ahead, it will be possible to probe beyond our solar system. Space science has benefited from the development of radio telescopes and, more recently, radar telescopes.

4

A

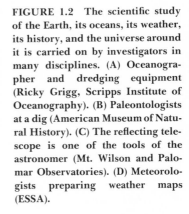

FIGURE 1.2 The scientific study of the Earth, its oceans, its weather, its history, and the universe around it is carried on by investigators in many disciplines. (A) Oceanographer and dredging equipment (Ricky Grigg, Scripps Institute of Oceanography). (B) Paleontologists at a dig (American Museum of Natural History). (C) The reflecting telescope is one of the tools of the astronomer (Mt. Wilson and Palomar Observatories). (D) Meteorologists preparing weather maps (ESSA).

B

C

D

E

FIGURE 1.2 (*cont.*) (E) Glacial investigation (Jerry Cooke).

A Look at the Beginnings of Geology

Geology, as is true of almost all the other sciences, started with the Greeks. For the Greeks, the Earth was a rather simple structure. It was composed of various combinations of four elements: earth, air, fire, and water.

Today, geologists recognize that there are elements that combine in various proportions to form the minerals and rocks in the Earth's crust. However, the elements of the Greeks are not found in the list of 92 naturally occurring elements in the modern periodic table of the elements.

Nevertheless, the Greeks had some excellent insights and concepts concerning the Earth. For example, Aristotle (384–322 B.C.) and the followers of Pythagoras conceived the idea that the Earth was round, and Eratosthenes, a Greek mathematician (approximately 276–195 B.C.), used geometric techniques to calculate the circumference of the Earth. This amazing mathematician achieved a measurement of about 25,000 miles for the circumference of the Earth—a figure that is less than 200 miles in error when compared to the currently accepted measurement.

The ancient Greeks also suggested explanations for the formation of mountains and valleys. Greece and the surrounding region is an area of great volcanic activity and many earthquakes. Thus, it was natural that the Greeks should believe that changes on the Earth's surface were brought about by sudden cataclysmic events of this type.

The Greek culture was succeeded by the Roman. The Romans were concerned with conquests, civil administrations, trade, and stabilization of their world. Their concentration on the realities of day-to-day life gave them little opportunity to operate as philosopher-scientists. But interest in science was nevertheless kept alive during the Roman period. In the first century A.D., Pliny the Elder included systematic observations and mineral studies in his philosophical writings. Pliny's dedication to science finally led to his death: he was killed during an attempt to observe the eruption of Mount Vesuvius.

As with the beginnings of every science, progress is made slowly from generation to generation. Great men separated by time and generations read, study, and draw new insights. The great Leonardo da Vinci (1452–1519) had many scientific interests. He recognized that running water was capable of removing soil and eroding rock. He also realized that some fossils found on land were the im-

A

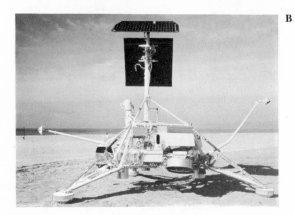

B

FIGURE 1.3   Robot spacecraft like (A) Mariner and (B) Surveyor are used to examine, in detail, the characteristics of other planets in the solar system (NASA).

FIGURE 1.4   Aristotle, one of the early, great scientist-philosophers, helped establish observation as an important aspect of scientific investigation (Alinari – Art Reference Bureau).

FIGURE 1.5 Leonardo da Vinci—a self-portrait. Da Vinci developed some of the earliest concepts regarding the proper origin of fossils (Bettmann Archive).

A

B

C

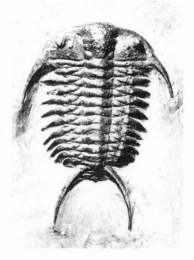

FIGURE 1.6 Animal and plant impressions left in rock allow for the reconstruction of the Earth's his-tory. (A) Fern (Dr. Wolf Starche). (B) Starfish (Dr. Wolf Starche). (C) Trilobite, an early marine organism (Yale Peabody Museum of Natural History).

prints of once living things that died in ancient seas, a fact that had been observed as early as 540 B.C. by Xenophanes, the first European paleontologist. Unfortunately, the observations of both these men had little impact on the thinking of the early geolo-gists, and their insights required new inter-pretations by later geologists.

The story of the Earth was unfolded slowly by men who studied its physical characteris-tics. Nicolaus Copernicus (1473–1543), a mathematician-priest, proposed the idea that

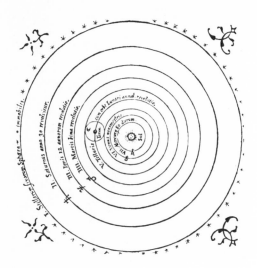

FIGURE 1.7 The Copernican view of the solar system placed the sun at the center. The planets, including the Earth, were seen orbiting the sun as its satellites (Bettmann Archive).

the sun was the center of the solar system, as shown in Figure 1.7. He made this proposal in spite of the fact that it was heresy to suggest that the Earth was not the center of the entire universe; his observations would not allow him to draw any other conclusion.

Copernicus' idea was upheld by Galileo Galilei (1564–1642), who was made to suffer greatly before the Inquisition for his beliefs. Galileo established the tradition of modern science, which calls for evidence to uphold theories. His observations of the heavens convinced him that the Earth is only one body out of countless bodies in the universe, all of which follow the same physical laws of motion. Although he was not honored for his ideas until nearly the end of his life, Galileo is now recognized as the founder of modern, experimental science.

Johannes Kepler's work, in 1619, produced the proof for the planetary motions suggested by Copernicus and Galileo. Kepler's work gave vindication to the stand taken by Galileo.

FIGURE 1.8 (*left*) Galileo Galilei, painted by Sustermans. Galileo was severely criticized and his work suppressed for his support of the Copernican view of the solar system (Bettmann Archive).

FIGURE 1.9 (*right*) Johannes Kepler founded modern astronomy as a quantitative science (New York Public Library, Picture Collection).

FIGURE 1.10 A contemporary cartoon sketch of James Hutton, the founder of modern evolutionary geology (Bettmann Archive).

Modern Geology as a Science

The ancient Greek and Roman philosopher-scientists were concerned with the Earth and its relationships to the universe. In fact, geology, like most of the other sciences, grew out of man's anxiety to understand his own place and that of the Earth in the scheme of things.

The first of the systematic, modern geologists was Abraham Gottlob Werner (1749–1817). Werner established mineralogy as a scientific study. He was also the founder of the Wernerian, or *Neptunist,* school of geology. He introduced the idea that rock layers, or strata, are built up from the precipitation and compaction of particles of eroded rock that were suspended in water.

Until Werner proposed his idea that geologic forms can develop from sediments deposited by water, the popular view was that volcanic action in the heat below the Earth's crust was primarily responsible for the changes and developments of Earth structures. This view was the *Vulcanist* theory of the Earth's development, which held that volcanic eruptions produced the formations we call igneous rocks. The name igneous literally means "from fire." The Vulcanists did

not recognize that the precipitation of materials from running water was a significant factor in the formation of rock strata.

The Neptunist view held sway for many decades because the theory fit in with the biblical story of Noah and the universal flood. The sedimentary formations, which were believed to have formed in a rather short period of time, were accepted as the remnants of the flood. Both the Neptunist and Vulcanist theories differed only in detail and emphasis from the catastrophism of the Greeks.

The modern view of geology was begun in the late eighteenth century by James Hutton (1726–1797). Originally a Vulcanist, Hutton developed a view of geology that has become known as *Uniformitarianism* in modern geology. Uniformitarianism is an evolutionary view of the Earth which recognizes that geologic activity is constantly taking place and that erosion and deposition of material require great periods of time in order to form rock strata.

The Modern View

Hutton's text, *Theory of the Earth,* laid the foundation for the theory of modern geology in 1795. In the text, Hutton recognized that molten rock, or lava, was forced onto the surface of the Earth, forming rocks of the type we call igneous. But he also realized that sedimentary rocks (Figure 1.11) were different in type and origin. These rocks were laid down over long periods of time, and not quickly and suddenly as in a universal flood. Hutton also believed that catastrophes like earthquakes, floods, and volcanoes were not the only processes that shaped the Earth's surface. Hutton proposed that examination of these processes and the landforms they produced would unlock much information about past events in the Earth's history. As Hutton put it, "The present is the key to the past."

For several reasons Hutton had great difficulty in establishing his ideas. Chief among these reasons was the great reputation and influence of Werner and his disciples. There was also a lack of evidence for Hutton's hypotheses.

Shortly after Hutton's death, his ideas were more strongly put forth by Charles Lyell in his classic book, *Principles of Geology*. Although Lyell himself made no great discoveries, he made Hutton's ideas more ac-

FIGURE 1.11 The realization that long periods of time were required for the formation of sedimentary rock strata was instrumental in the final acceptance of Hutton's view. Mummy Cliffs in Capitol Reefs National Monument are sedimentary strata (Union Pacific Railroad).

ceptable by producing a clearly written explanation of Hutton's theories coupled with evidence and examples of rock strata. In addition, the followers of Werner were beginning to find evidence which suggested that longer periods of time were necessary for the formation of geological structures than they had previously believed. Lyell's writings and his proofs of Hutton's theories helped to establish geology as an evolutionary science.

Studies carried on throughout the nineteenth century produced much quantitative data. Evidence accumulated rapidly from examinations of rivers, sedimentation rates in rivers and lakes, and glacial movement. This evidence helped to establish the Huttonian view more firmly in the minds of geologists.

Lyell's and Hutton's views of the evolving nature of the Earth's landforms even influenced Charles Darwin. In his writings, Darwin credited Lyell's ideas for inspiring his own research into the evolutionary nature of life.

Because of the work done by Hutton and Lyell, modern geology is an evolutionary science. Geologists recognize that catastrophes are responsible for landforms and changes. They also recognize that these processes have always taken place and are still taking place in the Earth's history. If we examine the rock strata and processes that are present today, we can understand the past processes and the history of the Earth.

Our Planet Earth

The extensive investigations that were carried on throughout the nineteenth century have continued on a greater scale in the twentieth century. Present scientific evidence indicates that the sun, Earth, and other members of the solar system are composed of the same group of 92 naturally occurring elements. Although the composition of each planet may vary, the structures of the solar system contain most of the same elements found on the Earth. Furthermore, the solar system and its members were formed at the same time, about five billion years ago.

The planet on which we live is now known

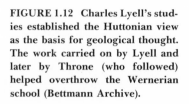

FIGURE 1.12 Charles Lyell's studies established the Huttonian view as the basis for geological thought. The work carried on by Lyell and later by Throne (who followed) helped overthrow the Wernerian school (Bettmann Archive).

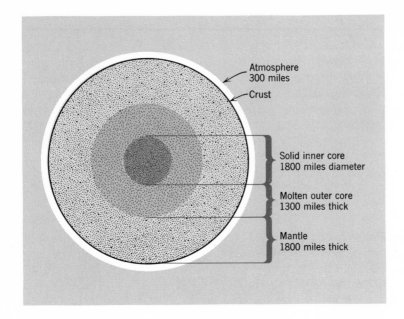

Atmosphere
300 miles

Crust

Solid inner core
1800 miles diameter

Molten outer core
1300 miles thick

Mantle
1800 miles thick

FIGURE 1.13 The layers of the Earth. The thickness of the crust is slightly exaggerated in this illustration. The crust may be as little as 3 miles thick in the ocean basins and as much as 25 to 30 miles thick in the continental mass.

to have two distinct layers beneath the outer crust. Closer to the surface is the mantle; below it is the core. The true composition of these inner layers is not known. However, geologists have developed hypotheses about their makeup that are based on much seismic evidence. It is probable that these hypotheses are pretty close to the truth. A diagrammatic view of the Earth is shown in Figure 1.13.

The Crust

The composition of the crust of the Earth has been determined with some accuracy. It is approximately as shown in Table 1.1.

The elements are rarely found in the pure state but are usually in combinations called minerals. Minerals are naturally occurring inorganic substances; that is, they are not laid down by plants or animals. These minerals, in turn, form the various rock types found in the Earth's layers.

Temperatures throughout the Earth's crust increase at a rate of 70° to 75°F per mile. Temperatures increase so rapidly that when miners excavate beyond a mile in depth, they cannot survive without special air-conditioning installations. This heat is partially because of the tremendous compression of rocks at great depths, but is attributed mostly to the radioactive decay of unstable elements.

The crust varies from about three miles in depth under the ocean basins to 25 to 30 miles under some of the mountain ranges. The continental masses on which we live are

Table 1.1

ELEMENT	PER CENT BY WEIGHT
Oxygen	46.60
Silicon	27.72
Aluminum	8.13
Iron	5.00
Calcium	3.63
Sodium	2.83
Potassium	2.59
Magnesium	1.41

All other elements, including important elements such as nitrogen, carbon, sulfur, and the precious metals—gold, silver, and platinum—make up only about three per cent of the crust.

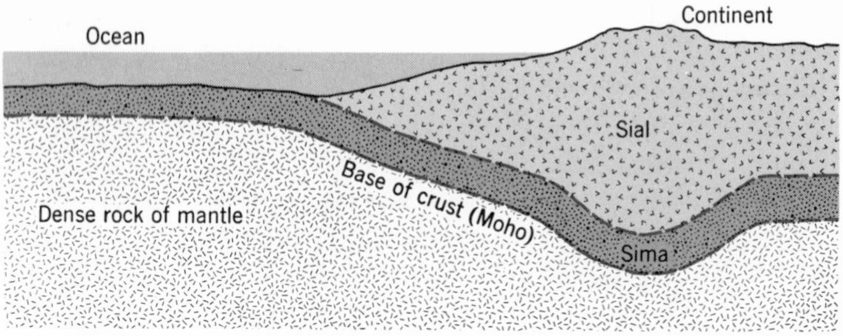

FIGURE 1.14 The location of the sial, the sima, and the moho. The ocean basins are not underlain by sial (After Longwell and Flint, 1962).

composed of granitic rock called *sial* (from silicon and aluminum, the major elements). These sialic rocks are, in turn, underlain by a layer of darker, heavier rock called *sima* (from silicon and magnesium, the major elements). As Figure 1.14 indicates, the ocean basins are not underlain by sial, but only by the bottom-most layer of sima. The absence of sial accounts for the relative thinness of the crust under the ocean basins.

The Mantle

Directly below the crust is the layer of dense rock known as the mantle. To date, our knowledge of this layer of the Earth comes from indirect measurement and inference. Analysis of earthquake waves and how they travel through the Earth's layers yields what knowledge geologists have managed to obtain about the mantle.

The boundary between the Earth's crust and the 1800-mile thick mantle was determined by a Yugoslavian geologist named Mohorovičić. This boundary is called the Mohorovičić discontinuity or Moho, after its discoverer.

Earthquake Waves

Mohorovićíc located the boundary by analyzing earthquake waves. These waves are one of our main sources of information about the interior of the Earth. Earthquakes produce vibrations in the form of wave motions. As these wave motions move through the Earth, they are affected differently by various rock layers. These effects give us our picture of the Earth's interior.

The two types of earthquake (or *seismic*) waves which are the most important for our purposes are the primary and secondary waves produced at the *focus,* or origin of the waves within the Earth. Primary waves, or P waves, are longitudinal waves produced by sudden movement of rock materials. As they move through the Earth, they are represented by a series of compressions and expansions of material. P waves are the fastest moving of all earthquake waves. The secondary, or S waves, are shear waves. Generated at the same time as the P waves, they are similar to the waves set up by flicking a rope that is held at both ends. Each S wave moves in undulations through the Earth's interior, as shown in Figure 1.15. It is significant for the analysis of the interior of the Earth that S waves do not move through liquids as do the P waves.

There are also two forms of *surface* waves, Rayleigh waves and Love waves, which are named for the men who discovered them. These waves, generated by the P waves, move through the upper crust from the *epicenter,* immediately above the focus on the surface, and may travel around the entire globe.

Seismograph stations are set up throughout the world to record the shock of earthquake waves that pass through their respective areas. Seismologists know the speed of each type of wave and calculate the length of time each wave should take to reach each station from the focus. Variations in the speed of a wave give seismologists a good idea of the types of material the wave passed through. S waves are never received by stations at opposite sides of the globe from the epicenter. As a result, scientists know that S waves must encounter a liquid somewhere near the center of the Earth. This is why it is presently believed that the core is at least par-

tially liquid. P waves are also refracted by the liquid portion of the core. There are stations, therefore, that receive neither S nor P waves. These stations are in the region known as the *shadow zone*, which is a belt some 3000 miles wide surrounding the epicenter at a distance from it of 7000 miles. It is this kind of information, together with laboratory testing, that gives us what knowledge we have of the Earth's interior.

Analysis of earthquake waves also reveals that the mantle varies in density. That is, it acts as though it were a structure made up of two principal layers. The outer layer is the prime focus of shallow earthquakes, which occur less than 40 miles deep. Intermediate and deep-focus earthquakes occur to depths

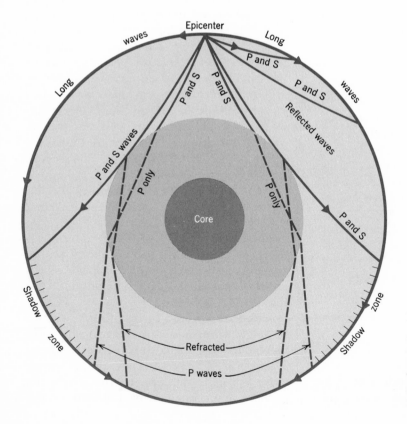

FIGURE 1.15   The effect of the various layers of the Earth on earthquake waves. The refraction of P and S waves creates a shadow zone in which no waves are recorded. The liquid core prevents S waves from passing directly through the Earth.

of approximately 400 miles. The rocks in this layer probably break under stress. In all likelihood, the rocks below a depth of 400 miles yield to pressure by flowing in slow, plasticlike movement.

The temperature increase observed in the crust does not continue throughout the mantle. If temperature underwent a constant rise with increasing depth throughout the Earth, the temperatures near the core would be tens of thousands of degrees Fahrenheit. Furthermore, in the core itself, the temperatures would increase to hundreds of thousands of degrees. These temperatures would be many times that of the sun, and it would not be possible for the Earth to reach this temperature without becoming gaseous.

The Core

The core of the Earth begins 1800 miles from the Earth's surface and is 4200 miles in diameter. Although its exact nature is unknown, the core seems to have two layers. The outer layer, which is 1300 miles thick, appears to be in a liquid state; the inner layer is made up of solid material. The entire core is thought to be predominantly iron and also to include some nickel. The density, or weight per unit volume, is about 5 times that of the crust and, at least, 11 to 12 times as dense as an equal volume of water.

The Earth's magnetic field is produced by the core acting as a gigantic dynamo (Figure 1.16). A dynamo produces currents of an electrical nature when a magnetic field is

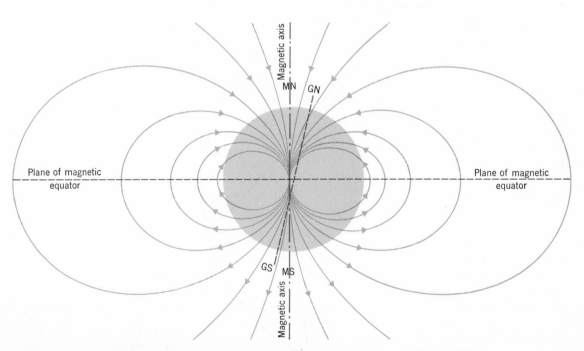

FIGURE 1.16 The Earth's magnetic field. Notice that the magnetic poles (MN and MS in the diagram) do not coincide with the geographic poles (GN and GS).

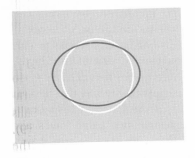

FIGURE 1.17 The relationship of an oblate spheroid (solid line) to a perfect sphere (white line).

"cut" by a moving magnet. As the Earth rotates, the core rotates with it, but the liquid nature of the outer core allows slipping. Thus, the core rotates more slowly than the surrounding solid mantle. As the metallic core rotates within the liquid core, the interaction produces a magnetic field that surrounds the Earth for many thousands of miles.

A View of the Earth

Although the interior of the Earth is inaccessible to geologists, the Earth's surface features are easily studied. One day, as our knowledge increases and our techniques improve, a better understanding of the Earth's interior will develop. For the present, there is much to learn about the external features of the Earth.

The Earth's Size and Shape

Although terrestrial globes represent our Earth as a perfect sphere, our planet is actually what scientists call an *oblate spheroid* (Figure 1.17). That is, the Earth is slightly flattened at the poles. The Earth's diameter at the equator is about 27 miles more than the diameter through the poles. In addition, the North Pole is about 80 miles farther from the exact center of the Earth than the South Pole, so that the Earth is slightly "pear-

shaped," a fact which became apparent through satellite observations.

Of course, a difference of 27 miles is very small when compared to the diameter of the Earth, which is nearly 8000 miles. Thus the flattening of the Earth is very slight. In fact, the Earth's shape deviates from a true sphere by only slightly more than 0.3 per cent.

The actual dimensions of the Earth are

Circumference	24,860 miles
Polar diameter	7,899.9 miles
Equatorial diameter	7,927.0 miles

Measuring the Earth's Mass

In addition to size, the Earth has mass. Mass is defined as the quantity of matter contained in an object. As a consequence of mass, the Earth exerts a pull, or gravitational attraction, on all objects on the Earth's surface. Scientists have been able to measure the mass of the Earth very successfully for over 200 years. Only slight variations in the early measurements have resulted from the increasing sophistication of measuring instruments as correspondingly more accurate measurements have been obtained.

The measurement of the Earth's mass is based on the fact that all objects exert a gravitational pull on each other. Thus, if a plumb bob is hung from a string, the Earth's mass exerts a gravitational pull on the bob which causes it to hang in a vertical position. If a large known mass of material such as a mountain is near the plumb bob, the mountain also exerts a pull on the plumb bob, causing it to deviate from a true vertical position (Figure 1.18). The amount of deviation of the

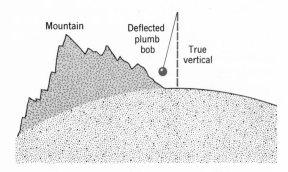

FIGURE 1.18 Large mountain masses exert a sufficient gravitational pull to deflect a plumb bob from the true vertical position.

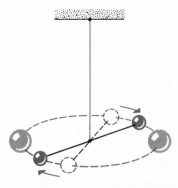

FIGURE 1.19 The Cavendish method of determining G—the constant of gravitation. The large lead balls exert a measurable gravitational force on the smaller lead balls.

plumb bob can be measured. By using the force exerted by the mountain on the known mass of the plumb bob, it is possible to calculate the unknown mass, that of the Earth.

In the late eighteenth century, Henry Cavendish, an English physicist, was the first to calculate the mass of the Earth in his laboratory. Cavendish set up two small lead balls at the end of a fine wire (see Figure 1.19). He placed two large lead balls near the smaller lead balls. The large lead balls exerted a "pull" on the small lead balls and caused them to deflect from their original position. Since Cavendish knew the masses involved, a microscopic measurement of the deflection allowed him to determine the force of gravity exerted by the Earth by using a formula derived by Sir Isaac Newton. Once the value for gravitational attraction had been calculated, it was a simple matter for Cavendish to use a similar principle in calculating an unknown mass, that of the Earth.

The accepted mass of the Earth today is 66×10^{20} tons. If this number were written out it would look like this:

$$6,600,000,000,000,000,000,000$$

This enormous mass exerts a gravitational attraction on all of the Earth. It is this attraction that causes the Earth's shape to assume that of an almost perfect sphere.

The Landforms of the Earth's Surface

The deepest point on the Earth's surface is the Mariana Trench, 1500 miles east of the Philippine Islands. This trench reaches a depth of more than 36,000 feet. The highest point on the Earth's surface is Mount Everest, which is 29,141 feet above sea level. Thus, on a sphere that is approximately 8000 miles in diameter, there is a total vertical variation, or *relief,* of only 12 miles between the highest point and the lowest point on the

Earth's crust. Although the Earth seems to be deeply scarred, it is only our small relative size that leaves us with this impression. In reality, the Earth is smoother in proportion to its size than the skin of an orange.

Our tiny size creates a sense of great heights when we examine the other points of elevation and depression between the two extreme points (Figure 1.20). The mountain masses and the ocean floors with their canyons and trenches are no more than slight wrinkles in the surface of the Earth. All these features have less than the height of Mount Everest or the depth of the Mariana Trench.

The land surfaces of the Earth have a wide variety of landforms such as mountain ranges, hills, plateaus, and ridges. These features are the result of several types of crustal movements that have been brought about by unstable conditions in the crust and mantle of the Earth.

The ocean basins descend from the continental shelves and slopes and display a variety of landforms. Plateaus, ridges, seamounts, and islands can be observed rising from the ocean floor. Deeps, trenches, and canyons are cut into the floor of the ocean. Curiously enough, the deep trenches and canyons are found, not in the center of the floor, but near the margins of the basins. This seems to suggest a possible common origin for all the surface features of the continents and ocean basins.

The Formation of the Earth's Crust

The crust of the Earth is a thin layer of material that is a zone of comparatively rigid and brittle rocks. We have observed that the continents are different in composition from the portion of the crust that forms the ocean basins. The continental crust is thicker and consists of rock material composed primarily of silicon and aluminum, whereas the ocean basins are mostly basalt. Furthermore, the continents are subjected to the processes of erosion, while the ocean basins are subjected largely to deposition of materials. The processes that shape and change the crust of the Earth cause the major differences in thickness between the two main parts of the crust. These processes also act to establish a condition of equilibrium known as isostasy.

Isostasy

The word isostasy means equal standing. In the Earth Sciences it refers to the hypothesis that, when weights of sediment are removed from an area, or deposited in other areas, the rocks beneath will move in a slow, plastic-

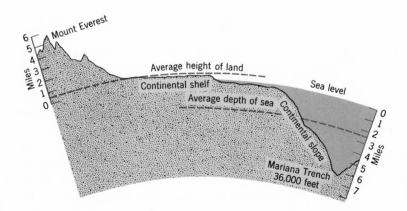

FIGURE 1.20 The total relief of the Earth's crust showing the greatest height on land and the deepest underwater trench.

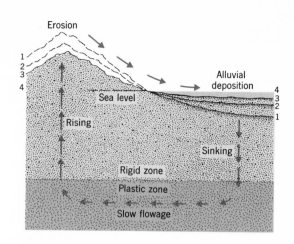

FIGURE 1.21 The theory of isostasy. The deposition of sediments causes the plastic rock beneath to flow away from the region of increasing pressure. The flow is toward an area of decreasing pressure such as the slowly eroding mountain mass, causing a slow rise in the area under the mountain. A corresponding sinking of the original valley floor occurs beneath the sediment layers.

like motion to accommodate the new weight above them.

The brittle, rigid rocks we call the crust rest on a thick, slowly flowing layer of rock that comprises the mantle. If great deposition of material takes place in a region, the underpinning of the crust will sink downward under the extra weight. The extra weight will cause the rock of the mantle zone to flow away from under the area of deposition toward that land mass where erosion of materials originally took place. The land mass so eroded, and accordingly lessened in weight, will rise in the slowly flowing mantle rock in a buoyant manner. Isostasy is diagrammed in Figure 1.21.

The deposition of sediments is a slow process that occurs on a large scale in producing isostatic adjustments. Huge downwarps of the Earth's crust, called *geosynclines,* are deepened by the weight of sediments carried from one region to another, for instance, from the Gulf Coast into the Gulf of Mexico. Pressures imposed on the crust by the slow deposition of materials over millions of years can be equalized by vertical movement. However, if deposition is too rapid, isostatic adjustment may take longer than the actual deposition of the sediments. When this occurs, continued vertical adjustment of the

materials causes the formation of huge, rounded geosynclines.

We have evidence of the process of isostasy from observations and measurements made in areas where the great weight of ancient glaciers caused a sinking of land masses. When the glaciers melted and receded the land masses did not immediately rise to their original levels. Certain land masses are still rising even today. The area around Scandinavia, once the center of a gigantic glacier, is slowly rising, one to three feet every century.

As sediments are carried down mountains and deposited in the ocean's margins, they sink slowly under the weight of the additional sediments. Although the sediments themselves make it seem that the level of the basin remains the same, measurements of the original ocean floor would reveal the sinking that is taking place. As land masses are worn away and ocean basins are filled, the general effect is one of smoothing off the Earth.

As material from the plastic zone of the mantle moves upward, it becomes part of the hard outer crust. When crustal rocks are forced into the mantle, they become plastic in response to heat and pressure. Thus, new material is constantly taking the place of the old in both the crust and the mantle.

20

To summarize, the continents, including the deep roots of the mountains, which form a part of the surface of the continents, are floating on a layer of plastic, yielding rock. The rock, like a thick putty, will flow away from areas of increased pressure and billow upward in areas of less pressure.

The Structure of the Continents

The continents, as we know them today, are not stable unchanging parts of the Earth's face. In the past, and in the present as well, the continents have yielded to pressure. At times, they have given way to slow isostatic movements. At other times, the forces that have been applied to the continents and the ocean basins have produced cataclysmic changes. Volcanic eruptions, earthquakes, slow folding of the crust, and the intrusion and retreat of seas, have produced far-reaching changes on the original crustal rocks.

All crustal rocks have taken one of three forms. The igneous rocks are those which have cooled from a melt, from lava on the Earth's surface or magma beneath it. The sedimentary rocks are those which have been compacted from sediments washed from surface masses or precipitated from solution and subjected to tremendous pressures. And the metamorphic rocks are those which have changed their crystalline composition or physical characteristics as a result of heat, pressure, and burial at great depths.

Although the continental masses are covered by a vast quantity of sedimentary rocks, we need not look very far to find evidence of other rock types. The oldest rocks known are igneous and metamorphic rock forms, which are approximately three billion years old. They conceivably formed under the heat and pressures that were exerted on the Earth after its birth pangs. Since then, the constant activity of erosion and deposition has formed the tremendous quantities of sedimentary rocks which presently cover the con-

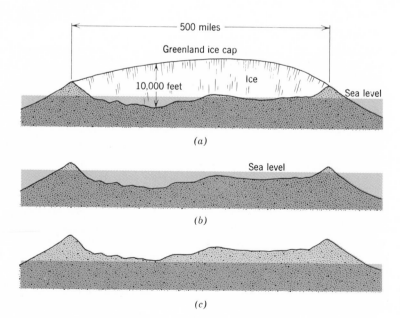

(a)

(b)

(c)

FIGURE 1.22 In areas that were once under the weights of gigantic ice sheets, depression of the crust occurred. Today, with the weight of the ice absent, a slow rise of the crust can be observed (After Zumberge, 1963).

tinents. But what events occurred to form today's continental masses? This is the question that presently needs much investigation.

One of the most interesting theories which has been proposed to date is that of *Continental Drift*. This theory first proposed that

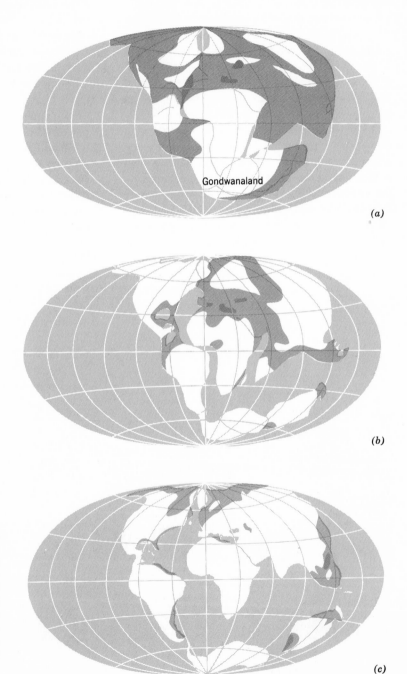

(a)

(b)

FIGURE 1.23 The present continents originated from a supercontinent (Pangaea) that broke apart into continental masses. Following this break the masses of rock slowly drifted across the face of our planet to their present positions (After Wegener, from a drawing by A. L. Do Tolt, 1937 — in *Our Wandering Continents*, Oliver and Boyd, Edinburgh).

(c)

the continents were once one large super-continent known as Pangaea. If you were to examine a globe of the Earth, you would notice that the east coast of South America and the west coast of Africa seem to fit together as though they were once part of some gigantic jigsaw puzzle that has come apart.

The Theory of Continental Drift suggests that the supercontinent Pangaea broke apart, and the continents have drifted slowly apart ever since. Further evidence has been gathered from the comparison of continental formations along the east coast of North America and the west coast of Europe. The past attachment of the continents, of course, has been lost. Present evidence suggests, in fact, that the ocean basins are spreading outward toward the continents. This process,

known as sea-floor spreading, has as its most visible evidence the Mid-Atlantic Ridge System (Fig. 1.24).

One of the chief objections to the Continental Drift theory was the difficulty in explaining exactly how solid, rocky structures like the continents have been able to move across a seemingly rigid basement of rock. In 1968, LePichon suggested that the earth consists of six large blocks or plates, which move relative to each other over a weaker layer. This discovery offered a new and credible explanation, and has satisfied most geologists.

Another theory which also has many adherents among modern geologists is that forces are produced by convection currents in the upper part of the mantle (Figure 1.25). This theory proposes that there are move-

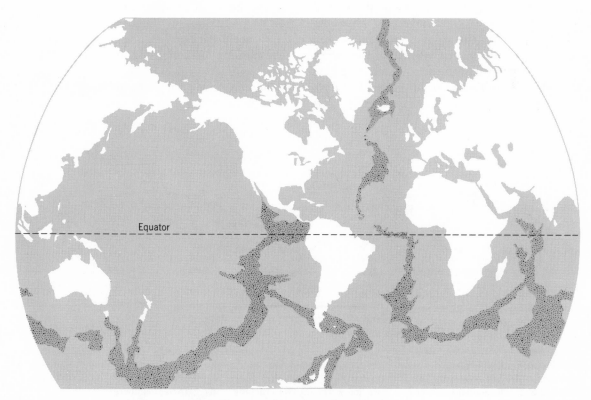

FIGURE 1.24   The location of the ridge-rift system throughout the ocean floors.

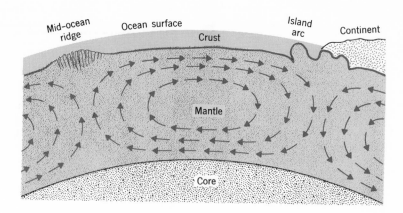

FIGURE 1.25 Large-scale convection currents below the Earth's crust produce folds and movement of the continental masses. The mid-ocean ridge represents the upwelling of these currents (After Navarra and Strahler).

ments of material within the mantle in response to upward heat flow. The heated rock rises upward, and when it reaches the rigid crustal rock, is forced sideward. Convection currents, which may be moving slowly beneath the crust, may be a factor in producing the crack on the Earth's crust recently discovered in the great Mid-Ocean Ridge system. As the rising convection currents separate under the ridge system, this area of the ocean basin might represent a weakened region of the ocean floor. The flow toward the continents may well be the effective push required by the Continental Drift theory for the spreading of the sea floor and the movement of the continents.

That continental structures are, in fact, still in motion has become evident from measurements made in the area of the Red Sea. These measurements show that there is a slow separation of the continent of Africa from the Arabian peninsula as Africa rotates in a slow, clockwise direction.

Other theories also have been proposed to explain the formation of the continents as we are familiar with them today. One theory holds that the original Earth was covered by a thin layer of granite, and that it contracted

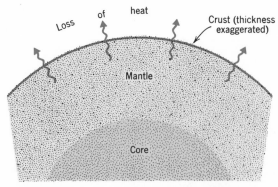

FIGURE 1.26 One older theory which attempts to explain the origin of the present continents holds that the Earth cooled and contracted. As contraction continued, the crust buckled and collapsed, forming the continental masses as we know them. The cracking and folding that took place gave rise to the mountain ranges and ocean basins that mark the Earth's surface (After Zumberge, 1963).

24

(Figure 1.26). As it did so, the continents were formed by the cracking and buckling of these granitic or sialic masses. According to this theory, the outermost crust cooled first, hardened, and then formed the continental masses and ocean basins in response to the collapse of the underpinnings of the crust as the underpinnings cooled, shrank, and moved away from the hardened, thin covering.

Another holds that the Earth has been expanding since its original formation, and that the ocean basins and continents formed as the thin granite covering cracked because of the expansion.

The oldest rocks known to man, the igneous and metamorphic rocks, seem to form the nuclei of the continents. These nuclei, known as the *continental shields* (Figure 1.27),

are covered and surrounded by sedimentary rocks. Another theory suggests that the continental shields are the remains of the original continents, and that additional material has been added by accumulation or accretion to build the continents to their present sizes and shapes.

At present, nobody is quite sure how the continents were formed. As evidence continues to accumulate, a new theory will probably evolve that has elements of all of these hypotheses.

Today's continents slope off into the ocean in several stages. The first of the underwater areas adjacent to the land are the continental shelves. The shelves generally range outward in a gentle descent to depths of 500 to 600 feet, where they end at an abrupt margin known as the *shelf break*. Beyond this is the

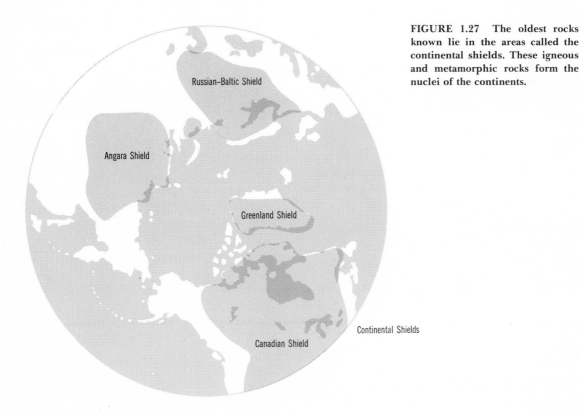

FIGURE 1.27   The oldest rocks known lie in the areas called the continental shields. These igneous and metamorphic rocks form the nuclei of the continents.

Russian–Baltic Shield

Angara Shield

Greenland Shield

Continental Shields

Canadian Shield

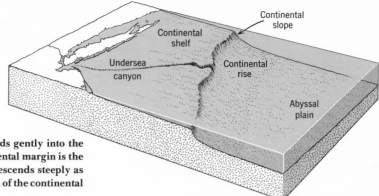

FIGURE 1.28 The continent descends gently into the ocean. The first portion of the continental margin is the continental shelf. The margin then descends steeply as the continental slope. The last portion of the continental slope meets the ocean basin.

continental slope, a comparatively steep slope, which is the outer portion of the actual continental structure. The base of the slope abuts the ocean basin.

Geologists have determined that as the Earth's crust has risen and fallen and the ocean has moved outward or impinged on the continents, the shelves have varied in size. The state of Florida and the east coasts of the southern states were, at one time, part of the shelf and were completely submerged.

Recent oceanographic explorations, as we shall see in Unit 3, indicate that far from being a flat plain, the ocean floor has even more topographic features than the continents.

Plateaus, ridges, seamounts, and islands rise from the ocean floor and basins, and deeps, trenches, and canyons are cut into it. We have observed something of the evolution of the science that is Geology. From man's first primitive questioning about his environment, through countless generations of investigations, a nearly exact science has developed. It is a science that is beginning to unfold the past history of the Earth and its formations.

In the following chapters we shall examine the landforms of the Earth in more detail. Their composition, structure, and the processes by which they have formed will be studied.

CHAPTER 2

The Composition of the Earth

Minerals

WE HAVE DISCUSSED SOME OF THE general characteristics of the Earth and its structural formations. At this point we shall examine, in detail, some of the minerals are found in the Earth. There is a huge variety of mineral combinations in the Earth's crust. These combinations make up the various types of rocks that, in turn, form the Earth's topographic features.

General Mineral Characteristics

A rock-forming mineral is a naturally occurring substance with fairly definite physical and chemical properties. The atomic structure of each mineral has a certain symmetry which gives it a characteristic external shape. That is, the atoms making up the specific minerals are arranged in a certain definite pattern. It is this crystalline structure that is the most reliable identification of a mineral.

The symmetry of a crystal is relative to a series of imaginary axes drawn through the crystal faces. The atoms are arranged around these axes in identical, repetitive patterns. Although the crystalline structure may not be apparent to the naked eye, special techniques such as microscopic and X-ray analysis will reveal the atomic pattern. The X rays will be diffracted at specific angles by the crystalline pattern within the mineral, and the angles of the axes may be measured.

All minerals fall into one or another of six general classes of symmetry determined by the axes drawn through the mineral faces. These classes are:

1. Tetragonal — Three axes which are at right angles to one another; two are equal in length and the other is of a different length. The mineral zircon is an example.

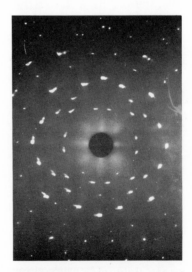

FIGURE 2.1 Laue photograph of salt (NaCl). The deflection of X-rays by the atoms within the mineral yields a characteristic pattern that is indicative of the crystalline structure (Professor C. Kerr).

2. Isometric	Three axes of equal length; all are at right angles to one another. Garnet is an example.
3. Hexagonal	Four axes; three axes are of equal length and one is at right angles to the others. Quartz is an example.
4. Orthorhombic	Three axes of unequal length all at right angles to one another. Topaz is an example.
5. Monoclinic	Three unequal axes; two are at right angles to one another, the other is at another angle to the first two. Gypsum is an example.
6. Triclinic	Three unequal axes all at oblique angles to one another. Plagioclase is an example.

These six classes are illustrated in Figure 2.2.

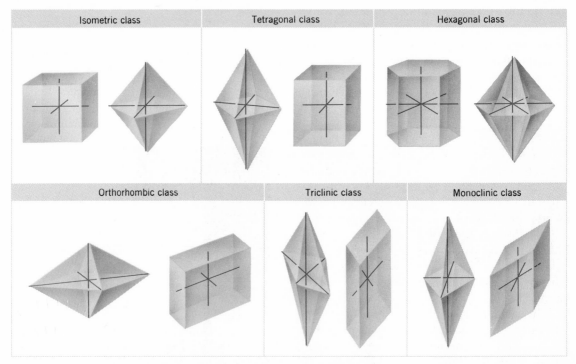

FIGURE 2.2 **The six crystal systems. Imaginary lines through the crystal faces are the axes that represent the configuration of the crystal.**

1. *Isometric class*. Three equal axes at right angles to one another. Example, garnet.
2. *Tetragonal class*. Three axes at right angles, two of equal length, the third of another length. Example, zircon.
3. *Hexagonal system*. Four axes, three equal axes at 60° to one another, the fourth at a right angle to the first three and of a different length. Example, quartz.
4. *Orthorhombic class*. Three unequal axes at right angles to one another. Example, topaz.
5. *Triclinic class*. Three unequal axes at oblique angles to one another (none at right angles). Example, plagioclase.
6. *Monoclinic class*. Three unequal axes, two at right angles to each other, one at an oblique angle. Example, gypsum.

There are a few minerals which are not crystalline in nature. Opal is an example. Minerals which are not crystalline have no regular form and are therefore called *amorphous*.

Mineral Formation

Minerals form in a variety of ways. One way is the evaporation of a solution. For example, consider lake water which contains dissolved salt. The water (the solvent) dissolves the salt as it filters through salt-bearing minerals. The salt (the solute) is held in the water as dissolved particles. A *saturated* solution holds as much salt as possible at that temperature. As the water evaporates, there is less room for the molecules of salt, and thus the water can hold less salt. The extra salt settles out of solution and forms a salt

bed. The slower the evaporation, the larger the salt crystals formed.

Crystals may also develop from a melt. When a solid is heated, as by volcanic action, it becomes molten. When subsequent cooling takes place, the melt forms crystals. For example, when sulfur is heated, it melts and then crystallizes when cooled.

Similarly, when one heats a solution such as a salt solution, it holds more than normal amounts of solute and becomes a *supersaturated* solution. That is, more salt—or any soluble material—can be dissolved in the hot solvent. When cooling takes place, crystals form. Slow cooling aids in the growth of large crystals. Dissolved minerals forced up by geysers through fissures in the Earth crystallize in this manner.

The process of sublimation—condensation of a gas directly to a solid—also results in

FIGURE 2.3 Salt beds such as the Morton mine in Grand Saline, Texas, are the result of evaporation from inland seas, which left the salt as huge deposits on the land surface (Morton Salt Company).

crystal formation. Gases forced out of volcanic fissures often recrystallize directly from the vapor to form solid particles. Sulfur and iodine, when heated to a vapor, will crystallize directly onto a cool surface.

FIGURE 2.4   Sulfur is obtained by creating a hot solution that is then evaporated. The sulfur crystals are recovered (Freeport Sulphur).

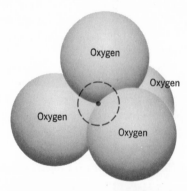

FIGURE 2.5   In the silicon tetrahedron, the silicon atom is linked to four oxygen atoms.

Mineral Classification

A bewildering variety of mineral combinations is found in nature. In attempting to deal with this seeming disorder, scientists have developed a scheme for identifying minerals that places this variety in some coherent order based on similar physical and chemical properties. One of the ways by which minerals are classified is according to similar chemical composition. There are several major groups in this system of classification.

The Silicates. These minerals include silicon and oxygen in their composition. They are the most common minerals. Silicates are characterized by a silicon atom surrounded by four oxygen atoms (Figure 2.5). This arrangement is known as the silicon tetrahedron. The silicates that also contain iron and magnesium are dark-colored and are known as *ferromagnesians*. The lighter-colored silicates are known as *felsic*. Quartz and garnet are two examples of the silicates.

The Carbonates. The carbonates include carbon and oxygen in their composition. They consist of three oxygen atoms and one carbon atom in addition to atoms of other elements which may be present. The carbonates are important constituents of many crustal rocks.

The Oxides. Oxides are produced when oxygen combines with another element. Quartz, which has the formula SiO_2, is a mineral which can be listed as either a silicate or an oxide since it satisfies the requirements of both groups. There are several iron oxides of importance, such as hematite—Fe_2O_3 and magnetite—Fe_3O_4.

The Sulfides and Sulfates. The sulfides and sulfates are produced by the different combinations of sulfur and oxygen. The sulfide known as fool's gold is actually iron pyrite, FeS_2. Galena is composed of lead and sulfur —PbS. Gypsum is a sulfate with sulfur in

FIGURE 2.6 Mineral materials. (A) Quartz crystal. (B) Muscovite. (C) Iron pyrite (fool's gold). (D) Halite salt crystals (American Museum of Natural History).

There are other groups of minerals, such as the halides (common salt), phosphates, borates, nitrates, hydroxides, and native elements. The elements (those minerals composed of one type of atom) such as gold, silver, copper, and graphite (carbon) are also considered minerals, since they not only satisfy the definition of a mineral but also have great economic importance. Although the list of minerals is long and varied, the most abundant minerals found in the Earth's crust are included in the groups just discussed.

Physical Properties of Minerals

Although crystalline structure and chemical composition are the most reliable bases for identification of minerals, the various

combination with four oxygen atoms plus calcium. In the case of gypsum, the chemical formula is $CaSO_4 \cdot 2H_2O$. Notice that molecules of water are attached to the gypsum structure, although this is not true of all sulfates.

physical properties of minerals are also important guides. Minerals are classified according to specific gravity, color, streak, luster, hardness, and other special physical characteristics. Many of the tests for these properties must be used together for reliable identification, although a single test may serve as a good indication of mineral type, particularly when a geologist is working in the field and a laboratory is not available for immediate testing of the specimen.

Specific Gravity

Specific gravity is defined as the weight of a quantity of the mineral compared to the weight of an equal volume of water. The relationship is a ratio, or a comparison between two substances, and does not carry a unit of measurement. The formula is:

Specific gravity

$$= \frac{\text{weight of the specimen}}{\text{weight of an equal volume of water}}$$

Color, Streak, and Luster

Although pure minerals have specific colors, the presence of impurities produces a wide variety of colors in minerals found in the field. However, when a specimen is rubbed against unglazed porcelain tile, the powdery streak left behind on the tile is always the same for a particular mineral. Thus, the original color is not as reliable a test as the streak test.

Luster is simply the manner in which light is reflected from the mineral surface. Glass has a vitreous or glassy luster, chalk has a dull luster, others have silky, pearly, or earthy lusters. Luster, then, varies somewhat according to the observer, but it is a good descriptive term for tentative identification.

Hardness

Minerals resist abrasion to varying degrees. As a result, a hardness scale called the Moh scale has been devised. The Moh scale is an arrangement of typical minerals set up in or-

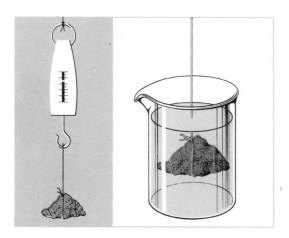

FIGURE 2.7 The determination of specific gravity. The same specimen is weighed in air and when submerged in water. The loss of weight in water is related to the buoyant effect of the water on the specimen.

FIGURE 2.8 The streak test — striking a mineral across an unglazed porcelain tile — yields a characteristic powdery, colored streak for many minerals. This test is a much more reliable test than color (Ward's Natural Science Establishment, Inc.).

FIGURE 2.9 Magnetite exhibits magnetism, the characteristic which causes it to attract iron particles. Lodestone, a form of magnetite, is used in early compasses (B. M. Shaub).

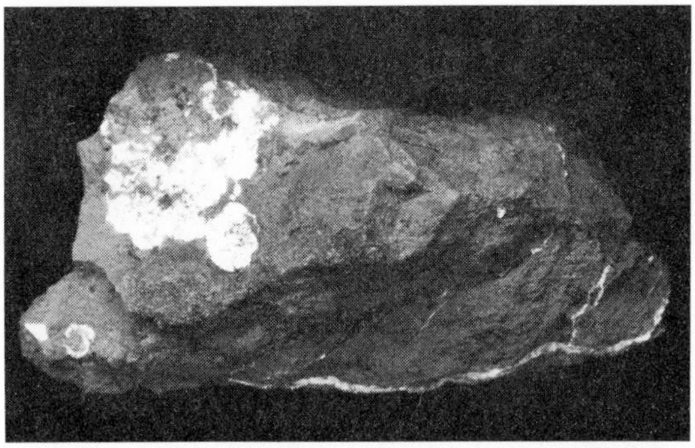

FIGURE 2.10 Ultraviolet light causes minerals such as this Andersonite to glow or fluoresce with brilliant colors other than the ones seen with the naked eye (Russ Kinne, Photo Researchers).

der of increasing hardness, from 1 through 10. They are:

1. Talc
2. Gypsum
3. Calcite
4. Fluorite
5. Apatite
6. Feldspar
7. Quartz
8. Topaz
9. Corundum
10. Diamond

Each type will scratch the ones preceding it on the numerical scale. For example, quartz will scratch feldspar.

As a comparison, a steel knife is about 5.5 on the scale of hardness, a fingernail is on the order of 2 to 2.5, and a penny is 3 to 3.5. These are useful devices for rapid estimation of the hardness of a specimen during field studies.

Cleavage

A mineral breaks along the surfaces that bound the atoms in the crystal structure. Thus, a crystal will break in a characteristic plane that is determined by its atomic symmetry. The smooth surfaces, or *cleavage planes,* reveal the crystalline structure of the mineral. The direction of cleavage parallels the direction of growth of most crystals.

Other Special Characteristics

Among the other characteristics used for identification are fracture and tenacity. *Fracture* is the manner in which a mineral breaks in an uneven fashion. *Tenacity* is the resistance to a change in form. For example, glass fractures in a series of concentric circles called conchoidal fractures; metals are malleable and can be hammered into different shapes. Substances which cannot be hammered are brittle, elastic, etc. *Magnetism* is a special characteristic of some iron ores.

Some minerals glow under ultraviolet light with colors other than their original ones. This is a property known as *fluorescence.* If they continue to glow after the ultraviolet light is removed, they are termed *phosphorescent.*

33

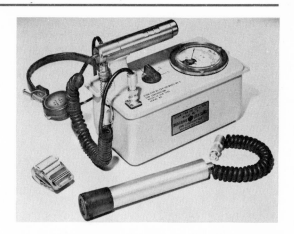

FIGURE 2.11 The Geiger counter yields a count of the amount and intensity of the radioactivity in rocks bearing radioactive minerals (Eon Corporation).

Chemical Properties of Minerals

Often, a single physical property is not sufficient for a reliable mineral identification. Chemical tests based on composition frequently must be considered in addition to several physical properties.

Minerals may be either single elements or combinations of elements in some fixed proportion. The constituent elements in a mineral govern its physical as well as its chemical properties. A series of tests to determine the chemical composition will firmly establish the identification of a mineral.

Radioactivity

Certain elements are unstable, and as they break down spontaneously they emit radioactive particles. A geiger counter, which is sensitive to these emissions, is used to detect radioactive elements within a mineral (Figure 2.11). Photographic film is also used to detect the presence of radioactivity. However, photographic film shows only the presence of radioactive emissions, whereas the geiger counter produces an accurate count of the intensity and amount of radiation.

Heat Tests

The application of heat to minerals will reveal the presence of any metal within the mineral. One test which utilizes heat is the blowpipe test (Figure 2.12). A small portion of the mineral is placed on a charcoal block in a depression gouged out for the purpose. A nearly horizontal flame is played over the mineral by blowing through a blowpipe. If metals are present in the mineral, they will be left behind as a residue. Copper remains as a red ball of metal; other metals leave balls of characteristic color, such as gray (lead) and white (zinc).

When metals are placed directly into a flame (Figure 2.13), the presence of the metal is indicated by a characteristic color within the flame as the metal oxidizes. Copper burns with a green flame, sodium with a yellow flame, and potassium with a purple flame.

Acid Tests

A small drop of acid will quickly reveal the presence of carbonate minerals within a specimen. Under the attack of hydrochloric acid, for example, certain carbonates will fizz or effervesce.

The Rocks of the Earth's Crust

A rock is usually defined as an aggregate of two or more minerals. Rocks in the Earth's crust vary from extremely hard forms to ones that are soft enough to be easily broken in your hand. At times, the mineral composition is very difficult to determine. For example, obsidian, a volcanic natural glass, contains no specific minerals that can easily be discerned. In some cases, large specimens of minerals may be called rocks, although this is not technically correct.

The Origin of Rocks

Rocks are classified according to the manner by which they are formed, the three major types being igneous, sedimentary, and metamorphic. Although igneous rocks are by far the most abundant rocks in the Earth's crust, they are subjected to erosion when exposed on the surface. As a result, sedimentary rocks are more commonly seen on the surface and are, therefore, most easily studied. Igneous rocks form as a result of the cooling of a melt of molten magma. Some igneous rocks cool and solidify below the Earth's surface, others cool on the Earth's surface.

Sedimentary rocks are composed of particles that have been worn away from other rocks, carried by a transporting agent such as water or wind to new locations, and there deposited. A compact mass is formed when the particles are compressed and cemented together with some other substance acting as the bonding agent.

Metamorphic rocks are derived from other rock types because of changes brought about by heat and pressure. These changes may be the result of rocks being buried and squeezed below the surface of the Earth, or may be because of the action of movements in the crust.

Since the origins of the Earth's first rocks are lost, no one knows which type of rock appeared on the Earth first, although we can reasonably assume that it was the igneous. Any one of the three types can be derived from either of the other two. That is, igneous and sedimentary rocks, when put under pressure, or exposed to extreme heat, can become metamorphic rocks. Metamorphic and igneous rocks, in turn, can be eroded to form sedimentary types, and metamorphic and sedimentary rocks that have become molten may produce rocks of the igneous type. Instead of fruitlessly debating on which type came first, geologists think in terms of a *rock cycle* (Figure 2.14), with no beginning and no end. Each of the three types of rock is constantly being developed, destroyed, and changed into another type.

Igneous Rocks

All molten material that is located below the Earth's surface is termed *magma*. On reaching the surface it is known by the more familiar term *lava*. When magma cools beneath the surface of the Earth, it cools slowly, and the minerals within it form large crystals. These large-grained, coarse-textured

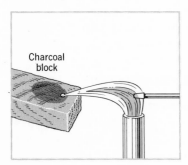

Charcoal block

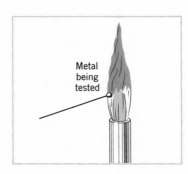

Metal being tested

FIGURE 2.12 (*left*) The blowpipe test will indicate the presence of metal in certain minerals.

FIGURE 2.13 (*right*) The presence of a metal in a mineral will result in a specific color when the mineral is subjected to a flame.

rocks produce a variety of *intrusive* rock bodies. Lava on or near the surface of the Earth
cools swiftly, and the crystal forms are quite
small. These fine-grained rocks are known as
extrusive rock structures. Although erosion
may eventually expose intrusive rock struc-

tures, the rock texture will usually reveal
its intrusive origin.

Intrusive rock formations vary in size, but
they are all characterized by the large-grained
texture indicative of slow cooling. The largest of these formations, and the largest of

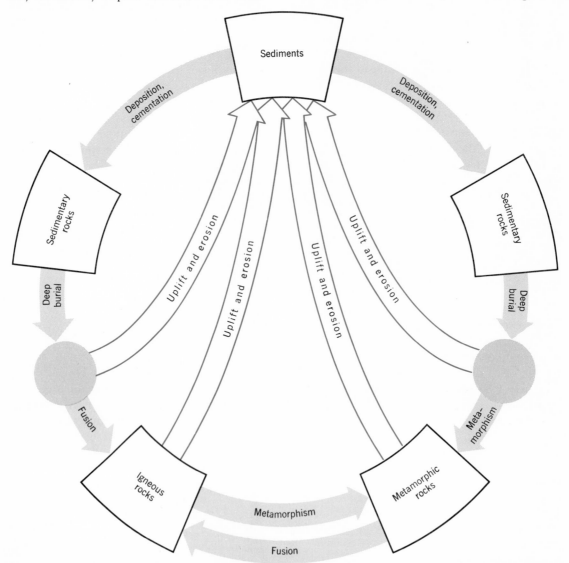

FIGURE 2.14 The rock cycle. The
materials that comprise the rocks of
the Earth's crust undergo repeated
and continuous changes. A rock
may be subjected to heat, pressure,
or erosional effects. Thus, any rock
may be changed by various processes to become one of the other
two types. The process is cyclical;
it has no beginning and no end. The
agents of change are various and
continually active (After Longwell,
Flint and Sanders).

FIGURE 2.15 An igneous intrusion formed this dike (dark column) cutting through metamorphic rock in this cliff in Ascutney, Vermont (B. M. Shaub).

the crustal rock forms, is the *batholith.* Batholiths extend beneath the Earth's surface for several miles and may be as large in area as an entire state. The Idaho batholith in the central portion of the state is nearly 100 miles wide and 300 miles long. The Sierra Nevada Mountain range is composed of huge batholiths. A smaller intrusive structure, similar to a batholith, is a form known as a *stock.* Stocks are less than 40 square miles in size. These vertical, domelike masses are found in great numbers in the mineral-rich areas of the southern Rocky Mountains, among other places.

Batholiths and stocks are masses of magma that cut through or push aside the preexisting rock structures in their respective areas. Structures that cut through rock strata are said to produce *discordant contacts* with the existing rock.

When an opening or fissure occurs in the crust, magma forces its way upward in a sharply inclined tabular structure known as a *dike*. Dikes, which may be rich in mineral deposits, also have discordant contacts with the preexisting rock. However, material from a dike may force its way between the preexisting rock layers, forming a shelflike structure known as a *sill*. Sills produce a *concordant contact* with the preexisting rock because they force strata apart instead of cutting through them like dikes. If the magma flow is very great, the sill may form a dome which pushes the existing rock upward, in which case the structure is known as a *laccolith* (Figure 2.17).

The larger intrusive structures cool rather slowly, and are characterized by large crystal structures. The extrusive rocks cool rather quickly, and are characterized by small crystal formations. This structural characteristic is especially important because it reveals the origin of the mass.

Extrusive rocks are formed from lava that has flowed out of fissures in the Earth's crust and, thus, they are volcanic in origin. Volcanic rocks contain particles that vary in size from dust, ash, and cinders to bombs and blocks several feet in width. These rock forms

FIGURE 2.16 This intrusive diabase sill along the Yellowstone River in Wyoming is of igneous origin. Notice the columnar jointing of the rock (B. M. Shaub).

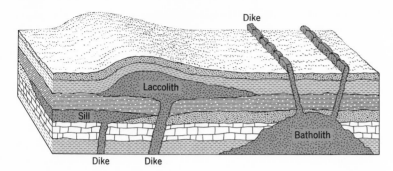

FIGURE 2.17 Igneous intrusive masses occur below the Earth's surface. The batholith and laccolith are larger intrusive masses. The batholith is cut through native strata; the smaller laccolith displaces the strata. The dike, which gave rise to the shelflike sill, was also formed below the surface and has been exposed by erosion.

FIGURE 2.18 Navajo Mountain, part of the Henry Mountains on the Arizona-Utah border, is a laccolith dome pushed up below previously existing rock. Erosion has exposed the dome as a broad, sloping mountain mass (Joseph Muench).

are known by the general term *pyroclastics.*

Extrusive rock oozes out of cracks in the Earth and may form layers of fine-grained masses hundreds of feet thick. The gases escaping from the lava leave the rock perforated with holes much like a sponge. This type of rock (found in all areas that possess lava flows) is termed *scoria.*

In some of the larger structures formed on the Earth's surface, cooling and shrinking causes vertical cracks called *joints.* Occasionally, these structures take the form of gigantic columns standing side by side. Called columnar joints, they are a rather spectacu-

FIGURE 2.19 Volcanic bomb blown out of an erupting volcano. Such bombs may be a foot or more in length (National Park Service).

FIGURE 2.20 Jointing in Devonian shale in Canyon Lake, Portland Point, New York. The joints are angular and rather straight (B. M. Shaub).

lar sight. Such formations can be observed along the banks of the Columbia River in Washington, and the Devils Post Pile National Monument in the Sierra Nevada mountains is a series of columnar joints.

Classification

The classification of igneous rock depends on several factors. The size of the crystal grains within the rock indicate the rate at which the rock cooled; thus, there are coarse-grained and fine-grained igneous rocks. In addition, the dark colored ferromagnesians will cool and form crystals before the lighter colored crystals of quartz and orthoclase. As a result, the identification and classification of igneous rocks depends on their chemical composition and their related color, in addition to texture. The igneous rocks fall into five general classes which are related to texture and subdivided according to composition.

1. The *coarse-grained* igneous rocks are those in which the grains are visible to the naked eye. The lighter colored members are granite, which is found in many mountain ranges, and diorite. Granite is composed mostly of quartz, feldspar, and biotite in addition to other minerals. Diorite is slightly darker, with feldspar, hornblende, and plagioclase dominant. The darker members of this class are gabbro, which has a high iron and magnesium content, and peridotite, which is composed of dark olivine and pyroxene with no quartz present.

2. The *fine-grained* and *felsitic* igneous rocks possess crystals which cannot be distinguished with the naked eye. Andesite is a lighter-colored rock with many feldspars present, while basalt is a gray to black form with feldspars combined with olivine and pyroxene.

3. The *porphyritic* rocks contain large crystals imbedded within a groundmass of fine-grained crystals. Varying rates of cooling produce these different sizes of crystals within the same rock mass. The large crystals, or phenocrysts, form slowly, within the bed of smaller crystals which take shape rapidly at higher temperatures. The rocks vary from light to dark and are described according to composition by the addition of a

FIGURE 2.21 Columnar basalt in the Devil's Postpile National Monument in the Sierra Nevada Mountain range. These are a spectacular example of columnar jointing (National Park Service).

prefix such as: felsite porphyry, basalt porphyry, andesite porphyry, etc.

4. The *cellular* igneous rocks are the group to which scoria belongs. These forms are fine-grained in texture and are composed of volcanic glass. Pumice is another volcanic rock in this class.

5. The *glassy* rocks result from rapid cooling. In these rocks, the atoms do not have time to aggregate and form crystals, although crystallization will take place later in the rock. They have many impurities and, thus, show a variety of colors.

A

B

FIGURE 2.22 Igneous rocks. (A) (*above right*) Obsidian showing conchoidal fracture (American Museum of Natural History). (B) (*left*) Coarse granite. The constituent minerals—quartz, feldspar, hornblende, and mica—are shown above the rock (Fundamental Photos).

Sedimentary Rocks

Sedimentary rock beds are often characterized by a layering effect of gradation. Larger particles settle out of the water first; each succeeding layer is composed of finer materials with smaller size individual particles.

The particles in sedimentary rocks vary greatly in size. They range from clay less than $1/1000$ of an inch in diameter, to sand which is up to $1/10$ of an inch in size. Pebbles are $1/4$ to two inches large, cobbles range up to nearly a foot, and the largest boulders are anywhere from one foot up.

The particles of sedimentary rock are cemented by materials which settle or precipitate out of solution. The precipitate is most commonly calcium carbonate ($CaCO_3$), which forms a weak, gray colored cement; silica, a hard glassy rock; or iron oxide, which forms a fairly weak bond of rust-red color. (See Table 2.2.)

Sedimentary rock solidifies by one of several processes. The weight of overlying sediments may compact the sediments and squeeze water out of spaces between particles, or precipitates may cement particles together.

Table 2.1   Common Igneous Rocks

	PREDOMINANT MINERALS	COARSE-GRAINED ⟶ PORPHYRITIC		FINE-GRAINED → GLASSY	
	Quartz				
		Granite		Rhyolite	Obsidian
	Plagioclase				Pumice
	Biotite or Hornblende	Syenite		Trachyte	Pitchstone
	Plagioclase Hornblende or Biotite	Diorite	Granite, Syenite, Diorite,	Rhyolite Trachyte,	Andesite
	Plagioclase Pyroxene	Gabbro	Gabbro, Pyroxenite, Peridotite, and Dunite	Andesite, and Basalt Porphyries	Basalt	Tachylite
	Pyroxene Plagioclase Olivine	Pyroxenite	Porphyries		
	Olivine Pyroxene	Peridotite			
	Olivine	Dunite			

Ferromagnesians to Felsites — *Light Color* — *to* — *Dark Color*

Increase in Cooling Rate ⟶ Decrease in Grain Size ⟶

FIGURE 2.23 Stratification layers (these are in Monument Valley) are typical formations found in many sedimentary rocks (Union Pacific Railroad).

A

B

FIGURE 2.24 Sedimentary rocks. (A) Conglomerate composed of pebbles cemented by limestone. The rounded pebble forms are characteristic of water erosion. (B) Coquina is a limestone cement containing shell remains and other debris (Fundamental Photos).

Table 2.2   Common Sedimentary Rocks

	ROCK		SEDIMENT
Clastic	Conglomerate	*Fine – Medium – Coarse*	Gravel and sand
	Sandstone		Sand
	Shale		Mud (clay)
Chemical	Limestone (calcium carbonate)	*Dense Composition*	Chemical and/or organic precipitates
	Dolomite (calcium magnesium carbonate)		
	Salt (halite) Gypsum		Chemical precipitates
Organic	Coquina		Shell fragments and sand
	Chalk		Soft, minute shells

Dissolved chemicals may precipitate out of solution to form a solid crystal mass, such as gypsum and halite. And chemical action between substances may change the original mass into a new compound that forms a new solid stratum.

Classification

The classes of sedimentary rocks are governed by origin and by the size of the particles found within them. *Clastic* rock types are comprised of particles that have been transported by water and later cemented or compacted together. (See Table 2.3.) The particles may be all the same size or of various sizes. Shale is a type in which clay-sized particles predominate. As a result, shales have fine textures in which particles are

Table 2.3   Definition of Clastic Particles and Aggregates

NAME OF PARTICLE	LIMITING DIAMETERS		NAME OF LOOSE AGGREGATE	CONSOLIDATED INTO ROCK
	MM.	IN. (APPROX.)		
Boulder	More than 256	More than 10	Gravel	Conglomerate and breccia
Cobble	64 to 256	2.5 to 10	Gravel	
Pebble	2 to 64	0.09 to 2.5	Gravel	
Sand grain	$\frac{1}{16}$ to 2	(Not of practical use)	Sand	Sandstone
Silt particle	$\frac{1}{256}$ to $\frac{1}{16}$		Silt	Siltstone
Clay particle less than $\frac{1}{256}$ mm			Clay	Shale

(From Wentworth, C. K., Jour. Geol., vol. 30, p. 377–, 1922)

not distinguishable. Sandstone is composed generally of sand-sized grains of quartz. Conglomerate consists of a fine cement matrix in which various sand-sized and larger rock particles are cemented. Breccia is similar to conglomerate, but the clastic particles are sharp and angular and not rounded as in conglomerate.

The class of sedimentary rocks that originate from precipitates and have fine, dense structures fall into the *chemical* class. Limestone, in which calcite ($CaCO_3$) dominates, is one form of chemical sedimentary rock. Evaporites like gypsum and salt from dried sea beds are also found in this group. Dolomite is similar to limestone but has a magnesium content in addition to calcium.

Sedimentary rocks arising from plant and animal remains are known as *organic* forms.

In general, calcium carbonate is commonly found in high proportion in this type of rock. It is the residue of deposited shells of sea life. Coquina is composed of large shell fragments bound by a fine matrix of calcium carbonate. It is similar to conglomerate except it is organic in origin. Chalk is composed of calcium shells of one-celled sea life called diatoms. Diatomaceous earth is a deposit of similar shell remains and is usually an indication of nearby oil deposits. Coal is rich in carbon and some clay. It is the remains of ancient fern forests. Other organic forms are lignite and peat.

Other Characteristics

Sedimentary rocks often display structures other than stratification. In addition to

FIGURE 2.25 Ripple marks left in sandstone cliffs in Zion National Park, Utah. The marks are evidence of waterborne deposits of silts (Frank Jensen).

FIGURE 2.26 Blocks of dried clay shown after evaporation of rain water in Arizona. The blocks are 12 to 16 inches across (Fred Taylor).

Table 2.4   Common Metamorphic Rocks

METAMORPHIC FORM	CHARACTERISTICS	ORIGINAL FORM
Slate	Fine grain, without banding Near-perfect cleavage	Shale
Marble	Crystalline, usually coarse Foliation revealed only by streaks of impurities	Limestone Dolomite
Quartzite and quartzite conglomerate	Compact, massive	Sandstone or conglomerate
Anthracite		Bituminous coal
Graphite		Anthracite
Serpentinite	Massive	Peridotite Pyroxenite
Schist and phyllite	Fine banding Pronounced foliation	Rhyolite, andesite, and basalt Shale and shaly limestone
Gneiss	Coarse banding Imperfect foliation	Granite and other coarse-grained rock Shale, shaly limestone, conglomerate

graded bedding, ripple marks left by water and wind may be found in hardened sedimentary rock surfaces. Fossil imprints of plants and animals are also common. Mud cracks left by the shrinking and drying of mud beds, and raindrop impressions, are also often found in sedimentary beds.

A rather interesting feature of some rock types is the geode. This is a cavity found in rock in which crystals grow inward. Often calcite or quartz, these crystals make a dramatic appearance when the rock is split.

Metamorphic Rocks

Metamorphic rocks are produced by the heat and pressure resulting from the compaction of rocks within the Earth. (See Table 2.4.) Heat and pressure bring about changes in the texture of the rock, or alter the mineral composition of previously existing rock. These changes occur because minerals recrystallize into different grain sizes, or the elements recombine and form new minerals.

Formation

There are several processes that cause metamorphic changes in rock. *Geothermal* action is the result of heat and pressure on buried rock forms. The pressures produce heat which brings about further changes in mineral composition. *Contact* metamorphism is produced by the invasion of native rock by

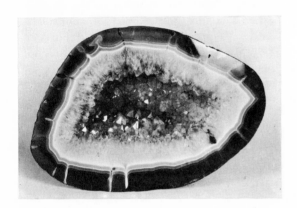

FIGURE 2.27 A quartz geode in which crystals have formed within a rock cavity (American Museum of Natural History).

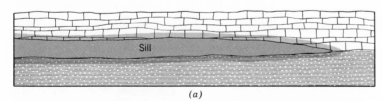

Sill

(a)

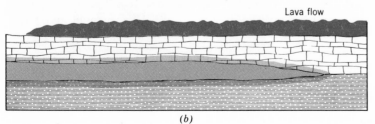

Lava flow

(b)

FIGURE 2.28 As magma intrudes into native rock, the rock bounding the intrusion will be altered by the presence of heat. Lava flows on the surface have a similar effect on the rock below. Some of the native rock will also be incorporated in the flow (After Zumberge, 1963).

FIGURE 2.29 Intense pressures create gigantic folds in rock strata that were originally laid down at a horizontal angle of repose. The effects can be observed in these strata (U.S. Geological Survey).

magma. The tremendous heat and pressure give rise to changes in the rock adjacent to the area of intrusion. (See Figure 2.28.) *Hydrothermal* action of very hot water and gases invading older rocks also causes metamorphism. This action usually removes some constituent minerals and replaces them with others. *Dynamic* or *kinetic* metamorphism is due to intense deformation of rock materials and usually takes place in shallow areas of the Earth's crust. The tremendous forces brought about by folding and faulting crush and shear the rocks into the various metamorphic rocks.

Classification

Metamorphic rocks fall into two major groups. One group possesses a distinctly banded, or parallel, arrangement of crystals within the rock. These forms, called *foliated* metamorphic rocks, generally cleave along planes following the bands, or folia. The foliated rock types include the schists in which foliated bands of quartz, hornblende, and mica predominate. They are termed quartz schist, mica schist, and hornblende schist.

The gneisses are the most abundant foliated rock types and are composed of very obvious, coarse bands of quartz, feldspars, and hornblende as well as other dark-colored minerals. They are called granite gneiss, hornblende gneiss, and muscovite gneiss depending on composition. The slates are fine-textured, metamorphosed shales in which individual minerals cannot be discerned. They cleave in flat, parallel sheets known as slaty cleavage.

FIGURE 2.30 This quartz vein in gneiss shows the foliated structure typical of this type of metamorphic rock (American Museum of Natural History).

The *nonfoliated* metamorphic rocks have interlocking crystal formations and, as a result, the folia are not obvious as in the foliated forms. The nonfoliated types do not break readily. Marbles are recrystallized limestone and dolomites in which calcite and dolomite dominate. The crystals form interlocking patterns, and the color varies from white to black. Quartzite is a fine-grained metamorphosed quartz sandstone. It is so rigid that it will break across grains instead of around them. Silica is often present as the cement. Although coal was listed under sedimentary rocks, when it is compressed and squeezed into anthracite, or hard coal, it too is classified as a metamorphic type. Anthracite is a very hard coal in which only carbon remains. It has very few impurities.

Some metamorphic rocks can be easily identified. In most cases, the changes that have taken place are so marked that the metamorphic forms are completely different from the original rock. In general, metamorphic rocks are more compact and there-fore harder than the rocks from which they originated.

Metamorphic rocks, when exposed at the surface, are also affected by the processes of erosion, mountain building, and faulting. A rather interesting process known as granitization produces some granitelike metamorphic rocks. Most granite comes from cooled magmas and, in some instances, materials dissolve and recrystallize in rocks containing high percentages of feldspars. This larger-grained rock is very much like a granite of the igneous form, and it is difficult to distinguish between the forms unless one is an expert geologist.

The occurrence of rock types is used as an indicator of the forces that have operated in a particular locality. On occasion, reconstruction of past events is very difficult to determine since many changes have taken place in rock strata. At the present time, Earth scientists are only beginning really to understand the forces that have shaped and reworked the crust of the Earth.

CHAPTER 3

Agents of Landscape Formation

THE LANDSCAPE OF THE EARTH IS AL-tered by agents of change such as water, ice, and wind. These agents of change all play major roles in scouring out and building up the different parts of the Earth's surface. The rocks and minerals of the Earth's crust are cleaved, crushed, and broken into those minute particles on which plant life and, ultimately, animal life depend: the soil of the crust. Although the effects of these forces may be modified by climate and topography, there is no place in the world that escapes their activity.

Surface Water

Water moves through an endless cycle known as the Hydrologic Cycle. Molecules of water evaporate from land and water surfaces. In addition, water is lost from plants by a process called transpiration. Water then condenses from the air, and falls back to the Earth's surface as precipitation. Some of this water filters into the ground and becomes ground water. The rest of it remains on the surface of the land. Under the influence of gravity, surface water moves downhill and finds its way into streams—brooks, creeks, rivers—and ultimately into the sea. About one third of the precipitation that falls to the Earth runs along the surface of the land under the influence of gravity. We are concerned here with this surface drainage, known as *runoff*, and the various land forms and water bodies it produces.

When water falls on the land surface as heavy rain, it first forms a sheet of water known as sheet wash. The flowing water erodes the land, carrying away particles from the softer soil and rock. It starts to cut small channels known as rills, which widen and consolidate with adjacent rills to produce gullies. Eventually, the gullies are cut deeper and wider to form the channels that contain creeks, streams, and rivers. An example of severe erosion of the land surface is shown in Figure 3.2.

Stream Types

Streams are generally classified according to the patterns in which they drain the land (Figure 3.3). Streams that are governed solely by the slope of the ground over which they flow are termed *consequent* streams. In areas where the ground slopes uniformly and the rock material is homogeneous, consequent streams will form a randomly branched pattern that is known as *dendritic* flow—meaning many branched, like a tree. In areas that are up-arched in domelike formations, the streams will flow downward from the peak in all directions, in a *radial* pattern.

Eventually, a stream erodes the ground covering the original slope, and the flow becomes adjusted to the slope of the underly-

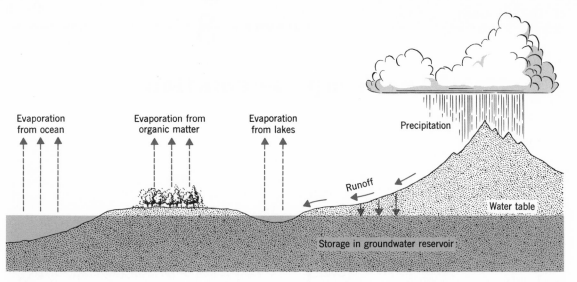

FIGURE 3.1 The elements of the hydrologic cycle. As water evaporates from the ocean, freshwater lakes, or other bodies of water, it enters the atmosphere. Precipitation onto the land returns to bodies of water as runoff. Some water filters through the soil to enter underground springs. Evaporation also occurs from the land and as waste from living organisms.

FIGURE 3.2 Death Valley shows severe gullying, the effects of unchecked erosion in a barren land. Water and wind have created a complex series of hills and gullies throughout this entire area (National Park Service).

ing rock strata in the area. These streams are termed *subsequent* streams. In areas where the rock strata are tilted in parallel formations, the stream will flow in a rectangular pattern known as a *trellis* pattern. In dome-shaped areas, where the stream flow has become adjusted to different resistances of rock, the flow is *circular*.

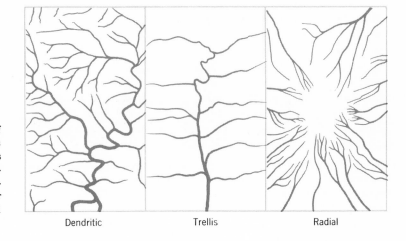

FIGURE 3.3 Different patterns of stream drainage. Dendritic streams form in flat terrain. Trellis patterns form when water flow follows parallel, tilted rock strata. Radial patterns originate on hills with water flow moving downward in all directions.

Dendritic Trellis Radial

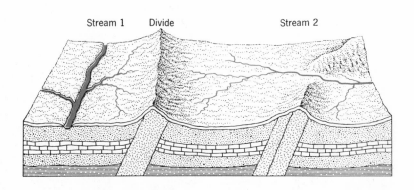

Stream 1 Divide Stream 2

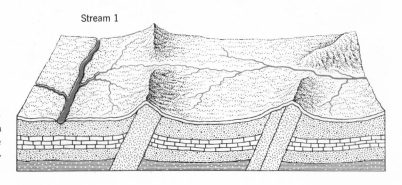

Stream 1

FIGURE 3.4 Stream piracy occurs as one stream cuts through a divide separating two streams. The water from Stream 1 is then diverted to Stream 2.

In some areas, one stream may become dominant, cutting through the bedrock more deeply and rapidly than nearby streams. If a stream cuts through a ridge separating it from a nearby stream, as shown in Figure 3.4, the water of the second stream will be diverted into the first. This process of beheading a stream's headwaters and stealing its source of supply is called *stream piracy*.

The amount of water supplied to a stream often varies, depending on local topography and climate. In humid regions, the upper surface of the groundwater (the water table) is high, and the groundwater feeds into the local streams. Streams fed by the groundwater table are known as *effluent* streams (Figure 3.5). When this condition persists throughout the year, a perennial stream flows in the area.

In dry arid regions, water from the streams quickly seeps into the ground and then works its way into the groundwater, well below the stream channels. Streams of this type are *influent* streams (Figure 3.6). Because the supply of water may vary during the year, the stream may periodically dry up. These intermittent streams are typical of arid regions that are completely dependent on precipitation to supply them with water.

Stream Cycles

There are three stages of stream development—youth, maturity, and old age (Figure 3.7). In youth, the major action of the stream is erosion—the removal, transport, and deposition of rock materials. Transport is accomplished by *traction*, the dragging of large boulders and particles along the channel bottom; by *suspension* of materials that will eventually settle out; and by *solution*, which is the process by which material is actually dissolved in the water. (Figure 3.8 illustrates this action of the stream.) As the stream approaches maturity, the channel cross section approaches the most efficient shape—a semicircular trough, which offers less resistance to flow. During the stream's last stage—old

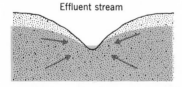

Effluent stream

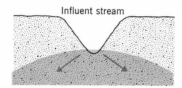

Influent stream

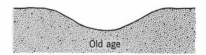

FIGURE 3.5 (*left*) Effluent streams are supplied with water by the ground in wet regions (After Zumberge).

FIGURE 3.6 (*right*) Influent streams lose water as it seeps into the ground in dry regions (After Zumberge).

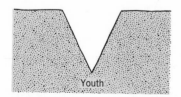

Youth

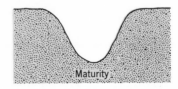

Maturity

Old age

FIGURE 3.7 Stream erosion changes the shape of the channel throughout its history. In youth, the channel cross section is V-shaped; as the stream enters maturity, the channel becomes rounded and U-shaped; and in maturity a gentle, rounded U-shaped channel is formed.

age—deposition increases and erosion diminishes as an important consequence of stream action.

Youth. Youthful streams are always marked by a steep longitudinal profile, or gradient (Figure 3.9). Gradient is the degree of descent downhill from the beginning to the end of the stream and is usually expressed in feet per mile. The major action of the youthful stream is erosion, which cuts downward into the underlying bedrock (Figure 3.10). Turbulence is especially common in the swiftly flowing water, since impediments have not

yet been removed or eroded by the water. Boulders and outcrops of rock produce rapids and a variety of swirls and eddies in the water.

Youthful streams are rather straight and narrow with V-shaped valleys and comparatively few tributaries. When the stream encounters rocks of differing resistance, a fall is produced. This phenomenon occurs because the water descends from a more resistant rock and rapidly cuts into a less resistant rock on the downstream side, producing a waterfall.

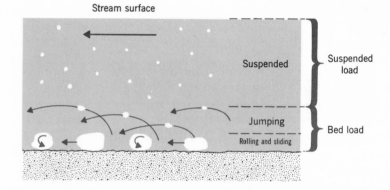

FIGURE 3.8 The stream load consists of materials in solution, in suspension and as bed load (materials dragged along close to the bottom).

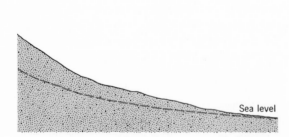

FIGURE 3.9 As a stream follows its original channel (solid line) it cuts downward toward a lower base line (dashed line). This process gradually lowers the gradient, or descent downhill.

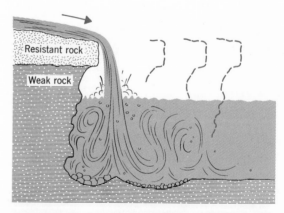

FIGURE 3.10 The erosion of weaker, underlying rocks results in backward migration of a waterfall. As undercutting progresses, the stronger rocks collapse as the weak underpinnings are worn away.

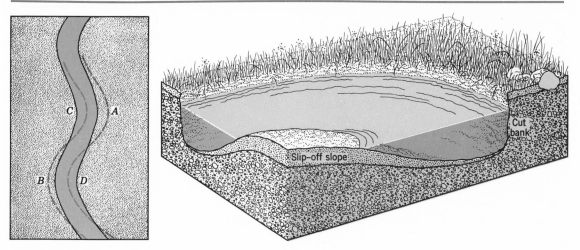

FIGURE 3.11 The development of stream curves. As the water en- counters resistant material, it is deflected to the opposite side of the channel. A slip-off slope and cut bank are formed.

Drainage of the land surrounding the youthful stream is poor, and swamps and marshes are usually located near its upper reaches. The Colorado River, which has been cutting into the Grand Canyon for at least 10,000,000 years, is an example of a youthful stream. As you can see, age has nothing to do with the classification of a river, which instead depends on the stage of development of the stream.

Maturity. In its mature stages, a stream valley is deep. Resistant materials are partially removed, and most of the erosional effects are directed sideward. The gradient is less in these stages of the stream's development, and therefore the velocity or speed of the water is less. The velocity is also related to the *discharge,* or quantity of water carried by the stream in a unit period of time. This in turn determines the size of the stream (the mean width times the mean depth of the water in the channel).

A mature stream cannot easily cut resistant material and, thus, it begins cutting from side to side around resistant materials and impediments. This action forms wide swinging curves called *meanders* (Figure 3.12). Floodplains, which are large, broad areas of deposition on either side of the stream channel, also form. In this stage, drainage is good and the tributaries that empty into the main stream are numerous. The lower Mississippi, with its wide, swinging channels, is a prime example of a mature stream.

Old Age. By the time old age occurs, the stream has nearly reached its *base level,* or the lowest level to which the land can be eroded by the stream drainage system. As a result, the velocity of the water is very slow and deposition takes place more rapidly. The meanders are exaggerated in an old stream, and some may cut through at the looped end. When this occurs, the meander curve is cut off from the main part of the stream and forms an *oxbow lake.* These lakes eventually become swampy areas on either side of the river. The general topography of the region is flattened, and broad floodplains are located on either side of the river. In addition, mounds of earth known as natural levees are formed on the banks of the stream. Such a region is often referred to as a peneplain.

(The lower Mississippi River is, for the most part, a stream in old age. Figure 3.14 diagrams some of its recent history.)

On reaching the sea, the speed of a river is suddenly checked and the suspended load dropped. The floodplains of rivers of this type are usually very fertile. These large fan-like deposits are the *deltas* that are found at the mouths of older streams such as the Mississippi and the Nile.

Streams also deposit materials in the channel through which they are flowing. These materials impede water flow and produce branches that become intertwined, creating a *braided stream.*

The levees laid down alongside the stream banks often prevent tributaries from entering the main stream in old age. The tributaries are then forced to run parallel to the main stream, often for many miles, before a junction can be achieved. Such streams are known as *Yazoo Streams,* after the Yazoo River which parallels the Mississippi River. The lower region of the Mississippi is one of the few examples of a stream in old age since, in many cases, rivers are uplifted, or raised during crustal movements causing the river to start the cycle all over again. The river is said to be *rejuvenated* in such cases.

Figure 3.17 summarizes the three stages of stream development.

Stream Energy

The water of a stream actually does not flow through the stream channel as a solid structure, but slides across itself in a series of separate flows called *laminar flow.* Because of the great friction between the water and the sides of the channel, the water in the center of a stream flows more rapidly and the water nearest the channel sides moves more slowly. (See Figure 3.18.) The fastest moving part of the stream is a point about ⅓ from the surface in the midpoint of the stream. Thus,

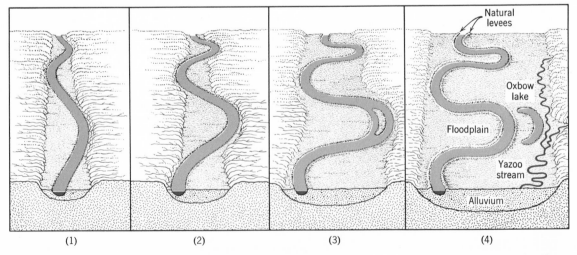

(1)   (2)   (3)   (4)

FIGURE 3.12 The formation of river meanders. As the meander is cut through at a bend, an oxbow lake is formed. In old age, Yazoo streams run alongside the main stream for many miles before entering it. During periods of flooding, deposits of material form along the main stream channel. These levees sometimes become high ridges along the river.

when a hydrologist measures the stream velocity by means of a current meter, he actually reports the mean velocity, since the flow of water is not uniform throughout the stream.

Groundwater

The second type of water is that which is found between particles of soil and rock in the crust of the Earth. This water is referred

A

B

FIGURE 3.13 (A) (*above*) As stream energy lessens, alluvial deposits like these sand bars are laid down (Alaska Pictorial Service). (B) (*left*) Broad, wide-swinging meanders appear as a river expends its energy in sideward erosion. The inside of the bend is the slip-off face, and the outer portion is the cutbank (Robert Perron).

to as the subsurface or groundwater. Just as surface water flows across land surfaces, groundwater flows through spaces in the soil and rock, dissolving materials, loosening the earth, and producing both erosional and depositional formations. Groundwater, too, flows from higher areas to lower ones under the influence of gravity. During its movement, groundwater carries on a great deal of work that is hidden beneath the surface of the Earth.

Groundwater is derived from three sources. First, volcanic activity releases *juvenile* water. This is water that is assumed not to have been part of the hydrologic cycle. Second, the water that was trapped in sedimentary rocks during their formation may be released. This *connate* water, as it is called, is usually salty because it originated in ancient seas. However, the most important source of groundwater is *meteoric* water, derived from precipitation that infiltrates the ground.

The Soil

The upper layer of the soil is known as the *zone of aeration*. This is the area of the soil that has high porosity, that is, a large number of air spaces or pores between particles. These spaces are filled partly with water and partly with air. Young plants are particularly de-

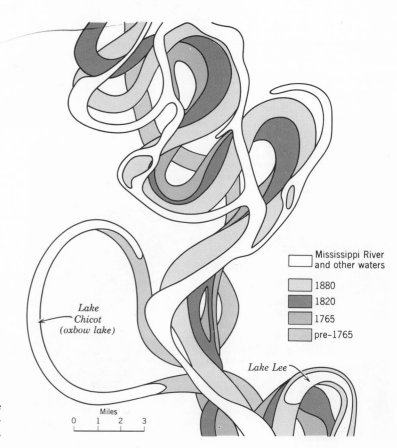

Lake
Chicot
(oxbow lake)

Lake Lee

Mississippi River
and other waters

1880
1820
1765
pre-1765

Miles
0   1   2   3

FIGURE 3.14 A diagram of the changes in the course of the Mississippi River from 1765 to 1880.

pendent on this zone for their water supply. In this zone, the portion just above the water table is the *capillary fringe*. Here, water is drawn up between the particles by capillary

action and held in place against the downward force of gravity.

As water permeates the soil, beneath the zone of aeration an area eventually develops

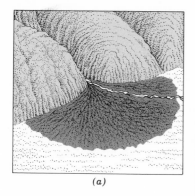

(a)

(b)   Delta

FIGURE 3.15 Here, a tidal delta has formed in Moriches Inlet, Fire Island, New York (Aero Service Corporation).

FIGURE 3.16 As a result of channel fill and varying resistance of a material, a braided stream forms. In these streams, the water flows around impediments and an intertwined pattern develops (V. C. Browne).

	Youth	Maturity	Old Age
Stream Profile	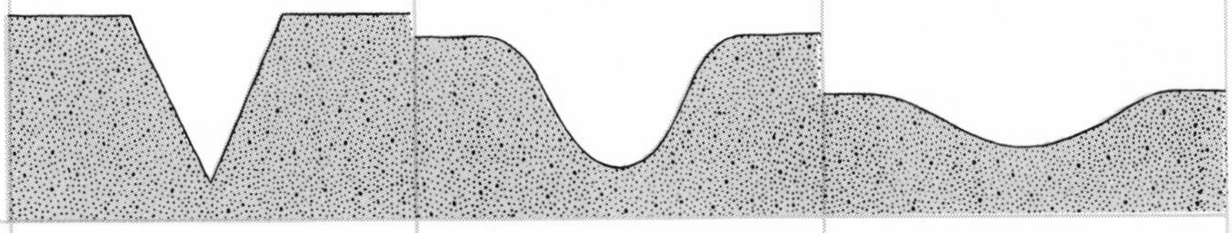		
Course of stream	Rather straight	Meandering	Very meandering
Width of stream	Narrow	Medium	Broad
Topography	Falls and rapids. Flat upland with V-valleys	Rounded; falls and rapids uncommon	Flat; no falls or rapids
Drainage	Poor; swamps and lakes in uplands	Very good; no swamps or lakes	Poor on floodplain
Vegetation	Some swamps	Often forested	Swampy
Floodplain	None	Medium	Wide. Levees and oxbow lakes
Tributaries	Few	Very many	Few

FIGURE 3.17   A summary of the three stages of stream erosion.

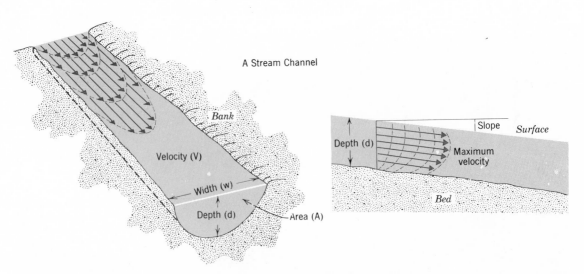

FIGURE 3.18 A typical stream flow. Because of friction between the water and the channel, the water moves fastest in the center of the stream and about ¹/₃ below the surface (After Strahler, *Intro-*duction to *Physical Geography,* Wiley, 1970).

in which all the pore spaces between soil particles are filled with water. This is the *zone of saturation* and is the zone one must reach when digging a well in order to have a steady water supply. The upper surface of this zone is called the *water table*. When a well is pumped too rapidly, or an area has too many wells in it, the water table drops rapidly and the wells run dry. The height of the water table also fluctuates with seasonal changes in precipitation. Therefore, it is important that a well be dug deep enough into the zone of saturation so that seasonal differences do not affect the source of supply.

In some areas, the zone of aeration is missing and the zone of saturation coincides with the surface of the land. This forms a swamp or a marsh, where all the soil is completely saturated. As a channel or gully is cut down into the saturated zone, an effluent stream or lake is formed.

Groundwater Action

As groundwater percolates through soil and rocks, it carries on considerable erosion and deposition. Salts and other minerals are dissolved in the water and produce "hard water." In addition, carbon dioxide dissolves easily in water and forms a weak carbonic acid. This hastens the solution of materials as the acidic water passes through underground formations. Solution is particu-

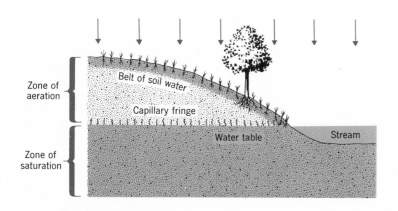

FIGURE 3.19 The location of the zone of saturation and zone of aeration. The zone of aeration includes the capillary fringe (just above the water table) and the region of soil water that is the main supply for most plants (After Zumberge).

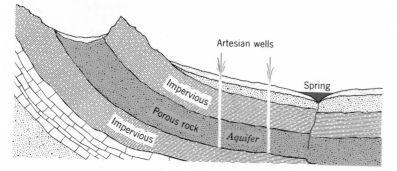

FIGURE 3.20 Artesian wells are formed when man drills through impervious rock covering an aquifer, or waterbearing rock. This water may also find its way to the surface through natural fissures and spaces between the rocks, or when the rock is eroded to the aquifer.

larly rapid in carbonate rocks, such as limestone formations, which are readily attacked by acids.

Regions of the Earth that are underlain by carbonate rocks are profoundly affected by meteoric water (Figure 3.21). These are areas of *karst topography,* so called because they resemble a region in Yugoslavia where this type of topography predominates.

In karst regions the solution of carbonate rocks produces underground caverns. The groundwater dissolves the easily removable

FIGURE 3.21 (*right*) **Karst topography occurs in areas underlain by easily dissolved limestone rock. These regions contain numerous underground caverns and springs. As portions of the surface collapse, sinkholes are formed.**

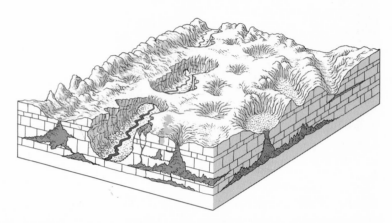

FIGURE 3.22 (*below*) **Karst topography is typical of regions underlain by groundwater. The solution of materials weakens the terrain, resulting in collapsed caverns that form sinkholes and filled-in lakes (W. H. Freeman).**

FIGURE 3.23 A sinkhole near Yucatan, Mexico. This region was sacred to the ancient Indians, as the temple in the background suggests (Mexicana Airlines).

limestone, creating large empty areas. As water drops from the cave ceilings, some dissolved carbonate is left behind as iciclelike stalactites hanging from the ceiling. As the water strikes the floor, mounds of calcium carbonate build up as stalagmites. These structures grow toward one another, and finally meet to form columns of carbonate rock.

As the caves and caverns are being enlarged and weakened in karst regions, they eventually collapse and form sinkholes, or open depressions in the land surface. These sinkholes may fill with water and form lakes throughout the region. Underground streams and rivers are also common.

Karst topography is found in Indiana, Kentucky, Tennessee, Virginia, and parts of Florida. Mammoth Cave and the Carlsbad, Howe, and Luray Caverns are some of the most famous structures in the United States produced by groundwater.

In addition to underground streams and rivers, groundwater also manifests itself at the surface, sometimes in the form of *hot springs* and *geysers*. Hot springs are believed to derive their heat from groundwater that has reached a hot body of igneous material underground. The hot water then travels to the surface through a fissure or crack in the rock material. Many health spas are built near thermal springs and hot mud accumulations of this type. The spas are built to take advantage of the supposed beneficial effects on certain types of disorders, such as arthritis and rheumatism.

The formation of a geyser is rather complex (Figure 3.25). A series of fissures interconnects near a source of heat. Water at the bottom of the geyser is heated, but is prevented from boiling off by the pressure of the water above it. However, when the hot water develops enough pressure, some of the accumulated water is forced out. The

FIGURE 3.24 Dissolved materials precipitate from groundwater to form spectacular underground cavern formations. These columns of calcium carbonate in Carlsbad Cavern are made up of stalactites and stalagmites that have joined (George Grant, National Park Service).

resulting drop in the water level now reduces the overlying pressure on the remaining water and, as a result, it becomes superheated and boils rapidly upward. Steam erupts violently out of the fissure. After eruption, water seeps back into the fissures, and the process periodically repeats itself. Old Faithful, at Yellowstone National Park, is a famous geyser of this type.

The geodes mentioned during the discussion of rocks are formed by groundwater that carries dissolved minerals. When the minerals precipitate out into cavities within the rock mass, crystals form, producing the geodes. On other occasions, precipitates of

minerals form around some nucleus of foreign material in rock masses. As layers of mineral matter build up, a formation known as a *concretion* develops.

The cements of sedimentary rocks form in somewhat the same manner as concretion rocks. Groundwater leaves deposits of minerals on grains of sediment. These compounds, such as silicas, oxides, or carbonates, settle out between grains and form the groundmass of sedimentary rock.

Among the most interesting forms created by groundwater are petrified structures. They are produced when organic materials are dissolved and minerals are deposited in their places. Buried materials such as wood, bone, shells, and even inorganic structures are sometimes replaced almost molecule by molecule by precipitates from groundwater. Eventually, the original material is completely replaced by the mineral precipitate, forming sedimentary rock. Thus are formed tree trunks which appear to be made entirely of silicas, as in the Petrified Forest in Arizona (Figure 3.26). Petrification is a slow process of replacement.

Aquifers

Rocks, like soils, have different *porosities*. Some have a high porosity and hold water very well. Others have a low porosity. In addition, the pore spaces may lack interconnections, so that the rock will have poor *permeability*. That is, water will not travel through the rock even though there are many pore spaces within it.

The fine particles of clay soils, when consolidated into a shale, do not produce a very permeable rock. The larger interparticle spaces of sandstone, and the gas pockets in some scoria, are good carriers of water. The permeable rocks are known as *aquifers* or water-bearing rocks.

Sometimes an aquifer, such as sandstone, is located between 2 nonaquifers, such as shale. Then groundwater is trapped within

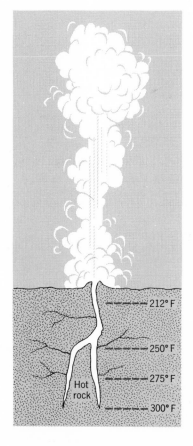

FIGURE 3.25 The geyser above is produced by super-heated rocks below ground that superheat the steam erupting from the geyser (as shown in the diagram at right) (B. M. Shaub).

212° F
250° F
275° F
300° F

Hot rock

the aquifer, and the aquifer acts as a natural pipeline. If the aquifer reaches the surface, the water flows out as a spring. In some areas, an aquifer may be found within the zone of aeration, above the water table. The aquifer may then form a *perched water table* (Figure 3.27) and may serve as an important source of water where the regional water table is particularly low, as in a desert.

Ice

Our planet still has an extensive covering of ice. Although not as extensive as in the past (Figure 3.28), great sheets of ice cover nearly 10 per cent of the Earth's surface. In the Antarctic region, ice masses on the order of 8000 feet thick cover the land area. Actively moving glaciers may account for as much transport of material as running water does, although glacial motion is much slower.

The Ice Age

It is commonly assumed that, in at least one period of the Earth's history, gigantic ice sheets covered a large part of the Earth's surface. In reality, ice sheets or glaciers are constantly recurring phenomena that have sculptured the Earth's surface again and again throughout geologic time. Glacial activity has been traced back at least 800 mil-

FIGURE 3.26 Petrified wood is the result of mineral replacement of the cellulose in the original wood fiber. The Petrified Forest in Arizona was formed from once buried trees that were acted on by silica-bearing water (National Park Service).

FIGURE 3.27 Perched water tables form in regions where impervious rock layers are situated above the water table. These rocks act as receptacles for water filtering through the soil.

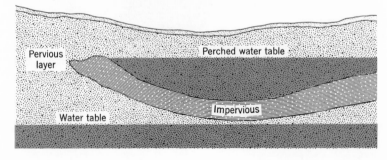

Pervious layer

Perched water table

Water table

Impervious

lion years in the Earth's past. The Earth has been partially covered by ice several times since.

We are living in a period of history in which the ice sheets are at their low point. When we use the term Ice Age, however, we are really referring to the last series of glaciers which began about two million years

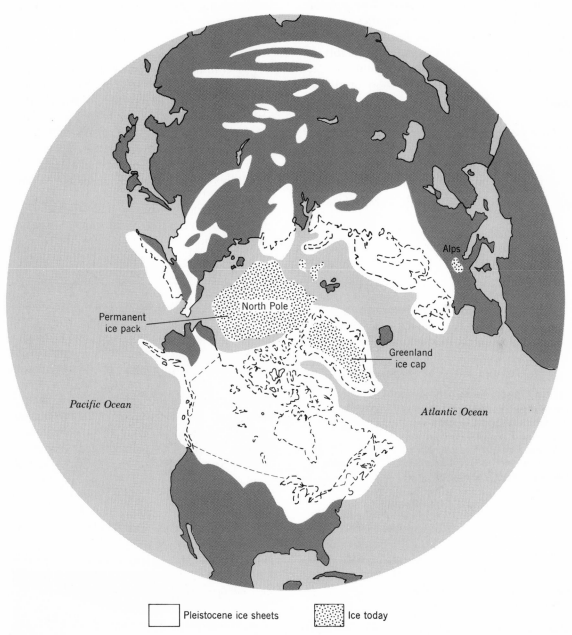

FIGURE 3.28   The extent of the Pleistocene ice sheets compared to the present-day ice sheets (After Navarra and Strahler).

ago during the Pleistocene Epoch of the Earth's history. Man's immediate ancestors had already evolved by the start of this last glacial period.

The last Ice Age consisted of four separate periods of ice advance. During these periods of time, nearly 1/3 of the Earth's surface was covered. The Nebraskan Age began with an advancing ice sheet about two million years ago and ended with the retreat of the ice about 1,750,000 years ago. A second advance began about 1,700,000 years ago and ended about 1,400,000 years ago. This was the Kansan Age. The Illinoian Age began about 750,000 years ago and ended about 350,000 years ago. The sea level during this age was lower than present by approximately 500 feet; a vast amount of water was tied up in the formation of the ice sheet. The last age, the Wisconsin Age, began about 270,000 years ago and ended about 7000 years ago. This extended the Ice Age into the era of modern man.

During this series of advances and retreats, which we call the Ice Age, many animals were forced to move southward. The Ice Age also caused the extinction of the giant mastodon, the saber-toothed tiger, and the North American camel. Great deposits of sediment were made in the upper Mississippi Valley, and Niagara Falls was created.

The Causes of Glaciers

Why glaciers form and recede is a question that no one has ever answered satisfactorily. Many scientists think that fluctuations in the Earth's temperature are the immediate cause, but no one knows why such fluctuations should occur. Many theories are current about the causes of glaciers; a number of them suggest explanations of the temperature changes as well. In general, most theories agree that there must be a worldwide drop in temperature; an increase in Arctic precipitation must occur, and a land mass must be present to receive precipitation from ocean winds. The Arctic regions satisfy the requirements once the temperature drop occurs. Some of these theories are as follows.

1. Variations in Ocean Circulation. Water circulates over a narrow rock sill between the Atlantic Ocean and the colder Arctic Ocean. As warm Atlantic Ocean water enters the Arctic and melts the ice cover, evaporation from the surface of the Arctic increases and precipitation in the form of snow increases over the surrounding land masses. The snow builds a broad ice sheet on the land and the glacier slowly moves southward. However, the precipitation that increases the size of the glaciers is water removed from the world's oceans, and the continued growth of the glacier causes a worldwide drop in sea level. The rock sill between the Arctic and Atlantic now prevents the warm Atlantic water from entering the Arctic and the Arctic freezes over. The frozen ocean can no longer supply great amounts of moisture to the air, precipitation decreases, and the ice sheet begins to recede. The melting ice enters streams and rivers and finds its way back to the oceans, causing a rise in sea level and the beginning of a new cycle. This rather recent theory has been proposed by Maurice Ewing and William Donn.

2. Carbon Dioxide Content in the Air. If the carbon dioxide in the air, which helps insulate the Earth and retain heat, were to decrease suddenly, temperatures would drop. This temperature drop would cause an increase in the Arctic snowfall. This increase in the accumulation of snow would cause the glacier to build and advance.

3. Fluctuations in Solar Radiation. According to this theory, the sun's energy fluctuates, causing periodic drops in Earth's atmospheric temperatures. This would lead to increased snowfall and glacial buildup.

4. Variations in the Earth's Orbit. Still another theory holds that a worldwide temperature decrease might be due to variations in the Earth's orbit around the sun. This would result in less solar energy being received by the Earth, and the chain of events already outlined would begin.

These are a few of the theories that have been developed to explain periods of glaciation in the Earth's past. Although most have some logical basis in fact, there are major objections to all of them. The growth of

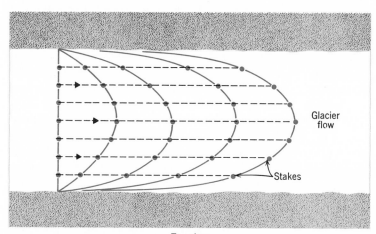

Top view

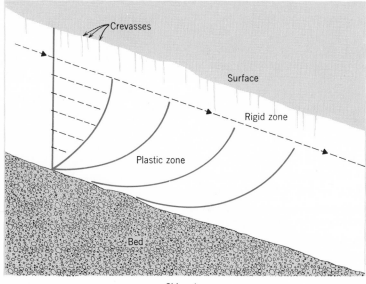

Side view

FIGURE 3.29 Laminar flow of glacial ice. Evidence from stakes placed equidistantly across the surface of glaciers shows that glacial flow is laminar in nature. Cross-section analysis shows that movement is most rapid just below the rigid zone (After Navarra and Strahler).

glaciers remains a mystery to glaciologists, and new theories are constantly being developed.

Glacial Movements and Formations

A glacier begins as an extensive snowfield where more snow accumulates throughout the year than melts or evaporates. This accumulation takes place at altitudes above the permanent snow line, since below the snow line snow melts completely during the warm season. As the snow accumulates, it is compressed by the weight of additional snow on top of it and becomes *néve,* a granular ice also called *firn.* The firn is denser than the surface snow and eventually recrystallizes into solid ice. (This process is similar to what occurs when snow is packed into a snowball.) As the snow increases in thickness, it begins to move under the influence of gravity. In mountains, the portion of the glacier above the snow line is called the zone of accumulation, and the portion below the snow line the zone of wastage.

Depending on terrain and snowfall accumulation, glaciers move at speeds anywhere from a few inches to well over 100 feet per day. It has been found that glaciers do not move uniformly throughout (Figure 3.29). They move most rapidly near the center and slowest near the edges, a fact that has been determined by planting a straight row of stakes across the glacier and by observing

FIGURE 3.30 Glacial till—materials carried along by the glacier—is a loose conglomeration of soil, pebbles, and boulders (U.S. Geological Survey).

FIGURE 3.31 Glacial markings are often left on rocks over which the ice has moved. These chatter marks on the lower rocks show the slow movement of the ice sheet that once covered much of the region (Charles Phelps Cushing).

their movement over a long period. The upper, rigid surface does not move smoothly, but is fractured and broken into deep crevasses. Below the surface, the tremendous pressures exerted on the layers of ice cause slippages. This results in the most rapid movement near the center of the glacial ice sheet.

As the glacier builds in size and advances along its path, it picks up whatever rock, soil, or other material is in its way. This debris, or *glacial till,* is dragged along beneath or within the ice, abrading bedrock and plucking up other materials that are dragged along with the glacier. The till eventually is deposited as the glacier recedes. Striations or scratches left on the bedrock are indications of glacial movement over the bedrock.

Valley glaciers form their typical U-shaped valleys by plucking and abrading the valley sides. As glacial ice moves over cliffs, it carves out hanging valleys located far above the main valley floor. At the origin of the snowfield, snow builds up in a bowl-shaped *cirque* or depression. The ice may be separated from a nearby cliff face by a deep crevasse known as a *bergschrund.* They are rather dangerous areas for explorers and mountain climbers, since the crevasses are often concealed by a light snow.

As the glacier advances, it deposits till along its boundaries on either side. The resulting mounds, called lateral *moraines,* are used by geologists to determine how far the vanished glacier extended. The end or terminal point of the glacier is marked by a mound of till known as a terminal moraine. These moraines generally curve in a downhill direction. When the ice retreats, a series of end moraines build up, one behind the other. These *recessional moraines*, as they are called, denote the withdrawal of the glacial ice. When two glaciers meet, they leave a deposit at the point of junction known as a *medial moraine*. Figure 3.34 diagrams the development of a typical glacier.

As a glacier cuts through a valley, the cliff walls are worn away. Steep tooth-shaped peaks called *horns* are left standing above the ice. Sharp ridges called *arêtes* are carved out

FIGURE 3.32 Valley glaciers move through a confined region between the sides of mountains. The resulting terrain is much more rugged than that left by an ice sheet (Spence Air Photos).

FIGURE 3.33 A valley eroded into a nearly perfect U-shape by glacial advance (F. E. Matthes, U.S. Geological Survey).

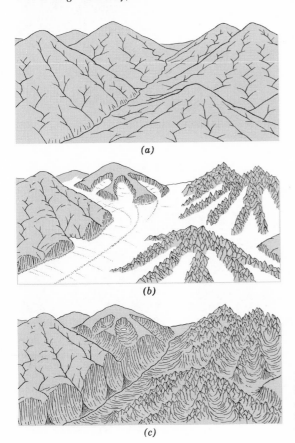

(a)

(b)

(c)

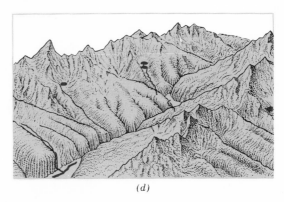

(d)

FIGURE 3.34 The development of a typical glacial area, including development of moraines. In a typical erosion cycle due to glacier formation, the terrain is eroded to become more rugged and pointed in appearance. The various formations typical of such an area are in evidence.

of the mountain sides. The arêtes denote lines of division between two or more glaciers.

When glaciers reach the sea, large blocks of ice break off in the water. These are the ice-bergs, which have only ⅑ of their masses exposed above the surface of the water, and are threats to shipping. Ships may be sunk when they ram the hidden portions of the icebergs.

FIGURE 3.35 (*right*) As valley glaciers descend from higher terrain to the foot of mountain masses, a piedmont glacier forms, fanning outward at the mouth of the valley (Steve & Dolores McCutcheon, Alaska Pictorial Service).

FIGURE 3.36 (*below*) The Great Ross Ice Sheet at the edge of open water. This photograph was taken in Little America, Antarctica (U.S. Coast Guard).

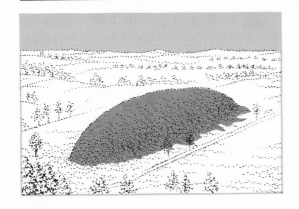

FIGURE 3.37 Drumlins are elongated hills formed by the progress of the ice over a heavily glaciated terrain.

Valley glacier activity tends to form sharp, peaked terrain. Deposits of till forming moraines are left to mark the boundaries of the glacier. The rocks in the area are rounded and smoothed—like knobs.

Continental glaciers, which are the largest of all ice sheets, are not confined to valleys and, thus, erode bedrock to a greater extent. Whereas valley glaciers tend to deepen the valleys between mountain ridges, continental glaciers remove almost all high relief in their paths. As they recede, they leave relatively flat terrain behind them.

Whatever the type of glacier, streams may flow out from under the ice sheet carrying sediments. When the sediments are deposited beyond the end of the glacier, they form *outwash plains.* As these streams flow under the ice, they also leave deposits called *eskers.* These ridges of material are built up

in tunnels carved through the ice by flowing water. Large *drumlins* are left behind as elongated hills. Other features found under the ice are pits known as *kettles,* and smaller mounds of earth called *kames.* Some bedrock is smoothed into round hills known as *roches moutonnées,* or "sheep rocks."

Another interesting glacial deposit is the glacial *varve.* These varves are fine sediments laid down in lakes in a series of layers, with thick layers of coarse particles alternating with thin layers of fine particles. It is assumed that the thicker layers were laid down in the summer during rapid melting of ice, and that the thin layers were laid down during the winter when little melting occurred. Estimates of glacial ages and activity are made by analysis of the number and the thickness of the varves.

Some of the rather marked changes

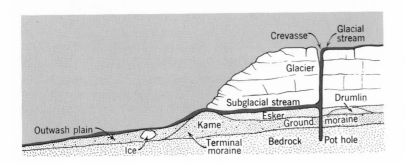

FIGURE 3.38 A typical glaciated landscape is marked by several types of erosional and depositional features.

brought about by glaciers are familiar to us all. The Great Lakes were carved out in the last ice age, as were many of the smaller lakes found in the same region of the United States (Figure 3.41). Long Island, New York, is for the most part glacial moraines. The famous Matterhorn was formed by valley glaciers. Glacial ridges and valleys are found in

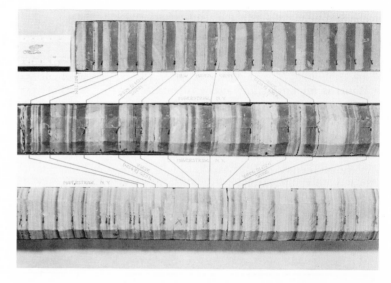

FIGURE 3.39 Varved glacial clay from Haverstraw, New York, shows alternate layering representing seasonal changes that affected the glacial ice. Each alternate dark and light band represents a deposit of one year (American Museum of Natural History).

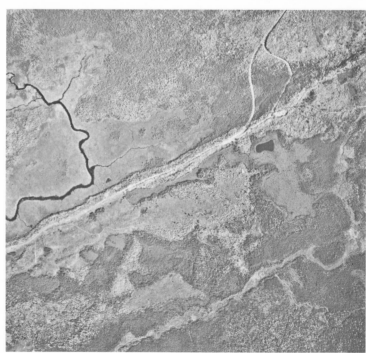

FIGURE 3.40 Eskers are longitudinal mounds of material deposited by steams flowing under the glacial ice (James W. Sewall Company).

Yosemite Valley in California, in Utah, and in Grand Teton Park, Wyoming. Glacial activity, past and present, is still obvious in almost all parts of the world, even the equatorial regions. Alaska, Greenland, Canada, Northern Europe, and Antarctic regions abound with evidence of glacial activity.

Winds and Its Effects

Wind is an important agent of landscape formation, although the effects of wind are less than those of the various forms of water. Where water is not present to any great extent, as in arid climates, the effects of wind are quite pronounced. Winds are present everywhere, and are bound to affect all parts of the globe.

The Work of the Wind

Wind accomplishes its work in two major ways, deflation and abrasion. Both are more pronounced where the Earth is dry and barren.

Deflation occurs when particles are picked up by the wind and carried for great distances. Huge dust storms, which were common in the American Midwest Dust Bowl in the 1930's, are examples of deflation on a large scale (Figures 3.42 and 3.43). Millions of tons of earth are moved in this manner. Coarser, heavier particles are also bounced along the surface by a process known as saltation (Figure 3.44). These are processes of removal. Depressions known as blowouts are also found in areas affected by

FIGURE 3.41 The Great Lakes were gouged out by glacial activity during the last ice age (Litton Industries-Aero Service Division).

deflation. Deflation occurs most readily in an area where particles are loosely packed and where there are no barriers to the wind.

The second major action of the wind is abrasion. During abrasion, hard tough grains of sand—which are usually quartz—and other particles carried by the wind are driven against stationary objects. The particles wear away the object much as sandpaper would. (See Figure 3.45.) In fact,

sandblasting is an artificial technique similar to natural abrasion. In sandblasting, a fast-moving stream of air drives sand particles against a stone building to wear away a thin layer of the stone. This is the most common method by which stone is cleaned of stains or defacing marks. Wind has the same effect on stone, although it is slower than man's sandblasting techniques.

Some rocks found in regions attacked by

FIGURE 3.42 A grove and buildings buried by drifting soil in South Dakota in 1935 (ESSA).

FIGURE 3.43 A wall of black dust being wind-driven in the midwest (Philip Gendreau).

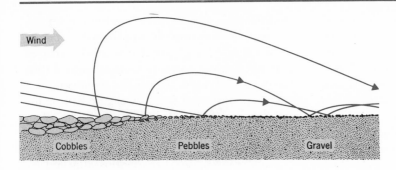

FIGURE 3.44 Saltation: the jumping of rock particles of various sizes in a current of air or water (After Zumberge).

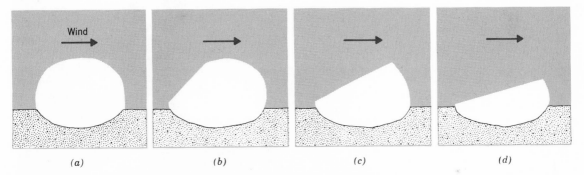

FIGURE 3.45 Ventifacts are typical of the effects of wind. The facets are gradually worn flat by abrasion.

FIGURE 3.46 Arches National Monument, Utah, is a natural bridge resulting from differential abrasion of rock strata (National Park Service).

FIGURE 3.47 Goblet rock in Natural Bridges National Monument, Utah, is a superb example of a pedestal of rock worn more rapidly at the base than near the top (National Park Service).

wind have facets, or flat faces, worn onto them. Rocks of this type are known as *ventifacts.* In some regions natural bridges are formed where wind wears away a cliff at the base, but not at the top. Rocks known as pedestals are common. These are large outcrops of rock which have been worn away more rapidly at the base than at the top, so that the resulting rock has a top larger than its supporting base.

Desert Formations

Most major deserts of the world are found between 35°N and 35°S latitudes. These areas generally have less than ten inches of yearly rainfall and a high rate of evaporation. Although deserts are thought to be rather barren, an abundance of highly specialized life is present. However, animal life is not as obvious during the day as at night.

The desert surface consists of rock and sand, and water is not a constant agent of change. However, alluvium can be found where mountain streams enter the desert. These streams form short-lived lakes called *playas.* The lakes leave salt beds behind them after evaporation takes place.

Large desert regions are found on nearly every continent. Africa claims the Kalahari Desert in the south and the Sahara and the Arabian Deserts in the north. South America has the Atacama and Australia, the Victorian Desert. North America claims the Sonoran Desert in the American Southwest as well as the Mojave Desert in California.

Although winds are active everywhere on the globe, their effects are most notable on deserts. In deserts, deposits of sand known as dunes are the best example of wind deposits and the transitory formations found in arid regions.

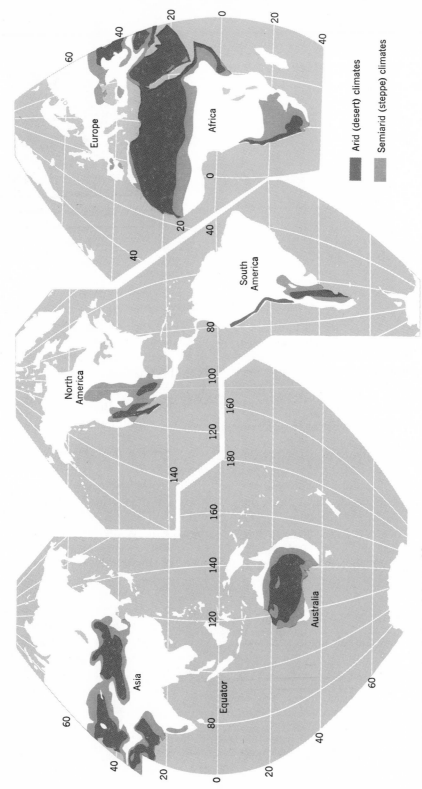

FIGURE 3.48 The major deserts of the world.

Arid (desert) climates

Semiarid (steppe) climates

FIGURE 3.49 Gila Monsters are inhabitants of some desert regions (Isabelle Corant, National Audubon Society).

Sand dunes result from the process of deposition by wind. A sand dune is a mound or hill of relatively loose sand that rises to a summit. It is independent of any fixed structure or object. Dunes range in size from a few feet to hundreds of feet in height, and may be several miles long. The dunes move in the direction of wind movement.

A sand dune consists of one gently sloping side that faces the prevailing winds called the windward side, and a steep leeward side, or slip-off face. As sand particles are blown over the crest and fall on the leeward side, the dune slowly creeps in the direction of the wind.

There are several types of sand dunes. Transverse dunes are long ridges of sand running perpendicular to the wind. When several stand together, they appear to be waves in the sand, and give it an oceanlike appearance. Longitudinal dunes are ridges of sand running parallel to the wind direction. Because the winds that produce them are strong, some of these dunes are over 60 miles long and 100 feet high. Smaller, crescent-shaped dunes also occur, with the points of the crescents facing the direction the winds are blowing. These are known as *barchans.*

Many other types of dunes are found in the deserts of the world. They are much less common than the ones already listed. One of them is the *parabolic* dune, which is oriented so that its points face into the wind's direction. When winds are not severe, dunes may become covered with vegetation. These

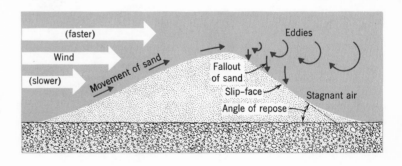

FIGURE 3.50 The movement of a sand dune occurs as sand is moved up the windward side and falls down the slip-off face (After Longwell and Flint).

dunes are less transitory, since the vegetation further slows the winds and the sand is not picked up and carried off very easily.

In areas where winds move from water onto a beach, long ridges called foredunes develop. These dunes run perpendicular to the winds and parallel to the beach. Dunes may be modified, and the various types be-

FIGURE 3.51 A longitudinal series of dunes in Saudi Arabia (Arabian American Oil Company).

FIGURE 3.52 Barchan sand dunes in Monument Valley, Arizona. Wind direction is from the right (Frank Jensen).

come mixed. Some areas have a complex series of dunes of many different types.

Sand drifts also appear in arid areas when sand piles up around an object or impediment and gradually forms a mound. These drifts are dependent on the impediment for their formation, whereas sand dunes are not. People living in the Dust Bowl in the 1930's found that sand drifts nearly buried homes and other large structures. Farmland was

FIGURE 3.53 Vegetation produces a stabilizing effect on sand dunes like these dunes on Chama River, New Mexico (John V. Young).

FIGURE 3.54 Loess deposits result from wind-borne silt dropped at a great distance from the point of origin. This material is easily eroded by water (E. W. Shaw, U.S. Geological Survey).

rapidly destroyed by the action of the winds. Extensive damage was done to what was once a great agricultural area.

Loess

Loess (lō'ĕs) is a form of depositional material carried by winds from desert regions and from river floodplain deposits. It consists of fine silt carried great distances from its source. Great deposits of loess are found ranging from the Mississippi Floodplain to the Far West. Deposits hundreds of feet thick are found in China. They are thought to have been carried from the Asian interior.

Soil

Soil is only a minute fraction of the Earth's crust, yet all life is dependent on it. Soil is a complex arrangement of minerals that depends on several factors for its formation. The parent material usually is bedrock, which is broken up and decomposed by physical and chemical weathering to give rise to the soil. The climate and topography of the land control the type of weathering processes that affect the bedrock, and the rate at which it is worn away. Biological activity determines the amount of organic litter or humus added to the mineral content of the soil. Finally, the amount of time the materials in the soil have been worked on by the forces of nature determines the extent of the soil covering in an area.

Pedology, the science of soil study, recognizes several layers or horizons in soil. The true soil, or *solum,* is classified into horizons. The solum rests on the subsoil.

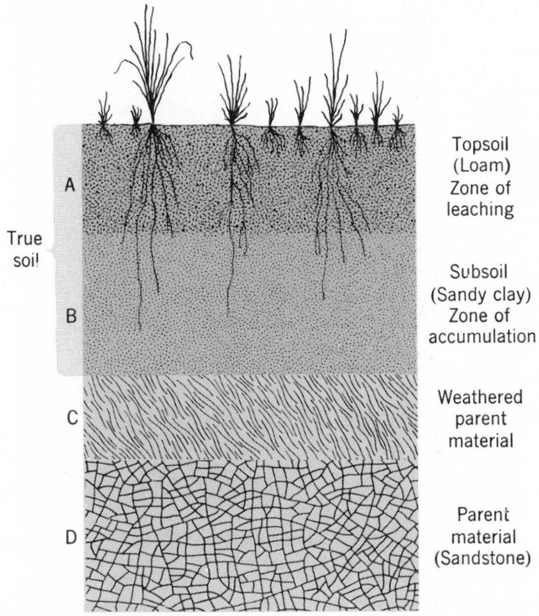

True soil

A — Topsoil (Loam) Zone of leaching

B — Subsoil (Sandy clay) Zone of accumulation

C — Weathered parent material

D — Parent material (Sandstone)

FIGURE 3.55 A cross section of an ideal soil profile showing the various soil horizons that are typical of a grassland region. The A horizon is the zone of leaching; horizon B represents the zone of accumulation for the leached materials.

FIGURE 3.56 A typical soil profile showing the four horizons. Much organic material can be observed in the A horizon (1) (Soil Conservation Service).

The Soil Horizons

There are, in general, four soil horizons. The *A horizon* is commonly called the topsoil. It is here that organic material (humus) is found. As rain filters through this horizon, soluble materials are carried downward through the soil. This is a process known as leaching.

The *B horizon* consists of the materials leached out of the A horizon plus compacted clays. Together with the A horizon, the B horizon composes the solum or true soil.

The *C horizon* consists of softer bedrock. It is not fertile as are A and B, but gives rise to the first two layers.

The *D horizon,* or bedrock, is the solid, unweathered material underlying the soil. It too, plays a role in the formation of the soil, but only A and B are important from an agricultural viewpoint. The C and D horizons compose the subsoil.

Soil Classification

Soil is classified in several ways. The organic and mineral content determines the fertility of the soil, its texture, and its color. As humus, or the organic content, becomes more and more dominant, the color ranges from brown to black and fertility increases. Iron oxides and iron and aluminum hydroxides produce red and yellow soils. Clay soils generally have a red color.

Sandy soils, which contain little clay or fine

A

B

C

FIGURE 3.57   Examples of clay soil (A), loam (B), and sand (C). Topsoil (B) is excellent for the propagation of plant roots, as can be observed from the photo. The other two soils have little organic material, closely packed grains, and poor plant growth (Soil Conservation System).

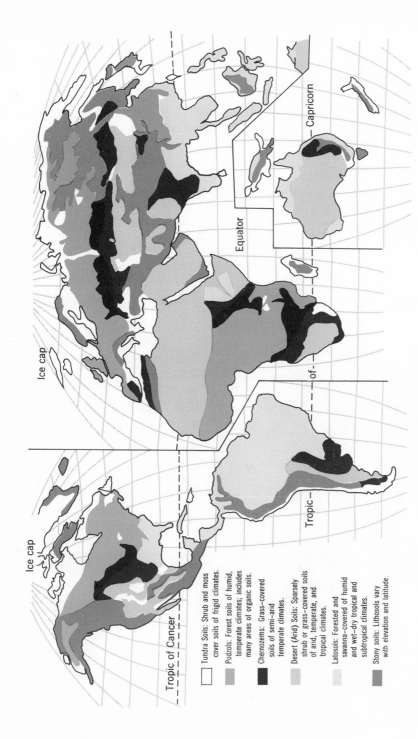

FIGURE 3.58 World distribution of the principal zonal soil groups. Latosols and lithosols are not as important as the soil types mentioned in the text. Important areas of organic soils and other intrazonals are omitted as well as very important bodies of alluvial soils along great rivers such as the Mississippi, Amazon, Nile, Niger, Ganges, Yangtze, and Yellow (adapted from the U.S. Department of Agriculture).

Ice cap

Ice cap

Tropic of Cancer

Equator

Capricorn

Tropic of

Tropic

Tundra Soils: Shrub and moss cover soils of frigid climates.

Podzols: Forest soils of humid, temperate climates; includes many areas of organic soils.

Chernozems: Grass–covered soils of semi–arid temperate climates.

Desert (Arid) Soils: Sparsely shrub or grass-covered soils of arid, temperate, and tropical climates.

Latosols: Forested and savanna–covered of humid and wet–dry tropical and subtropical climates.

Stony soils: Lithosols vary with elevation and latitude.

sediments, are coarse and do not hold water very well. *Loam* is a mixture of sand and clay in varying proportions and has an intermediate texture between that of sand and that of clay. *Clay* is a fine particled, dense material, through which water does not filter very easily. The loams are best for general agricultural purposes.

A

B

C

FIGURE 3.59 Soil Regions. (A) A typical desert region in Arizona (Moos-Hake Greenberg). (B) A beech forest from the podzolic region in the northeastern United States (P. Berger, National Audubon Society). (C) Chernozem soils are typical of the midwest United States (Joseph Muench). (D) Bog soils occur when the water table is near the surface and vegetation forms a compact mass (U.S. Department of Agriculture). (E) Tundra found in Alaska and other Arctic regions is a soil with poorly decayed vegetation (Alaska Pictorial Service).

D

E

Mineral content, texture, and color are the factors used in setting up standards for soil classification. Most of the major soil types are associated with specific climatic areas of the world. There are more than 100 types of soil; but most of the important soils of the United States can be listed by a few different types.

Desert soils are mostly silica and lack humus. They contain lighter colored, coarse sand grains. Many desert soils have a high salt or alkali content and do not support life even if water is available. The alkali soil and salt flats of the southwest are typical examples. However, with irrigation, some desert soils do produce well.

Podzols are associated with moist areas. There is some humus in podzol soils, and they range in color from ash gray to brown. Great leaching takes place in podzols and the B horizon is a dense, heavy clay. Because of its acid condition, podzol soil is well suited for evergreens and other coniferous forms. Extensive podzol regions are found in the northeastern United States and in the Great Lakes area.

Chernozems are found in temperate climates. These soils are rich in humus and are very fertile. The A horizon is deep, and there is little leached material in the B horizon. The color of chernozems varies from brown to black. These soils are particularly suited for grain crops and are common in the Midwest United States.

In addition, there are dozens of other specialized soil types. Some of them are widespread and others are the result of very localized conditions.

For example, *bog soils* develop where the water table is at the surface. Vegetation forms a compact mass and decays slowly, creating peat.

Laterite soils are found in very hot and humid climates. Heavy leaching removes silica and leaves iron and aluminum in the form of oxides and hydroxides so that the laterites are yellow to dark red in color and contain little humus. The name laterite comes from the Latin word for brick and refers to the fact that these soils can become very hard when baked in the sun. In parts of Asia, laterite soil is cut into blocks, dried, and used for buildings.

Tundra soil, found in moist polar regions, is a poorly developed soil consisting of several particle sizes. The alternate freezing and thawing that the area undergoes throughout the year creates a soil with poorly decayed vegetation. This climate forms a peatlike mass of vegetation within the soil.

Soils and their changes are important not only to the farmer but to the miner-geologist. They yield many natural resources. For example, potash and gypsum are important as fertilizers. Beds of these minerals found in soils are, therefore, of great economic importance. In other areas, certain portions of the less resistant soil are worn away leaving the more resistant iron and aluminum oxides behind. These deposits are actually enriched by weathering, since worthless material has been removed. Complex chemical reactions may also take place, enriching the soil in metallic salts. Thus, soils are changed by weathering processes as well as being the result of weathering.

CHAPTER 4

The Dynamics of the Earth's Crust

Movement of the Earth's Crust

YOU DO NOT HAVE TO BE A SCIENTIST TO realize that the Earth's crust is a dynamic structure. From its beginning, it has undergone many physical changes. Many evolutionary processes are constantly going on to build up the land in one place and tear it down in another.

Four major processes of crustal formation will concern us in this chapter: (1) the crustal movements called diastrophism, which include the folding and fracturing of the Earth's crust, (2) magma activity, or vulcanism, (3) weathering processes, and (4) erosion and deposition.

Diatrophism: Folding of the Crust

The rocks of the Earth's mantle and lower crust are plastic; that is, they are capable of flowing under pressure. They move away from areas of increased pressure, causing the crust above to subside and raise or uplift the area toward which the plastic rock is flowing. Great horizontal and confining pressures may cause giant folds in the rock strata of the Earth's crust, just as horizontal pressure on several layers of clay will produce folds. If you push in on either side, the clay buckles in the center, forming a large fold or upwarp. When sediments on the Earth's surface are carried down into the lower portions of folds such as these, isostatic adjustments cause a lowering of the valley and a raising of the upturned portion of the fold.

The high portion of a fold in the Earth's crust is termed the *anticline,* and the lower portions on either side are the *synclines.* As the fold becomes more distorted, the sides of the fold assume different positions with respect to one another. If the two sides of the fold are tilted in opposite directions at the same angles, the fold is *symmetrical;* if the different parts of the fold are tilted at different angles, the fold is *asymmetrical.* Folds may also be overturned so that they overlap one another, forming *overturned* folds.

As parts of the folded strata are subsequently eroded away, the geological history of the area becomes increasingly difficult to recreate.

To aid them in deciphering the past,

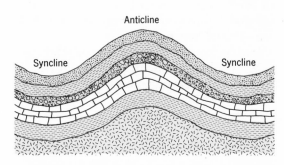

FIGURE 4.1 Lateral pressure on rock layers produces folded strata. The higher region of the fold is known as the anticline, the lower sides the syncline.

geologists have developed a system of measuring the distortion of strata relative to their original horizontal position. Thus, on geologic maps, a system of *strike* and *dip* symbols record the attitude of the rock strata.

Strike indicates the bearing or azimuth of a line of contact of the folded strata with a horizontal reference plane. It is recorded as a specific compass direction. Dip is the angle the slope of the strata makes with the hori-

FIGURE 4.2 Symmetrical folds are the ones that have a vertical axis and in which the attitudes of the strata on either side are similar. Overturned folds occur as the fold is inclined more steeply on one side than on the other.

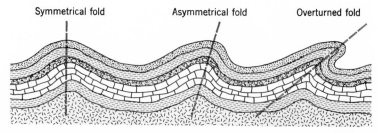

Symmetrical fold   Asymmetrical fold   Overturned fold

FIGURE 4.3 The higher portions of rock folds are termed anticlines. This anticline occurs near Kingston, New York (B. M. Shaub).

zontal plane. Thus, a figure such as 37°NW indicates that the rock stratum is tilted at an angle of 37° downward from the horizontal, and points in a northwest direction. (See Figure 4.5.)

The process of folding takes place on both large and small scales. When a small upwarp of the crust occurs, each end of the warp plunges into the Earth's surface at opposite sides, forming a small, elongated dome. A

FIGURE 4.4 (*above*) The lower arms of the rock fold are termed the syncline. This synclinal fold is in Maryland (C. D. Walcott, U.S. Geological Survey).

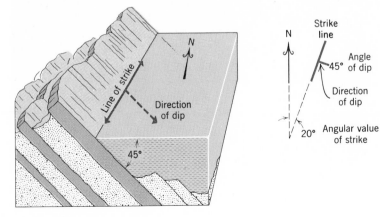

FIGURE 4.5 (*left*) The relationship of strike and dip symbols in rock beds. The angle of dip is in relation to the horizontal. The strike direction is at right angles to the dip.

downfold of this type will form a basin. Large-scale folding forms huge mountain ranges. The Jura and Alps mountains in Europe as well as the Appalachian and Rocky Mountains in this country were created in this way. They are thousands of feet above sea level. Enormous fold mountains, such as these, began when great thicknesses of sediment were deposited as horizontal rock beds. Subsidence occurred, forming a deep trough known as a *geosyncline*. Intense folding followed, and the rock strata were deformed. As compression, shortening of the crust, and faulting occurred, a portion of the geosyncline was uplifted. Soon after, erosion began to sculpture the mountain range.

Adjacent to the folded regions, sediment-filled troughs remain. The Gulf of Mexico and the Mediterranean Sea are two geosynclines which are still receiving much sediment even today. Most major mountains of the Earth were formed from the uplifting of geosynclines, resulting in deeprooted mountain ranges.

Diastrophism: Faulting of the Earth's Crust

Crustal rocks do not always yield to pressure by folding. Sometimes they fracture, and the masses of rock on either side of the fracture move upward or sideways with re-

(a)

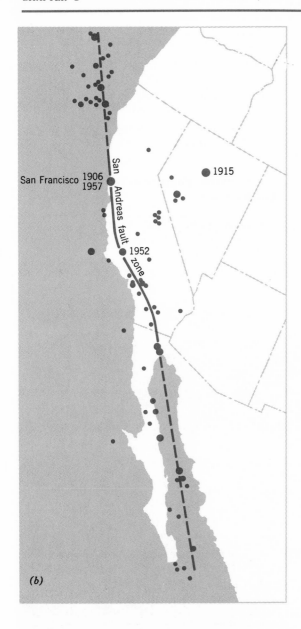

San Francisco 1906 1957

San Andreas fault zone

1915

1952

(b)

FIGURE 4.6 **(A)** *(left)* **The San Andreas fault is a strike-slip or lateral fault. The horizontal movement of rock on either side of the fault is evident from the offset of the rock masses (Spence Air Photos). (B)** *(above)* **In this diagram of the San Adreas fault, earthquakes are shown as black dots (After Zumberge).**

spect to one another. The slipping movement of the rock masses on either side of the fracture, or *fault,* may displace rock masses from a fraction of an inch to more than 100 miles. This movement of two rock masses past one another, if sufficiently rapid, may result in an *earthquake.* Earthquake-producing movements sometimes form mountains; the Sierra Nevadas, part of the Appalachians, and parts of the Rockies were raised by faulting processes. These mountains are termed *fault-block mountains.*

As the rock masses on either side of the fault move, the pressures that caused the movement force the two parts of the original rock mass to assume different attitudes depending on the direction of the movement. The motion of the rock masses may be either vertical or horizontal with respect to one another. The two rock masses produced by faulting movements are respectively termed: the *hanging wall,* which is the wall that forms on the overhanging surface, and the *foot wall,* which forms on the underside below the fault surface.

Faults are of four distinct types (Figure 4.8). A *normal fault* develops when the hanging wall moves downward with respect to the foot wall. A *reverse fault* or *thrust fault* occurs when the hanging wall moves upward and is thrust over the foot wall. This type of fault may form a cliff face that is above the surface of the downthrown block. In some cases, the hanging wall may actually slide over and settle on top of the foot wall to form an *overthrust fault.*

The fourth type of fault is one in which the movement is primarily horizontal instead of vertical. Here the two masses move sideways with respect to one another. This is a *strike-slip* or *lateral fault.* The San Andreas Fault in California is the best known example of a strike-slip fault. Here movement

FIGURE 4.7 A fault scarp, or cliff face resulting from a fault line. The scarp delineates the downthrown region from the higher standing formation. Other faults can also be seen in the photo (Sigurdur Thorarinsson).

along the fault line has involved more than 300 miles of its length as these great masses of rock slipped past one another.

Faulting usually occurs in a complex series of events in which several factors are involved. Two normal faults result in a large block being dropped down between the fault planes, resulting in a deep *rift valley*. The downthrown block, which may be several hundred miles long, is called a *graben*. In other situations, the block may be raised to form a *horst*, an uplifted area between reverse faults. This uplifted mass of rock will eventually become eroded, forming a complex mountain chain. Fault-block mountains are found in southeastern California, southern Oregon, and the Southwestern United States. The New Jersey lowlands and the Connecticut Valley are representative of grabens filled with sediment from uplifted ridges or fault blocks during the Triassic.

Earthquakes

Earthquakes are one of the most spectacular and disastrous natural processes known to man. They take place suddenly and cover vast regions and their effects can be observed on every continent and in all the ocean basins. At present, there is no foretelling

96

with any accuracy where or when earthquakes will occur. However, modern seismologists have constructed maps of the epicenters, or foci of specific earthquakes, that show regions of unusually great earthquake activity. These epicenters, when plotted accurately, overlap and show that there are large areas of the Earth that are underlain by stressed rock. With this information, some tentative predictions can be made about the occurrence and intensity of earthquakes. The evidence is slowly accumulating, and the possible locations of future earthquakes are becoming more predictable. Unfortunately, they cannot yet be predicted with sufficient accuracy for earthquake warnings to be given like storm warnings.

Earthquakes take place at various depths in the crust and mantle. Shallow earthquakes occur within the first few miles of the Earth's surface. Intermediate earthquakes occur at depths of 40 to 200 miles, and deep-focus earthquakes take place as deep as 400 miles below the surface.

The science of seismology, the study of earthquakes, has enabled scientists to collect a great deal of information about the interior of the Earth. As we learned in Chapter 1, our knowledge about the core and mantle has been primarily derived through analysis of the data from earthquake waves.

Earthquake Zones

The area of the Earth where earthquake activity occurs most frequently is the margin of the Pacific Ocean. The island arcs of Indonesia, Japan, southern Alaska, and the west coasts of North and South America have been the sites of spectacular earthquakes in modern times. India, the Mediterranean area of North Africa, Greece, and Turkey are also regions of greater than normal earthquake activity. They are all active, young areas of the Earth, and are underlain by great faults and fractures. As Figure 4.10 indicates, volcanic activity is closely associated with earthquake zones.

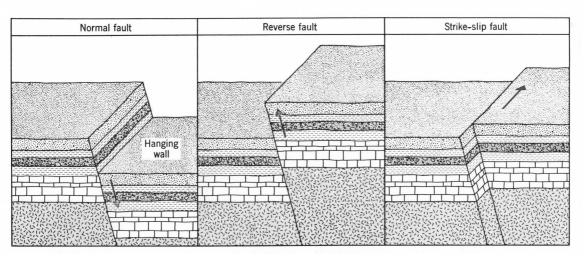

FIGURE 4.8 The attitudes of strata in various fault types. In a normal fault, the hanging wall moves down- **ward; in a reverse fault the hanging wall moves upward. A strike-slip fault occurs where rock displace-** **ment is horizontal rather than vertical.**

However, great earthquakes have also taken place in the continental interiors and in other parts of the world. As yet, no obvious sequence of earthquake occurrence has been detected. This is primarily why earthquake predictions are so difficult. The only pattern shown thus far is that they occur in close association with volcanic regions of the globe along zones in the Earth in which mountain building is still going on. The Pacific coast of

FIGURE 4.9 (A) (*right*) A rift valley or graben ("grave") is a downthrown block with higher portions of rock (horsts) remaining on either side (from A. N. Strahler, *Physical Geography*, 3rd ed., © 1969, John Wiley & Sons, New York). (B) (*below*) The graben in the photograph is Crag Lake, Idaho (U.S. Geological Survey).

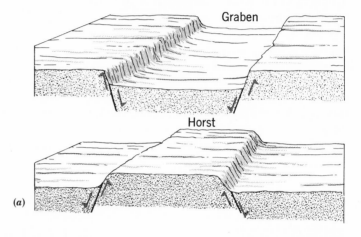

(*a*)

(*b*)

North and South America is one of the zones associated with earthquake activity around the Pacific Ocean basin.

Seismographs

Scientists use seismographs to record earthquake waves. In its simplest form the seismograph has only a few basic parts. A concrete base is firmly anchored to bedrock so that the base moves with the bedrock. Connected to the base is a support post, and a wire from the top of the post supports a weight. When the base and post move, the free-hanging weight remains stationary; its inertia keeps it motionless. A marking needle attached to the weight moves against a drum wheel which has a marking paper attached to it and makes a

permanent record of the earthquake's activity. This is a simplified explanation and does not go into the many problems inherent in the device. It is sufficient to say that this is the basic principle behind the device. Modern seismographs employ electronic aids and give a series of readings necessary for proper analysis.

Some Famous Earthquakes

In 1755, a severe earthquake struck Lisbon, Portugal. During this catastrophe, the waters drained out of the harbor and then returned in a gigantic tidal wave. More than 50,000 people were killed or washed out to sea, and the shock of the earthquake was felt as far away as Loch Lomond, Scotland. Tidal

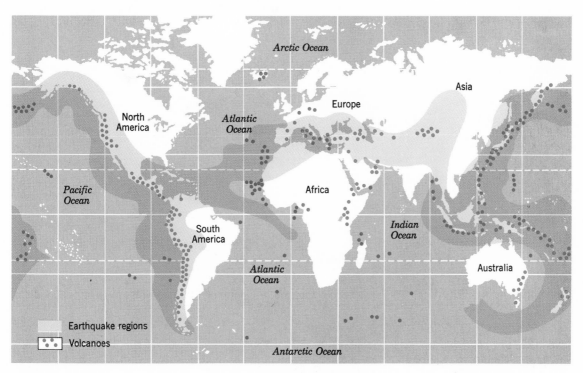

FIGURE 4.10 Major volcanic and earthquake regions of the world.

waves are common secondary effects in earthquakes that occur in or near ocean basins.

In late 1811 and early 1812, New Madrid, Missouri, was struck by a series of shocks for a period of several months. These earthquakes created new topographical features over thousands of square miles, and shocks were felt over most of the United States. Reelfoot Lake in Tennessee was created by the New Madrid earthquakes.

The great San Francisco earthquake in 1906 killed nearly 1000 people and destroyed a large part of the city. The city was later racked by fires caused by the earthquake.

In 1923, Yokohama and Tokyo, Japan, were very nearly destroyed by a shift of several hundred feet in the ocean bottom near the shoreline. Nearly 150,000 people were killed by fire and collapsing buildings. Gigantic tidal waves destroyed ships and harbors in the area.

In 1964, an earthquake off the south coast of Alaska created waves of water that caused great damage to the northwest coast of the United States. Land shifts in some places amounted to nearly 20 feet in a vertical direction.

Although earthquakes are usually rather local in their origins and effects, they are one of the most frightening of all geological occurrences. It has been estimated that 13 million people in the last 4000 years have lost their lives because of earthquakes. Earthquake activity is of great concern, especially in the United States. The areas that have been struck by earthquakes in the past, most notably in California, are now heavily populated. If new earthquakes were to occur, the tragedy might be of apocalyptic proportions.

Volcanoes

The study of volcanoes, called Volcanology after Vulcan, the Roman god of fire, is con-

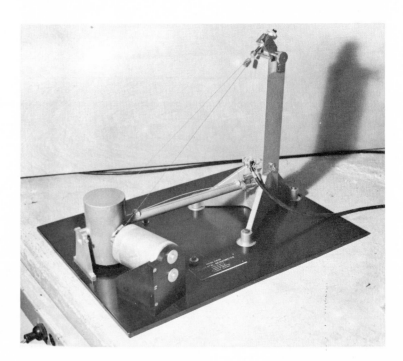

FIGURE 4.11 Seismographs record the energy vibrations resulting from earthquake activity. The instrument is set into bedrock, and a weighted arm reacts to the vibrations from the quake (Lamont Geological Observatory).

FIGURE 4.12   Craters of the Moon National Monument, Idaho, shows ropy lava flows that have cooled and hardened at the surface (Frank Jensen).

cerned with volcanic activity as well as volcanoes themselves. Volcanic activity is comparable to earthquakes in power and suddenness. Most continental volcanoes are isolated peaks, but at places on the ocean floor clusters of volcanoes occur, many of which are still active.

An active volcano ejects many types of materials. It releases large quantities of gases, including water vapor, carbon dioxide, carbon monoxide, and sulfurous fumes. In addition, it expels rock fragments, or pyroclastic material, such as ash, dust, cinders, and blocks as large as a foot or more in length.

These very large particles are called bombs. Volcanoes also spew forth much molten lava.

Volcano Types

Volcanoes are formed when lava flows forth from fissures and vents in the Earth's surface. As Figure 4.15 indicates, the manner in which the lava flows determines the type of volcano that is formed.

The quietest volcanoes are the ones in which lava flows slowly from fissures. These become broad, dome-shaped structures called *shield volcanoes*. The Hawaiian Islands are great shield volcanoes which have risen from the ocean floor until they extend far above the surface of the water.

Some volcanoes alternate between periods of quiet lava flow and fierce eruptions. These are steep-walled *stratovolcanoes* or composite cones in which the layering or stratification is very obvious. The famous Mount Fujiyama in Japan and Mount Rainier in the United States are examples of this type of volcano.

Small, steep mounds of cinder often build up around a series of vents in the surface. These vents spew forth a large amount of cinders and some lava, forming a *cinder cone.*

When the eruption is mostly lava and larger pyroclastics, *spatter cones* accumulate.

A secondary cone may arise on the side of a previously formed volcano when lava flows from the main vent flow into a smaller side vent. Thus, a small cone forms on the side of the earlier cone.

Volcanic Eruptions

Volcanic eruption is usually prefaced by rumbling and the emission of gases and steam, although advance warning does not always occur. When the volcano actually

FIGURE 4.13   San Lucas volcano in Guatemala shows a typical volcano cone formation (Verna Johnston, Photo Researchers).

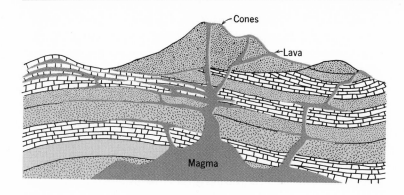

FIGURE 4.14 Fissures and lava flow typical of volcanic regions. Magma is forced upward by pressure.

erupts, it may vary from a quiet flow of lava through a series of small explosions up to gigantic explosions that tear the cone apart.

A curious formation sometimes occurs when a crater in a dead volcano becomes

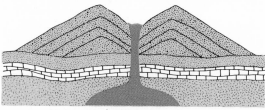

Cinder cone

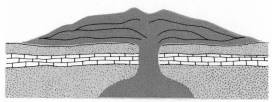

Shield cone

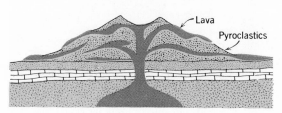

Composite cone or stratovolcano

FIGURE 4.15 Common types of volcanoes.

plugged with solidified lava. Erosion wears away the cone, but the more resistant plug remains. The plug mound is then the only obvious evidence remaining of the once active volcano.

Crater Lake in Oregon is a fine example of a *caldera,* or *volcanic cone* that has collapsed and filled with water. After the lake formed, a small cinder cone arose in the lake. Thus, Crater Lake is a large, water-filled dead volcanic cone, in which a second cone, Wizard Island, rises out of the water.

There are at least 300 active volcanoes throughout the world. They are most numerous where earthquakes are common in the circum-Pacific basin and the Mediterranean area. In fact, the circum-Pacific basin has been dubbed the "circle of fire" or "ring of fire" because of its volcanoes. East Africa, Japan, Indonesia, and parts of the Caribbean have also experienced volcanic activity in recent years.

Some Famous Volcanic Eruptions

Lassen Peak in Northern California is the only volcano in the continental United States that has erupted during this century. Perhaps the most notable volcanic activity in this century was the series of eruptions off the south coast of Iceland which, in 1963, gave birth to the Island of Surtsey. These eruptions have been the most carefully studied and thoroughly documented in history.

Krakatoa, a cinder cone, produced one of the most famous volcanic eruptions of the nineteenth century. Located between Sumatra and Java, it gave vent to a gigantic explosion in 1883. The shock produced huge waves in the sea that destroyed most of the island and killed more than 30,000 people. The eruption produced dust particles in numbers that completely blotted out the sun. The results of this explosion were still evident in the atmosphere months after the eruption.

Mount Vesuvius in Italy, a composite cone, was thought to be dormant until 79 A.D. when Pompeii and the surrounding area were destroyed by a tremendous eruption. The city was completely covered by molten lava and volcanic ash. Since then the city has been partially excavated and many artifacts as well as body molds of the victims have been found in the hardened lava and cinders. Vesuvius had another deadly eruption in 1631 and several smaller ones during this century.

Hawaii National Park has two active volcanoes, Kilauea and Mauna Loa. A government observatory located near them studies the volcanoes and the material they issue.

Although no one is sure of how earthquakes and volcanoes are related, both activities apparently take place in zones of weakness in the Earth's crust. Whereas earthquakes result from rocks snapping under stress, volcanoes push materials out of cracks and fissures in the Earth. Volcanoes may erupt catastrophically, or they may erupt quietly.

Erosional and Weathering Processes

Landforms are sculptured and produced by crustal movements, volcanic activity,

FIGURE 4.16   A volcano plug. The vent of the volcano became plugged with hardened lava. Erosion has removed the original sides of the volcano cone; the plug remains as a butte (U.S. Department of Agriculture).

FIGURE 4.17 Crater Lake is a supreme example of a caldera. The original crater has collapsed and a secondary cone has formed within the lake (Delano Aerial Surveys).

weathering, deposition, and erosion. The last two sculptural processes involve the wearing away of land and the deposition of materials. These processes are sometimes referred to as *degradation* and *aggradation* of the land. Many of the erosional processes of water, wind, and ice operate simultaneously.

Weathering, or the disintegration of materials, operates through physical and chemical changes. Physical weathering involves changes in the size, the shape, or the form of

FIGURE 4.18 Corrosion of an iron pipe resulting from a chemical reaction with ions in the water (Fundamental Photos).

rocks. For example, the disintegration of rock by frost action is a physical weathering process. Chemical weathering is a change in chemical composition. The weak carbonic acid in rain water, which attacks rock such as the carbonates, is a chemical weathering agent.

Physical Weathering

Ice is one agent of physical weathering. We have observed that it abrades and scours rocks, but smaller-scale formation of ice and frost can also create changes. As water seeps into rocks and soil, it may fill all the crevices it encounters. Then the water freezes and expands, forcing the cracks or pores to spread. The process occurs repeatedly. Eventually the soil pores are widened, the cracks in rocks become larger, and the rocks are split and broken. Piles of these broken particles are often found lying at the base of cliffs. This loose, jagged, and unsorted material found at the base of cliffs is called *talus.*

Repeated heating and cooling of rocks also cause disintegration. As the rocks expand and contract under alternately warm and cold conditions, cracks widen and the rocks split. Rocks also lose surface materials in layers. This process, known as *exfoliation,* is a form of "peeling" of rock surfaces.

These weathering processes, when continued over a long period of time, eventually reduce large rock masses into loose, fine-grained particles. These processes are among the first steps in soil formation.

Rock masses and soils are also affected by gravity through a process known as *mass wasting.* Talus formations form at the base of cliffs because of the pull of gravity. Soils become water-filled and flow downhill. This is a process known as *soil creep.* In mountain-

FIGURE 4.19 A bust of Constantine in Rome showing the effects of weathering on exposed artwork (Anderson—Art Reference Bureau).

ous terrain, earthslides are common. Huge masses of earth flow downhill much like avalanches of snow.

When mud dries, it contracts and breaks apart, leaving cracks in the resulting hardened material. The hardened mud has angular joints and forms weak rocks, which shatter easily when weathered.

Living things also act as agents of erosion. In addition to man's well-known processes of soil removal, plants and animals also cause Earth movements, but on a smaller scale. Plant roots and burrowing animals separate and loosen soil; tramping animals pack soil down.

Streams and rivers are the chief agents of soil erosion. Materials are carried in solution, plucked or picked up, and pushed along. Other materials are abraded with the help of the materials carried in the water. Winds can also carry on the same processes.

Some Depositional Forms

Many of the landforms mentioned in previous chapters are deposits of erosional material. Floodplains on either side of a river are formed by deposition, as is a delta at the mouth of a river. The levees of an old stream are deposits of waterborne silt. Silt and other materials deposited by water are called *alluvium*. As an intermittent stream transports material onto a flat plain after passing down a mountainside, a fan-shaped deposit of material forms. This is known as an *alluvial fan*.

Eskers and outwash plains of glaciers also originate as deposition. The structures are all formed by glacial streams and are made

FIGURE 4.20   **The erosional effects of water on statuary in San Francisco (George Daniell, Photo Researchers).**

up of glacial sediments. Drumlins are rounded hills of unsorted glacial till.

Chemical Weathering

Chemical reactions are most marked when water is the agent of change and take place most rapidly in moist, warm climates. There are several ways in which water is involved in chemical weathering: carbonation, hydration, solution, and oxidation. These reactions generally involve the addition of some substance to the minerals already present. This results in a change in form and structure of the rock material being acted on.

Chemical Weathering by Water

Carbonation. Rocks composed of calcium carbonate are most susceptible to the effects of water solutions. Carbon dioxide, which is easily dissolved in water, forms a weak carbonic acid. This acidic water reacts with calcium carbonate rock such as calcite, marble, or limestone. For example:

Carbon dioxide + water → Carbonic acid
$$CO_2 + H_2O \longrightarrow H_2CO_3$$

Carbonic acid + calcium carbonate →
                    Calcium bicarbonate
$$H_2CO_3 + CaCO_3 \longrightarrow Ca(HCO_3)_2$$

FIGURE 4.21 Talus formation is the result of rock wastage, chiefly as a result of mechanical weathering. The rock forms an apron at the base of the cliff face (B. M. Shaub).

FIGURE 4.22 Rock that shows the result of exfoliation. Royal Arch Lake in California shows the piles of debris that has "peeled" off the cliff face (National Park Service).

FIGURE 4.23 A huge mass of earth that has moved downslope in a gigantic mud flow (John Shelton).

Solution. Some minerals and salts, as calcium, sodium, and magnesium, are at least partially soluble in water. In fact, water has been called the universal solvent. This is true because, given enough time, water will dissolve nearly every material available.

Hydration

Some reactions add water to the mineral already present. Feldspars are particularly susceptible to hydration. The resultant materials are referred to as hydrates:

Feldspar + water + carbon dioxide →
Kaolin (clay) + silica + potassium carbonate
$$2KAlSi_3O_8 + 2H_2O + CO_2 →$$
$$Al_2Si_2O_5(OH)_4 + 4SiO_2 + K_2CO_3$$

Calcium sulfate is chemically written $CaSO_4$. When it is attacked by water, it becomes $CaSO_4 \cdot H_2O$, or gypsum. Here, it is perhaps easier to see that hydration is the simple addition of water.

Hydration produces materials that are softer than the original rock for the hard, compacted material expands and swells. The hydrates are more susceptible to erosion and other weathering processes.

Chemical Weathering by Oxygen

The process of chemically adding oxygen to a substance is called oxidation. This takes place most readily under moist conditions. For instance, iron rusts (oxidizes) when

109

damp. Hematite is a mineral which is an oxide of iron, that is, it is composed of iron plus oxygen. Iron pyrite oxidizes and forms sulfuric acid which will further attack bedrock.

$$2FeS_2 + 2H_2O + 7O_2 \rightarrow 2FeSO_4 + 2H_2SO_4$$

Pyrite $\longrightarrow$ Ferrous sulfate + sulfuric acid

It is interesting to note that the minerals

FIGURE 4.24 (*right*) Bowen's reaction series.

FIGURE 4.25 (*below*) Soil erosion in the United States.

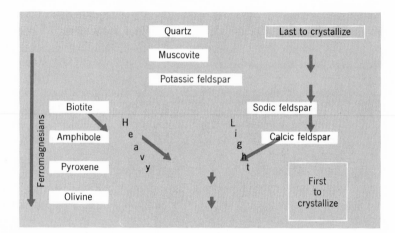

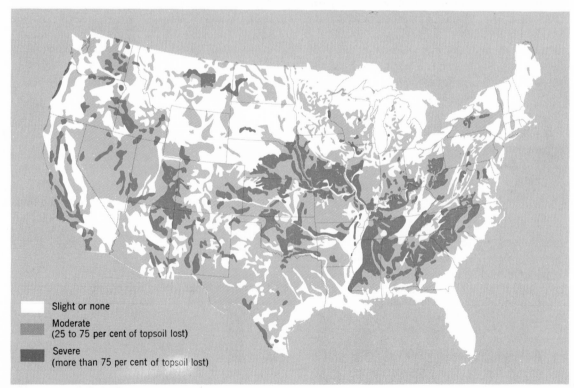

that first crystallize from molten material are the most susceptible to weathering (Figure 4.24). Those minerals that crystallize last from the magma are the last to be attacked and disintegrated by the agents of weathering. For example, the ferromagnesians—pyroxene and olivine—form first and are attacked and decomposed most rapidly. Quartz forms last and is the most resistant.

Man and His Environment

The natural resources of the Earth have taken millions of years to take form and to develop. They are the one crop that man cannot replace, regrow, or harvest a second time. There is no second cycle to the natural resources and scenic beauty with which the Earth was originally endowed.

One of man's chief problems is to prevent weathering and erosion from exacting their toll on agricultural lands. Weathering processes and the dynamic movements of the Earth's crust have shaped our land and formed our soils. Great rivers like the Mississippi and the Nile have produced great deposits of silt, which have developed into some

of the richest agricultural regions of the world.

But there are times when man wishes to prevent these processes from taking place. He has evolved many methods of doing so. For example, a gentle slope may experience a runoff of 5 per cent of the precipitation that strikes it. A steeper grade may lose 50 per cent or more as runoff. A simple planting of alfalfa may cut the runoff on a steep grade by one half or more.

To prevent sheet wash from carrying off soil and grasses on hilly terrain, bushy plants are planted. When city planners design water reservoirs, they specify planting in the uplands to slow the rate at which sediment is washed into the reservoirs. A rapid rate of sedimentation would eventually begin to fill the reservoir, effectively cutting its capacity.

To slow erosional process, farmers plow their fields in contour lines following the lay of the land. Winds are cut by the use of windbreaks—tall, closely planted trees placed along the edges of fields. Grasses grown on fallow land also act as windbreaks and slow runoff as well.

Experiments in South Africa, the Nether-

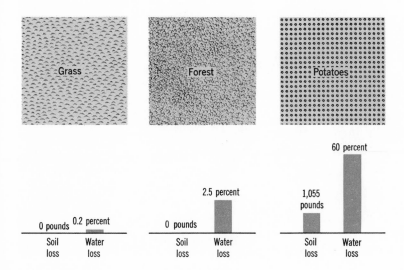

FIGURE 4.26 Relative water loss from land with different coverings.

A

B

112

C

D

FIGURE 4.27 Special procedures must be followed to slow erosion in agricultural regions. Contour plowing (A) follows the contour of the land. Terracing of hillsides to slow and hold water (B) is particularly suited for rice planting. Wind breaks and strip cropping (C) also aid in maintaining soil fertility. Wind can also be broken by dune fixation by using grasses and other fast-growing vegetation (D). [(A) U.S. Department of Agriculture, (B) FAO, (C) U.S. Dept. of Agriculture, and (D) FAO.

113

lands, and Israel have shown that erosion can be slowed by spraying a flexible film of synthetic rubber and oil on soil. This does not affect plant growth.

To prevent agricultural land from becoming barren, soil nutrients must be replaced. Plants require a supply of potassium, phosphorus, nitrogen, and traces of manganese, cobalt, sulfur and many other elements for proper growth.

In desert areas, man has had some success with irrigation and fertilization. Barren soil has been turned into productive farmland. But even these techniques must be used wisely. When irrigation is used unwisely, the rapid evaporation of the water leaves salt behind it in the soil. The soil becomes alkaline, and the result is the destruction, rather than the production, of rich farmland.

Fuels and minerals are one-crop harvests. Billions of tons of coal have been removed from the coal fields of West Virginia, Pennsylvania, Illinois, Kentucky, and other states. At the present rate of use, in 500 to 1000 years, this resource will become scarce. There is only enough oil in Texas, California, Louisiana, and Oklahoma for a few more decades of supply. Although the Near East oil fields seem to be a source of never-ending supply, these fields cover a very small area in comparison to the number of countries that are dependent on it.

Granites and other building stones are plentiful in the Earth's crust. But many of these rocks do not form outcrops, and their extraction from deep in the Earth is becoming a more expensive operation with every passing decade. Metallic ores are not found in the pure state and they, too, must be extracted at high cost, from poorer and poorer grades of ore. Although the last few years have seen new supplies of minerals discovered and synthetics developed at a rapid rate, we are facing the disappearance of many

minerals which have taken millions of years to be produced. They are irreplaceable.

In addition to being necessary for the support of life, water remains one of man's most valuable economic resources. It is required for industrial purposes and for drinking, and rivers are still prime highways of commerce. These uses are in danger because of a lack of planning in population centers.

Rivers receive sewage from cities, and carry industrial wastes as well as silt. Pesticides carried through the soil into rivers kill millions of fish each year. Sewage from cities upriver enters the water supply taken from these same rivers by cities downriver. The Hudson River in New York State, for example, has been almost hopelessly polluted in this fashion.

Wells, a valuable source of drinking water, have been affected by the many local increases in urban population. In Minnesota, some 50,000 to 75,000 wells recently were polluted by the sewer drainage of a nearby urban area. In coastal areas, the rapid pumping of wells can allow sea water to invade the fresh water in the wells.

Pesticides, industrial poisons, and human waste upset the ecology of wildlife populations. The poisons used to kill insects accumulate in plants and animals that are eaten by birds and other animals, and kill these organisms as well. Sewage ruins the streams on which animals depend for food. And, the poisons in the air upset the smells on which animals depend for obtaining food and even mates during the reproductive season.

The population of the United States may reach 300 million by the year 2000. Of these, 80 to 90 per cent will live in large cities. The accumulations of industrial smoke, automobile exhaust, and burning wastes will create even greater problems from smog than at present. Studies are now being conducted to determine how smog is related to the high incidence of emphysema, a serious respiratory disease, in the population of Los Angeles. In 1952, in London, more than 5000 deaths were recorded in a few weeks as a result of a temperature inversion (a layer of warm air settled over a layer of cold air), which prevented the dissipation of industrial smog. Similar situations have occurred on a large scale in New York in 1953 and 1963, and again in 1956, in London.

The removal of waste products from human civilization requires careful planning. Man has polluted the air with industrial waste gas, and the soil and water with metabolic waste and industrial poisons. New methods of disposal and even recovery are being sought to safely remove these wastes far from human and animal habitation. For example, the waste products of human civilization can be treated and turned into fertilizer instead of being lost when burned. But it is increasingly apparent that, if man continues to increase in number and if he does not take greater steps to prevent upsetting the balance of nature, future food, air, and water supplies will be in short supply partially because of contamination.

Questions

CHAPTER ONE

1. In what ways have the early Catastrophist views of the Earth evolved into our modern explanation of natural catastrophes?
2. What are minerals? Discuss the relationship of minerals to crustal rocks.
3. Describe the various parts of the Earth as it appears in cross section. How do we know the composition of each section?
4. How does the concept of isostasy explain the differing heights of various topographical features of the Earth?
5. What are some of the current theories that have been developed to explain the formation of the continents?
6. Compare and contrast the general appearance of the ocean floors with the surface of the continents.

CHAPTER TWO

1. Why is color a poor test for mineral identification? List and explain at least four other tests that can be used for mineral identification.
2. What is specific gravity and how is it obtained? What information does this test yield about a specimen?
3. What is meant by the rock cycle?
4. Explain how the various intrusive igneous rock structures form.

In what major way do intrusive rock masses differ from extrusive masses?

5. By what processes do sedimentary rocks form?

6. By what processes do metamorphic rocks form? For each class of metamorphic rock, list one specific form and its origin.

CHAPTER THREE

1. Describe the general changes found during the life cycle of a stream. What is the general effect of surface water on local topography?

2. What is turbulent flow? Why does this type of water movement occur?

3. List the water zones in the soil. Relate each to water content and plant growth.

4. Compare the theories that attempt to explain the onset of glacial periods. What do they have in common with one another?

5. What is soil? Explain its formation and the general characteristics of the soil profile.

6. In general, the Earth is buffeted between two major, opposing forces. What are these forces and what effects do they have on the Earth's surface?

CHAPTER FOUR

1. What relationship exists between the twin catastrophes of earthquakes and volcanoes?

2. Describe the various types of faulting that occur in the Earth's crust.

3. How do geosynclines form? What structures result as they continue to evolve?

4. Describe the various types of volcanoes and the ways in which they form.

5. In what way does analysis of earthquake waves produce information about the Earth's interior? What characteristics of the various types of waves allows us to collect this information?

6. Describe the differences between physical and chemical weathering. In what ways can man slow the effects of physical weathering?

Bibliography

BOOKS

Clark, Thomas H., and C. W. Stearn, *Geological Evolution of North America*, 2nd Ed., Ronald Press, New York, 1968.

Deer, William A., *Introduction to the Rock Forming Minerals*, John Wiley & Sons, New York, 1966.

Dennison, John H., *Analysis of Geologic Structure*, Norton, New York, 1968.

Ernst, W., *Earth Materials*, Prentice-Hall, Englewood Cliffs, New Jersey, 1969.

Hunt, C. B., *Physiograph of the United States*, W. H. Freeman, San Francisco, California, 1967.

Longwell, C. R., Flint, R. F., and Sanders, J. E., *Physical Geology*, John Wiley & Sons, New York, 1969.

McCullough, Edgar J., Jr., *Historical Geology*, Brown, W. C., Dubuque, Iowa, 1966.

Navarra, J. G., and Strahler, A. N., *Our Planet in Space*, Harper and Row, New York, 1966.

Ordway, Richard J., *Earth Science*, D. Van Nostrand, Princeton, New Jersey, 1966.

Ramberg, H., *Gravity Deformation and the Earth's Crust*, Academic Press, New York, 1967.

Shelton, J. S., *Geology Illustrated*, W. H. Freeman, San Francisco, California, 1967.

Sinkankas, J., *Mineralogy*, D. Van Nostrand, Princeton, New Jersey, 1966.

Thornbury, William D., *Principles of Geomorphology*, 2nd Ed., John Wiley & Sons, 1969.

Vening-Meinesz, F. A., *Earth's Crust and Mantle*, Am. Elsevier, New York, 1964.

Woodford, A. O., *Historical Geology*, W. H. Freeman, San Francisco, California, 1965.

Zumberge, J. H., *Elements of Geology*, 2nd Ed., John Wiley & Sons, Inc., 1963.

PERIODICALS

Frazier, K., "Turning to the Earth — Smaller Crustal Plates," *Science*, February, 1970.

Larsen, R. L., "Gulf of California: A Result of Ocean Floor Spreading and Transform Faulting," *Science*, August, 1968.

LePichon, X., "Sea Floor Spreading and Continental Drift," *J. Geophys. Res.,* **73,** pp. 3661–3697, 1968.

Mansinha, L., and Smylie, D. E., "Earthquakes and the Earth's Wobble," *Science,* September, 1968.

McDonald, G. J. I., "How Man Endangers the Climate," *Current,* January, 1970.

McElhinny, M. W., "Paleomagnetism and Gondwanaland," *Science,* May, 1970.

Meyerhoff, A. A., "Continental Drift," *Journal of Geology,* January, 1970.

Miller, F., "How to Predict an Earthquake," *Saturday Review,* November, 1968.

Mitchell, A. H., and H. G. Reading, "Continental Margins, Geosyncline and Ocean Floor Spreading," *Journal of Geology,* November, 1969.

Smith, S. D., "Study of the San Andreas Fault: Something Very Interesting Is Going on Deep Beneath the Earth's Surface," *Time Magazine,* January, 1970.

Van Houten, F. B., "Molasses Forces: Records of Worldwide Crustal Stresses," *Science,* December, 1969.

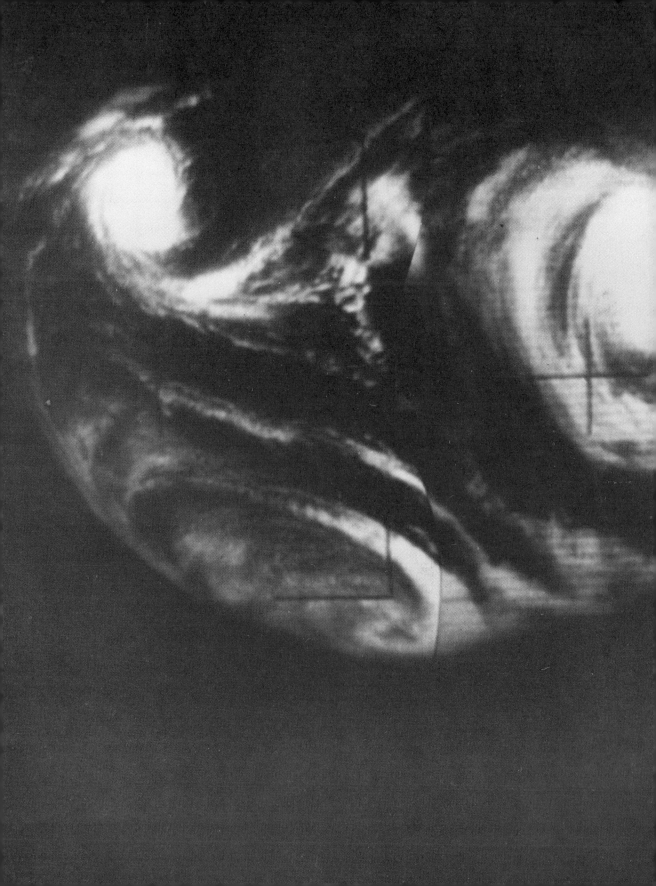

The Atmosphere

THE EARTH'S SURFACE IS SURROUNDED BY A GASEOUS envelope known as the atmosphere. This blanket of gas is constantly absorbing solar energy from space and from the surface of the Earth. The energy exchanges that take place between the atmosphere and the Earth's surface, between the atmosphere and space, and within the atmosphere produce physical and chemical changes within the atmosphere. These various exchanges of energy produce the effects we call weather. The energy-absorbing properties of the atmosphere have allowed conditions suitable for life to be achieved on our planet. The atmosphere supports and protects the multitude of life forms that exist on the surface of our planet. An understanding of weather and its associated phenomena is essential to man's existence on Earth.

CHAPTER 5

The Composition of the Atmosphere

THE ATMOSPHERE THAT SURROUNDS THE Earth today is very different from the atmosphere that first formed around the Earth, and is probably very different from the atmosphere that will exist a million years from now. It is constantly undergoing change.

As we ascend from the surface of the Earth to the outer limit of the atmosphere nearly 300 miles up, we find that the chemical and physical properties of the air do not remain the same. The dissimilarities among these properties have led man to identify and name several different layers of the atmosphere.

The Development of the Atmosphere

Every moment of every day, subtle changes take place within the atmosphere. What is the atmosphere like at this very moment? No one can answer this question in minute detail with any certainty. To do so would require that a vast amount of data be collected and analyzed instantaneously throughout the world. No such system of collection and analysis exists at present. Thus, if we build a picture of the present atmosphere, or the atmosphere long ago, it must be developed from scattered bits and pieces of information. These bits and pieces come from many sources. Some are exact, sophisticated measurements, and others are projections of what *might* have been or what *might* exist. Some of the projec-

tions are suggested by theories that relate to the Earth's history. Still others are guesses and predictions about directions in which the atmosphere might evolve in the future. But once we have accepted the uncertainty that exists in our knowledge, it is possible to put the bits together and to attempt to develop a picture of the atmosphere—past, present, and future.

The Primeval Atmosphere

What was the atmosphere like 5 billion years ago? It is rather obvious that we have no direct observational evidence, since man arrived on the scene, at most, 2 million years ago. Thus our idea of the primeval atmosphere must be largely speculative.

One of the first modern scientists to speculate about the primeval atmosphere was Louis Pasteur, the great nineteenth-century microbiologist. Pasteur asserted that the first life forms could not have developed in an atmosphere that contained as much oxygen as it does now. He was among the first to suggest that the primitive atmosphere had a chemical and physical composition other than that at the present time.

Since Pasteur's time, many other suggestions have been made. The most widely accepted proposal is that of A. I. Oparin, the noted Soviet biologist and theoretician. He has suggested that the primeval atmosphere

FIGURE 5.1 Louis Pasteur was one of the first to speculate about the origin of the atmosphere in relation to primitive life forms (Bettmann Archive).

was composed of methane (CH_4), ammonia (NH_4), hydrogen, and water vapor.

In the current view, volcanic activity played a major role in the formation of the atmosphere by expelling gases from the Earth. Among these gases were carbon dioxide (CO_2), nitrogen, and sulfurous gases, as well as methane, ammonia, hydrogen, and water vapor. Before the Earth began to cool, these gases were driven into space. Later, however, the Earth cooled sufficiently to allow the expelled gases to form an envelope around the Earth, and the formation of the atmosphere began.

As the primeval atmosphere began to cool, the hot water vapor began to condense. The ensuing precipitation removed some carbon dioxide and sulfur compounds from the air by a process of solution, so that the atmosphere was left rich in ammonia, hydrogen, and methane. As we shall learn in "The Biosphere" (p. 287), these chemicals are the important constituents from which primitive

life forms develop. (Figure 5.3 shows these important constituents.)

As everyone knows, oxygen is an important constituent of our present atmosphere. But oxygen itself was probably never in the atmosphere in any great quantity until after the development of primitive life forms. The very earliest life forms were probably anaerobic bacteria (Figure 5.4), which can exist without oxygen being present in their environment. The modern representatives of these bacteria are *Escherichia coli,* found in the colons of mammals. Some of the first oxygen was probably introduced into the atmosphere as a byproduct of the metabolic processes of anaerobic bacteria. More oxygen may have been added to the developing atmosphere as the action of solar energy split water molecules in the air.

Today, the atmosphere consists of nitrogen and oxygen, with traces of carbon dioxide, inert gases, and several other gases, as shown

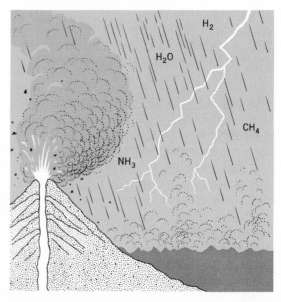

FIGURE 5.2 The ancient atmosphere. The primeval Earth was affected by much volcanic action that released great quantities of gas to the atmosphere.

in Figure 5.5. Evolution of the atmosphere, from its first primitive beginnings to its present form, has taken more than 4 billion years. But the story of its development is not complete. We have only small patches of information.

The Present Composition of the Atmosphere

The ocean of gases that surrounds the Earth is not bound together chemically. It is a mixture, in which each gas is distinct and separate from all the others. (See Table 5.1.)

The most abundant gas is nitrogen. It makes up approximately 78.1 per cent of the Earth's atmosphere. Man cannot use nitrogen in its gaseous form, but it is important for plant life. Certain bacteria in the soil and the roots of some plants convert gaseous nitrogen into nitrates, which are essential for protein synthesis in plants. Thus, when man eats green plants he receives proteins only because there is nitrogen in the atmosphere. The nitrogen cycle is shown in Figure 5.6.

Approximately 20.9 per cent of the air is composed of life-supporting oxygen. This second most abundant gas in the atmosphere is essential in converting food to energy. Oxygen must combine with another substance in a chemical reaction called oxidation before food can be converted to energy. Without this gas, the metabolic processes of

Name	Molecular Formula	Structural Formula	Models	
			Space Filling	Ball–and–Stick
Hydrogen	H_2	H — H		
Water	H_2O	H — O — H		
Ammonia	NH_3	H — N — H ￨ H		
Methane	CH_4	H ￨ H — C — H ￨ H		

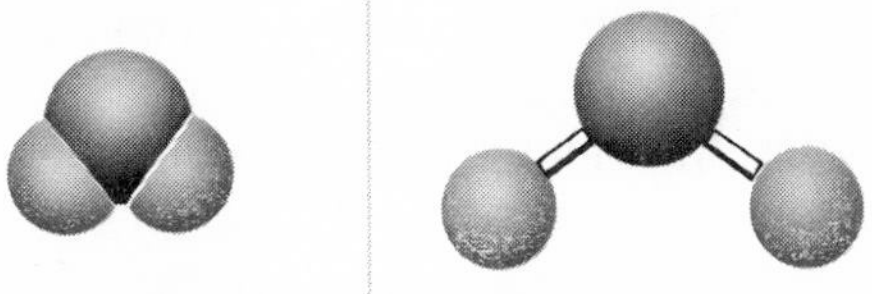

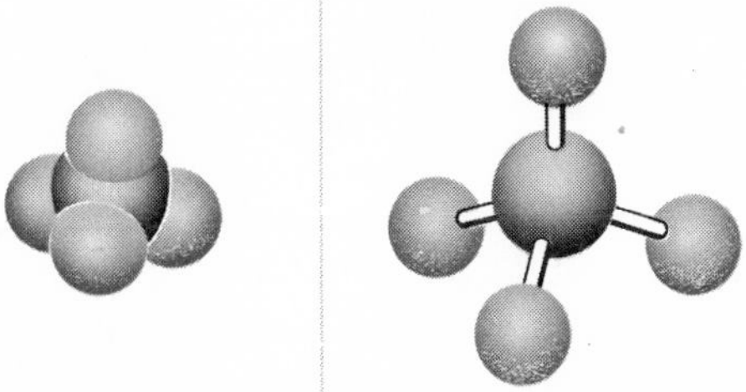

FIGURE 5.3   The formulas and models of the major components in the primeval atmosphere.

most living things would not occur and the organisms would cease to exist.

Nitrogen and oxygen together make up 99 per cent of the atmosphere. The remaining 1 per cent of the atmosphere includes a number of gases. About 9/10 of this 1 per cent is composed of the inert gases: argon, neon, krypton, helium, and xenon. These gases do not, under ordinary circumstances, combine with other substances as do oxygen and nitrogen. The inert gases apparently have no value for living things.

In the remaining 1/10 of 1 per cent of the air, about 3/100 is carbon dioxide. Although this is a very minute portion of the total atmosphere, carbon dioxide is an important gas. It is directly involved with the absorption of solar energy and the insulation of the Earth.

The remaining 0.07 per cent of the atmosphere is composed of methane, hydrogen, dust particles, water vapor, and a huge variety of industrial gases. As we shall learn later in our study of the science of meteorology, water vapor is an important constituent of the air.

The proportions of the gases in the atmosphere vary from place to place and from one day to the next. Carbon dioxide and water vapor are the most variable of all the gases. Carbon dioxide is utilized by plants in photosynthesis, and expelled by animals in respiration. Water vapor enters the air through evaporation and is removed as a result of condensation. Environmental conditions, numbers of plants and animals, and temperature affect the amount of these gases in the atmosphere. (See Figure 5.7.)

Man's Effect on the Atmosphere

Since the nineteenth-century Industrial Revolution, man's industrial processes have been producing great quantities of carbon dioxide (CO_2) and carbon monoxide (CO) gases as wastes. Because of this fact, much concern has arisen over man's effect on the atmosphere.

Changes in the carbon dioxide content of the air may affect the onset of glacial periods, because they affect the Earth's temperature. Although the percentage of carbon dioxide is only 0.03 per cent of the total atmosphere, a change of only a few hundredths of 1 per

Table 5.1   Present Composition of the Atmosphere

GAS	SYMBOL OR FORMULA	PER CENT (BY VOLUME)
Nitrogen	N_2	78.1
Oxygen	O_2	20.9
Argon	Ar	.934
Carbon dioxide	CO_2	.03
Neon	Ne	.001818
Helium	He	.000524
Methane	CH_4	.0002
Krypton	Kr	.000114
Hydrogen	H_2	.00005
Nitrous oxide	N_2O	.00005
Xenon	Xe	.0000087

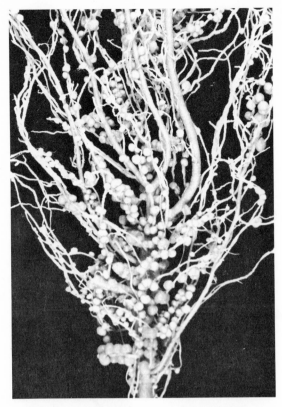

FIGURE 5.4 Bacteria-containing nodules on the roots of leguminous plants fix free nitrogen from the air (The Nitrogen Company).

cent might have very serious effects on the Earth's climate. No one really knows. Nevertheless, it has been estimated that sufficient amounts of carbon dioxide have been added to the air in the last 100 years to have raised the average temperature of the Earth 1°F. This situation is presently being studied in order to determine what climatic changes have already occurred and what changes might occur in the future.

Other atmospheric pollutants are suspected of being major causes of respiratory ailments in man. Carcinogenic, or cancer-causing, agents spewed forth by industrial processes may be one of the major causes of lung cancer in urban populations.

Sulfurous compounds resulting from industrial processes combine with water vapor in the air to produce a weak sulfuric acid. Rains containing low concentrations of sulfuric acid attack the façades of buildings and are presently obliterating detail on great works of art exposed to the elements in Greek and Italian cities.

Automobile exhaust fumes also release irritants into the air. Furthermore, the action of solar energy on waste gases produces chemical reactions that bring about further changes in the gases. Ozone, a poisonous gas, is produced when sunlight affects automobile exhaust. Ozone is otherwise rare in the lower atmosphere. Sulfur dioxide, as we have said, eventually forms a weak sulfuric acid when dissolved in rain water. These gases are particularly corrosive and irritating in large quantities.

The Layers of the Atmosphere

As sounding balloons and rockets reveal, the atmosphere changes in pressure, in temperature, and in composition with increasing

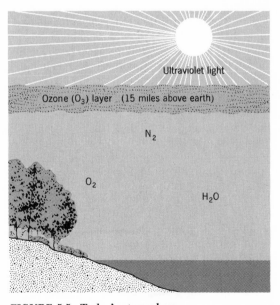

FIGURE 5.5 Today's atmosphere.

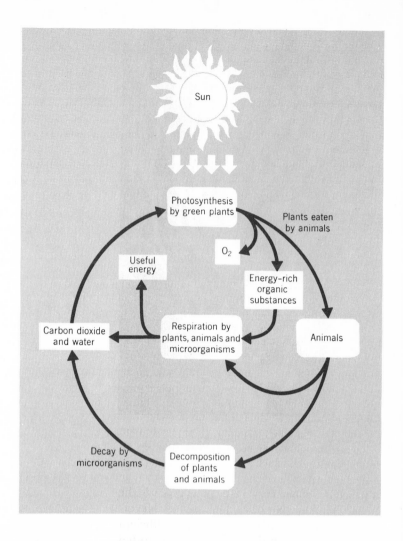

FIGURE 5.6   The nitrogen cycle.

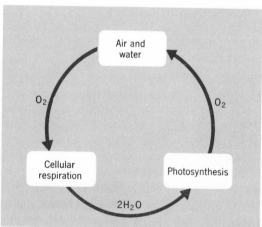

FIGURE 5.7   The oxygen cycle.

height. In fact, the atmosphere has several sharply defined layers and sublayers. (See Figure 5.10.)

The Troposphere

The layer nearest to the Earth is called the *troposphere*. It is comparatively dense, and is characterized by a great deal of turbulence. As different air masses circulate throughout this layer, they make contact with one another and begin to change as they mix. This mixing of air masses produce our daily weather.

Examination of the troposphere reveals that temperatures decrease at a rate of 3.5°F for every 1000 feet of height. This phenomenon is known as the *normal adiabatic lapse rate*. This is the average normal decrease in the temperature of air with height in the troposphere.

However, ascending air masses present a different set of circumstances. In rising dry air, the *dry adiabatic lapse rate* or temperature decrease is 5.5°F for every 1000 feet of height. In rising moist air, the *wet adiabatic lapse rate* is about 3.2°F for every 1000 feet of height. In short, ascending air masses in which condensation of water vapor takes place cool more slowly than those in which no condensation takes place. These decreases in temperature result from the expansion of air, not an actual loss of heat. As these air masses descend, they warm up by compression at about the same rate. (See Figures 5.11 and 5.12.)

The temperature drop continues to a minimum of about −70°F at the outer limit of the troposphere—these minimum temperatures are encountered at a height of six miles above the poles and ten miles above the equator. The troposphere tends to bulge like the solid

FIGURE 5.8 The Gateway Arch in St. Louis, Missouri, shrouded in smog from polluted industrial smoke and fog in 1966 (UPI).

Earth. The upper boundary of the troposphere is known as the *tropopause*.

The Stratosphere

Above the tropopause, at around 11 miles in the middle latitudes, the stratosphere begins. Temperatures in this layer of the atmosphere range from −70°F at the lower edge to −30°F at the upper limit, called the stratopause. One important feature of the stratosphere is its ozone layer. The energy of the sun causes atoms of oxygen within a molecule to separate and rejoin as a layer of ozone (O_3), a poisonous form of oxygen. This gas differs from "normal" oxygen (O_2) in that three atoms of oxygen join together instead of the usual two.

Although ozone is poisonous to human beings in sufficient concentration, it is indirectly essential to the support of life. The reaction involved in forming ozone uses a great amount of the potentially harmful ultraviolet frequencies of solar radiant energy before they reach the earth's surface. Therefore, the formation of ozone protects man from an excess of the ultraviolet rays produced by the sun.

A

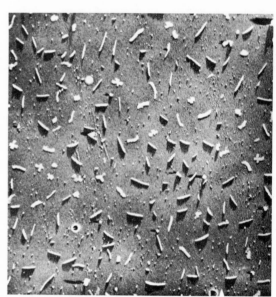

B

FIGURE 5.9   (A) A sequence showing Los Angeles, California, at different times on the same day. The top photograph was taken at 8 A.M., the middle at 9 A.M., and the bottom at 10 A.M. (B) Los Angeles air magnified many times (UPI).

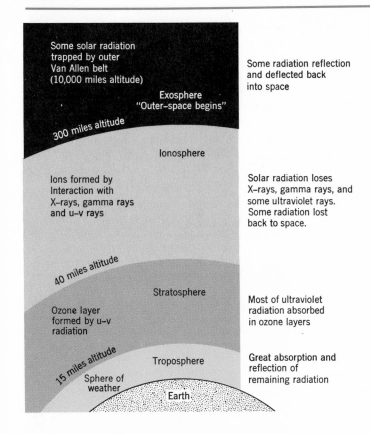

Some solar radiation
trapped by outer
Van Allen belt
(10,000 miles altitude)

Some radiation reflection
and deflected back
into space

Exosphere
"Outer-space begins"

300 miles altitude

Ionosphere

Ions formed by
Interaction with
X–rays, gamma rays
and u–v rays

Solar radiation loses
X–rays, gamma rays, and
some ultraviolet rays.
Some radiation lost
back to space.

40 miles altitude

Stratosphere

Ozone layer
formed by u–v
radiation

Most of ultraviolet
radiation absorbed
in ozone layers

15 miles altitude

Sphere of
weather

Troposphere

Earth

Great absorption and
reflection of
remaining radiation

FIGURE 5.10 Generalized cross section of the atmosphere.

The stratosphere is a relatively calm layer of the atmosphere, even though some interactions occur between it and the troposphere. Nowhere in this layer are there turbulence and sudden changes like the ones that occur in the troposphere.

The Ionosphere

The boundary between the stratosphere and the ionosphere is referred to as the stratopause. Beyond the stratopause, in the ionosphere, at a height of approximately 55 miles there is a strong increase in temperature. In the lower levels of the ionosphere the temperature climbs to 170°F (Fig. 5.10).

At the stratopause, there is a great concentration of electrically charged particles. These particles, called ions, give the *ionosphere* a special set of characteristics. As solar energy bombards this layer, some atoms of gases lose electrons, and other atoms pick up excess electrons. Those atoms that lose electrons are left with a positive charge; those that gain electrons are left with a negative charge. These interactions produce a marked increase in temperature. At the outer fringes of the ionosphere, the kinetic temperatures reach well over 1500°F.

The outer boundary line of the ionosphere, the *exosphere,* is nearly 300 miles from the surface of the Earth. This is the point that marks the beginnings of "outer space."

The Kennelly-Heaviside Layer

The Kennelly-Heaviside layer is a particularly interesting subdivision of the ionosphere. It is at an approximate height of 60 miles, within the lower limits of the ionosphere.

The Kennelly-Heaviside layer reflects radio waves as if it were a gigantic reflecting "mirror" (Figure 5.14). If it did not exist, radio stations would have to be in a straight line with one another in order to receive short waves. That is, if the receiving station were below the horizon of the sending station, the curvature of the Earth would prevent direct communications. But the Kennelly-Heaviside region and other similar regions of the atmosphere are used to extend the range of normal communications by hundreds of miles.

The waves reflected off the Kennelly-Heaviside layer can be picked up by stations below the horizon of the sending station because the atmosphere has "bent" the waves around the curve of the Earth. It might be added that this cannot be done with television waves, which are more powerful than radio waves, because they are able to penetrate the layer.

The Homosphere and the Heterosphere

When the atmosphere is considered in terms of differences in composition rather than in temperature, we see that it is made up of two different zones, the homosphere and the heterosphere.

The *homosphere* includes the troposphere, stratosphere, and ionosphere. It is composed of a relatively uniform mixture of oxygen, nitrogen, and other gases. In the *heterosphere*, on the other hand, the mixture of gases is not uniform.

In the heterosphere, temperatures increase steadily, and a distinct layering occurs. Each layer of the heterosphere is composed of different gases with the lower layers containing gases of higher atomic weight than the higher layers. The base layer of the heterosphere is

mostly molecular nitrogen and oxygen, with a smaller percentage of atomic oxygen (O). At a height of about 140 miles, a layer begins in which oxygen atoms become dominant. A small percentage of nitrogen is also present.

The two highest layers in the heterosphere are very different from anything encountered at lower levels. The third layer, which begins

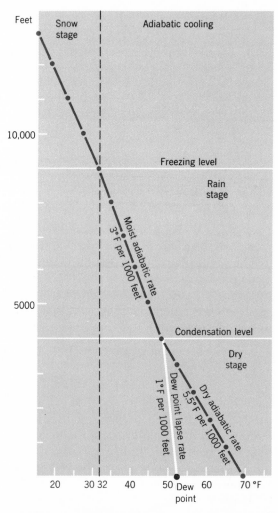

FIGURE 5.11 Moist and dry adiabatic rates of cooling. Adiabatic cooling is a consequence of the expansion of the gases, and not an actual loss of heat.

700 miles from the Earth and is 1500 miles thick, consists mainly of helium atoms. The fourth layer is mostly hydrogen. It is curious that both of these gases play minor roles in the homosphere, yet are abundant in these layers of the heterosphere.

The heterosphere is not very dense. In fact, it may be said that from an altitude of 100

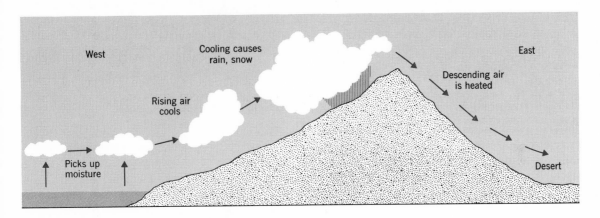

FIGURE 5.12 (*above*) Local topography affects the character of air masses. As an air mass is forced to rise over a mountain mass, moisture is lost because of adiabatic cooling. As the air descends the other side of the mountain, it is a heated, drying wind.

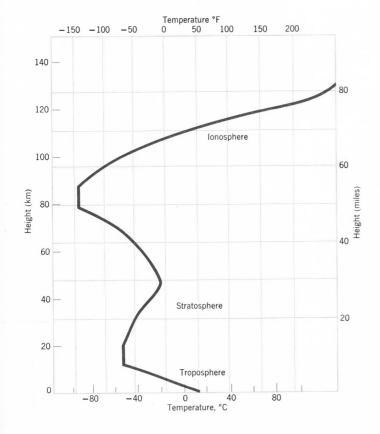

FIGURE 5.13 (*left*) Approximate temperature ranges of the three atmospheric layers. Notice the sudden and sharp rise in the stratosphere in the region of heavy ozone concentration.

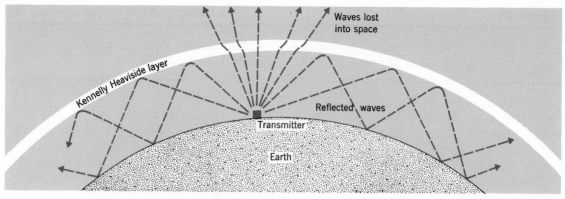

FIGURE 5.14 Radio waves are reflected from portions of the ionosphere around the curve of the Earth. This reflection is due to the electrical nature of the layer of gases found in this section of the atmosphere (After Navarra and Strahler).

miles, conditions similar to those of outer space exist. The low densities in these layers point to the weak gravitational attraction that holds this last portion of the atmosphere to the Earth.

The Jet Stream

Man has recently discovered a phenomenon located near the upper limits of the troposphere in the Temperate Zone of the Northern Hemisphere. This phenomenon, known as the jet stream, is a cylinder of rapidly moving currents of air several hundred miles wide. As shown in Figure 5.16, the currents travel one inside another, almost like a series of various-sized tubes.

The winds in the jet stream increase in speed from the outermost to the innermost regions of the column. In the outer regions, the wind is estimated to be 50 miles per hour. It gradually increases toward the center, where the wind speed is well over 200 miles per hour.

Jet stream winds move in an easterly direction, following an undulating horizontal path across the United States. Pilots use the jet stream to push their planes when flying in an easterly direction and avoid it when flying in a westerly direction.

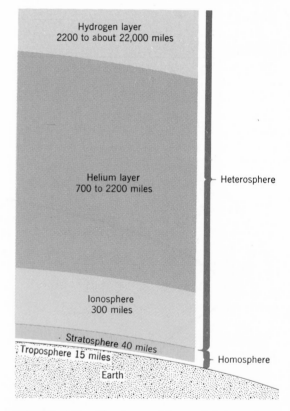

FIGURE 5.15 The homosphere and heterosphere classification of the layers of the atmosphere shown in relation to the previous classification system. This classification uses the gaseous composition of the atmosphere, whereas the previous classification used various physical characteristics as its basis.

134

The Atmosphere and Insolation

Many properties of the atmosphere are results of the interactions between solar radiation and the molecules of gas that make up the atmosphere. As well as producing radiant energy, the sun also releases large quantities of subatomic particles that bombard the Earth's atmosphere.

The Earth's atmosphere, like the Earth's surface, gains its heat by absorbing solar radiation and interacting with subatomic particles. It is primarily this heat that is responsible for the activity found in the atmosphere as it moves across the Earth's surface.

The solar radiation that reaches the Earth and is involved in heating the atmosphere is often referred to as *insolation*. The term for this energy is taken from the abbreviation for *In*coming *Sol*ar Radi*ation*.

The Aurorae

The aurora borealis, or "northern lights," and the aurora australis, or "southern lights," are ionospheric disturbances caused by solar energy. These aurorae are the result of solar particle emission, which increases and decreases in 11-year cycles that correspond to sunspot activity on the surface of the sun. When sunspot activity increases, the aurorae become active. In the Northern Hemisphere, the aurora borealis is visible above 70°N latitude. The aurora australis is visible below 70°S latitude in the Southern Hemisphere.

Auroral displays take place within the ionosphere. Since their activity corresponds

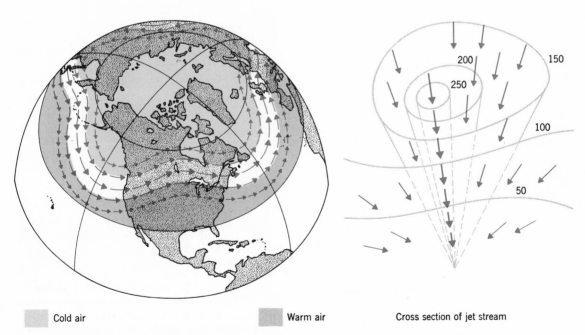

Cold air   Warm air   Cross section of jet stream

FIGURE 5.16 Typical movement of a jet stream across the face of the globe in the Northern Hemisphere. Jet streams vary in location and configuration.

to solar activity, it has been concluded that they are a result of atmospheric interchanges with particles and energy emitted by the sun. Storms and other unusual electrical disturbances on the sun's surface produce more particles than usually come from the sun, and these particles are thought to excite the atoms of gas in the ionosphere and thus produce the aurora. The excitation of the layers of the ionosphere apparently also causes disturbances in communications systems on the Earth.

Atmospheric Protection

The Earth is constantly bombarded with solar energy and particles carrying either negative or positive charges. These particles, which come not only from the sun but from other sources in space as well, are called cosmic rays.

Cosmic rays interact with the atmosphere in ways that prevent them from reaching the earth. Some particles are absorbed, and some are used in chemical reactions within the atmosphere. Thus, a large percentage of these particles are filtered out far above the Earth's surface. Because of the atmosphere, the amount of cosmic rays actually reaching the surface is safely within the limits man can tolerate.

The Van Allen Radiation Belts

In 1958 early exploratory satellites made a most exciting discovery in the upper atmosphere. They found that the Earth is surrounded by two doughnut-shaped belts of electric particles following the configuration of the Earth's magnetic field. These belts have been given the name *Van Allen radiation belts,* after the scientist who first proposed their existence. They are shown in Figure 5.18.

FIGURE 5.17 Various auroral displays, all taken in Alaska. The first display is a ray, the second a folded band, and the last converges to a point (Victor P. Hessler).

136

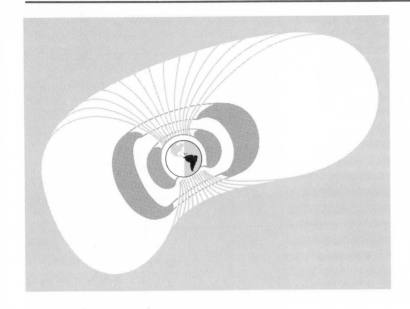

FIGURE 5.18 The Van Allen radiation belts were the first significant discovery accomplished by the early satellite program. They extend outward in two doughnut-shaped bands from a height of 500 miles to more than 10,000 miles.

The particles making up the Van Allen belts result from cosmic rays attacking the atmosphere. The Earth's magnetic field traps these charged particles permanently, so that they follow a tortuous, twisted, involved path of travel from pole to pole (Figure 5.19).

The lower limits of the belts of radiation are about 400 to 500 miles from the Earth's surface; they extend outward for at least 10,000 miles. Together they form one huge region of radiation. The Van Allen belts prevent much radiation from ever reaching the Earth's surface—radiation which might prevent the existence of life were it to reach the Earth's surface.

The Greenhouse Effect

The study of solar energy reveals several interesting facts. For example, consider some of the facts we know about the moon. The temperatures on the moon vary through a range of several hundred degrees in a single lunar day. On the Earth, temperatures vary no more than one tenth as much. The average

surface temperature on the moon is −40°F. On the Earth, the average is 40 to 50°F. Why should there be such a difference between the two celestial objects? The moon lacks one significant feature of the Earth: it has no atmosphere. This accounts not only for the great swings in lunar temperature but also

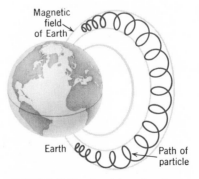

Magnetic field of Earth

Earth

Path of particle

FIGURE 5.19 Particles that become trapped in the magnetic field of the Earth spend their existence following a tortuous, twisted path between the poles of the Earth.

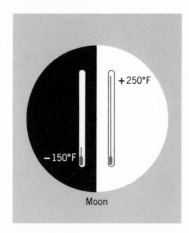

FIGURE 5.20 The temperatures on the moon yield a wide range throughout its "day" and "night." The absence of an atmosphere offers no protection like that found on Earth.

mains in the atmosphere or is absorbed by the Earth. This quantity of heat is called the *heat balance.* It is passed around the Earth by circulating winds and enables the entire Earth to maintain an equivalent distribution throughout.

In contrast with the moon, the Earth's atmosphere allows less than ½ of the entering solar energy to reach the Earth's surface. The rest is partially absorbed in the formation of ozone in the stratosphere, by carbon dioxide, and, most important, by water vapor in the troposphere. Some of the energy that reaches the Earth is reflected back by clouds, by other particles in the atmosphere and by the surface. The degree to which the Earth, or any surface, reflects solar energy is described by the term *Albedo.*

Solar energy is mostly in the range of visible light and shorter wave frequencies. Thus, a good proportion can penetrate the cloud and atmospheric covering of the Earth. However, when the Earth absorbs and reradiates solar energy, the energy is in a long-wave form that does not easily penetrate the atmosphere. Thus, the radiated energy is trapped by the atmospheric cloud covering and retained in the atmosphere as heat. The total effect is like that of putting clothes on the body. The clothes trap body heat, raising the inside temperature above that of the outside. In a similar way, the trapped energy

for the lack of water, the nonexistence of life, and the fact that a steady stream of meteors bombards the moon every day rather than being burnt up above the surface. The lack of an atmosphere is directly related to the fact that the moon possesses only a weak gravitational pull. It is insufficient to hold any great amount of atmospheric gases.

The atmosphere of the Earth traps heat in the form of solar radiation. Although a small quantity is radiated back to space, much re-

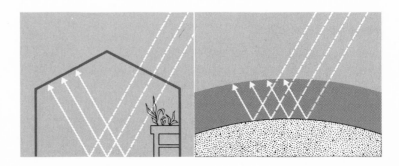

FIGURE 5.21 The Earth's atmosphere acts like the glass in a greenhouse. Short-wave radiation passes through the atmosphere, but long-wave radiation is reflected back toward the surface.

in the atmosphere raises the average temperature of the Earth.

To sum up, the atmosphere eliminates some of the solar energy, preventing the Earth from having high temperatures during the day. Conversely, radiation from the Earth is retained by the atmosphere, producing a higher night temperature. It is for this reason that the Earth's heat is not lost at night, as it is on the moon. Thus the temperatures do not drop suddenly and drastically. This entire process has been termed the *greenhouse effect* because of its similarity to the effects of glass in a greenhouse.

CHAPTER 6

Dynamics of the Atmosphere

AIR IN THE ATMOSPHERE IS IN CONSTANT movement. As it travels from one locality to another, its physical characteristics change rather drastically. Since these changes in large part determine the weather of the new location, the physical characteristics of air are a primary concern for the meteorologist.

Although the meteorologist studies the entire atmosphere, he concentrates on the troposphere because this is the layer in which weather changes take place. The troposphere is in constant motion, with air moving in all directions depending on local conditions. The meteorologist must study local conditions in order to make accurate weather predictions.

Winds

Air may move over a small area, or it may travel over thousands of square miles, encircling the entire globe. To simplify our discussion, we shall refer to horizontal air movements as *local breezes* or *winds*, and large-scale, relatively uniform parcels of air as *air masses*.

Any parcel of air moves when the pressure exerted on one side is greater than on the other. This difference in pressure, called the *pressure gradient*, forces the air toward an area of lower pressure. Air moves either horizontally or vertically.

From this simple description we might expect winds to travel in straight paths from one place to another. However, many factors come into play. As a result, winds in the Northern Hemisphere move toward the right side of what might be the expected path of movement, and in the Southern Hemisphere move toward the left of the observer. This motion toward one side or another results from what is called the Coriolis Effect.

Coriolis Effect

The Coriolis Effect is easy to conceive if we imagine a simple analogy: If the Earth were a stationary object and we were to stand at the North Pole and fire a rocket at some point directly south on the equator, the rocket would appear to travel in a straight line to the target. But the Earth is not a stationary object; it is rotating in a west-to-east direction. While the rocket is traveling in a southerly direction, the Earth will have rotated underneath the rocket toward the observer's left. When the rocket reaches the equator, it will strike a point far to the right of the original target. To the observer who is moving with the rotating Earth, the rocket appears to veer to the right under its own power. Actually, however, it is the Earth that is turning. (See Figure 6.1.)

This *apparent* shift of the object—called the Coriolis Effect—is the result of the Earth's rotation. The Coriolis Effect also plays a role in the movement of winds passing over the

Earth's surface. Winds appear to veer to one side of the observer because of the Earth's movement.

If we fired a rocket from the South Pole, the planet would be rotating toward our right and the rocket would appear to veer to the left. As before, the Earth actually moves beneath the rocket and the rocket only seems to move toward the left. Since the observer is rotating with the Earth, he interprets the Earth's motion as the motion of the rocket.

The Coriolis Effect is caused by a phenomenon known as angular momentum. Although this phenomenon can only be fully discussed in difficult mathematical terms, a few general remarks can be made. The curvature in the path of a moving object results from the fact that, as it moves from the North Pole in a southerly direction across the face of the Earth in the Northern Hemisphere, it crosses increasingly larger circles of latitude until the equator is reached. As the object crosses circles of greater circumference, its velocity slows and its speed lags behind that of the Earth's rotational motion. Thus, the object appears to veer to the right of an observer in the Northern Hemisphere. When the object moves north from the equator, it speeds up as it crosses parallel circles of latitude that decrease in size. Thus, the moving object increases its speed in relation to the rotating Earth and again appears to veer to the right of the observer.

Movement Around Pressure Systems

Air movement is influenced by the Coriolis Effect and by the pressure gradient. In the Northern Hemisphere, these two forces cause winds to move in a clockwise motion around an area of high pressure and in a counterclockwise direction around an area of low pressure. Low-pressure systems are known as *cyclones,* and the counterclockwise wind movement is called a *cyclonic wind flow.* Winds converge on the center of such a system. A high-pressure system, referred to as an *anticyclone,* has winds moving out of the center in an *anticyclonic,* clockwise flow. As winds from high-pressure systems sink and warm up, they impose an area of fair weather within the system. Within low-pressure systems, the wind rises, cools, and produces an area of turbulence often accompanied by precipitation.

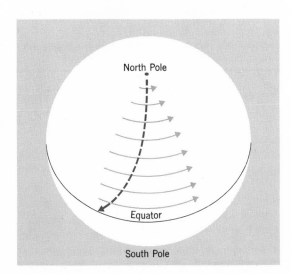

FIGURE 6.1 An illustration of the Coriolis effect acting on a missile moving southward from the North Pole. Relative speed is slowed by the increasing size of the circles of latitude, so that the missile lags behind the rotational velocity of the Earth.

Vertical Movement

Winds also move vertically. Several factors may produce this vertical movement, but in the majority of cases it is because of the differences in density of air masses. These differences arise from unequal heating. For example, if one heats a gas, its molecules move faster and require more room to move. Therefore, a particular volume of atmospheric gas contains fewer molecules when warmed than the same volume when it was cool (Figure 6.3). The heated gas weighs less per unit volume than it did when it was cooler.

Air receives heat from the Earth's surface. Thus, the air in contact with the surface tends to be warmer than the air above it. The warmer surface air weighs less than the surrounding, cooler air. This difference in density causes the cooler air to move into the area of low pressure, buoying up the warm air. In turn, there will be a corresponding upper-level movement of air from the region above the warm air toward the cool air. Two motions will then be present—a low-level motion of cool air toward the low-pressure region, and an upper-level motion in the oppo-

site direction. This circular motion of air is termed a *convection current.* It is one process by which heat travels through liquids and gases. (See Figure 6.4.)

General Circulation of the Atmosphere

The circulation of the atmosphere is controlled by convection currents and by the deflection that the Coriolis Effect produces. Figure 6.5 illustrates the general circulation in diagrammatic form.

We shall begin our discussion of atmospheric circulation with the equator. The equator receives a great deal of solar energy throughout the entire year. As air in this region is warmed and expands, a low-pressure

FIGURE 6.2 **Air movements around high- and low-pressure systems. In the Northern Hemisphere, winds move in a clockwise direction out of a high and in a counterclockwise direction into a low.**

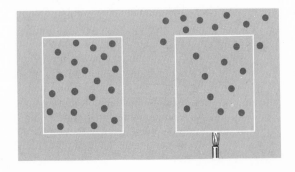

FIGURE 6.3 **A warm air mass contains fewer molecules than a cool air mass of equal size. Therefore, the warm air mass is less dense and will be buoyed upward by the colder.**

system is created. The air moves upward toward the upper level of the troposphere. As a result, the equatorial region is relatively calm during most of the year and is referred to as the *doldrums*. Winds move toward the equator from both the northeast and the southeast. Because winds from both hemispheres meet here, the equator is also known as the *intertropical convergence zone*.

A wind is named for the direction from which it comes, since its point of origin determines its character. Winds originating between the equator and the Tropics of Cancer and Capricorn are known as *prevailing easterlies*. These winds, commonly known as the trade winds, since they drove the ships of the early explorers to the New World, rise upward at the equator and move north and south until they reach approximately 30°N

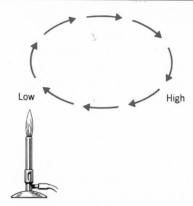

FIGURE 6.4 The warmer, less dense mass of air over a heat source will rise, leaving a low-pressure system behind. Cooler, surrounding air will move into the low-pressure system, forming a generally circular path as it is heated and rises.

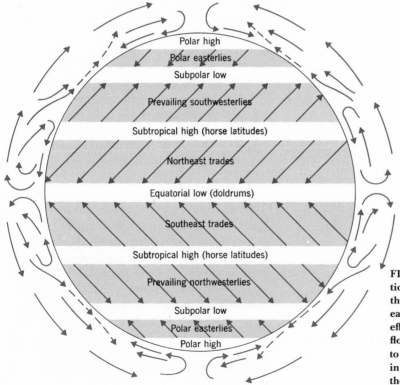

Polar high
Polar easterlies
Subpolar low
Prevailing southwesterlies
Subtropical high (horse latitudes)
Northeast trades
Equatorial low (doldrums)
Southeast trades
Subtropical high (horse latitudes)
Prevailing northwesterlies
Subpolar low
Polar easterlies
Polar high

FIGURE 6.5 The general circulation of the atmosphere, showing the three general cells of circulation in each hemisphere. The Coriolis effect on the direction of surface flow forces each air mass to move to the right from its point of *origin* in the Northern Hemisphere and to the left in the Southern Hemisphere.

latitude and 30°S latitude. This is the limit of the *first cell* (section) of general circulation. At 30°N latitude and 30°S latitude the upper atmospheric portion of this convectional circulation descends to the Earth's surface and sweeps toward the equator in an east-to-west direction. At 30°N latitude and 30°S latitude the *second cells* of circulation begin. These are regions of high pressure, since air settles toward the Earth's surface and accumulates there. Consequently, these regions are known as the *subtropical high-pressure belts*. Some of the accumulated air, the easterlies, moves toward the equator. The remainder moves from this high-pressure area along the surface from southwest to northeast to approximately 60°N latitude, and from northwest to southwest at 60°S latitude. These winds are called the *prevailing westerlies* in the Northern Hemisphere, and the "roaring 40's" in the Southern Hemisphere.

Wind-driven trading ships were often becalmed in the subtropical high-pressure belts for long periods of time. To conserve food and water, the horses on board the ships were thrown overboard. This happened so often that the sea was frequently littered with dead horses, and the region was given the unusual name of the horse latitudes.

The third cell of circulation lies between 60°N or 60°S latitude and the respective pole. Between the second and third circulation cells lie the *polar fronts*. At the Arctic Circles (66½° N and S), the air again rises, leaving low-pressure areas behind. This air moves toward the poles in the upper atmosphere and sinks at the poles. Thus, the poles are regions of high pressure. The surface winds move from the poles along the surface of the Earth in a northeast to southwest direction in the Northern Hemisphere and a southeast to northwest direction in the Southern Hemisphere. These winds are called the *polar easterlies*.

To summarize, each hemisphere is marked by three cells of circulating air. The first cell is located between the equator and 30° N and S latitude, the second between 30° and 60° N and S latitude and the third between 60° N and S latitude and the poles. Each cell is a general convection current modified on the surface by a deflection caused by the Coriolis Effect. If one stood at the origin of each surface wind and faced in the direction toward which the wind was moving, the deflection would be toward the observer's right in the Northern Hemisphere and to the left in the Southern Hemisphere.

The Earth's Heat Balance

The statement can be made that the motion of air is the result of solar energy. That is, heating and the beginnings of a convection current are the primary reasons for the general circulation of the atmosphere. The circulation is complicated by the Coriolis Effect and by friction between the air and the Earth, but they are not the driving forces behind the initial movement.

The circulation of the atmosphere distributes heat around the Earth. The equatorial regions of the Earth receive more heat through radiation than they lose during the course of a year. If there were no circulation, there would be a continuous buildup of heat in the equatorial regions. Similarly, the polar regions lose more heat than they are able to gain, so that, if there were no circulation, the poles would constantly grow colder.

But fortunately the atmosphere is continuously in circulation. Instead of the polar regions always growing colder while the equatorial area gets hotter, heat is constantly transferred from the equator toward the poles by atmospheric circulation. A heat balance is maintained, so that each area of the Earth keeps a fairly constant temperature throughout the year. This air circulation, combined with the greenhouse effect, pro-

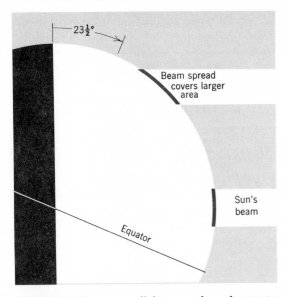

FIGURE 6.6 Because sunlight passes through a greater portion of the atmosphere at higher latitudes, the energy is diffused over a larger portion of the Earth's surface. Therefore, the same amount of energy covers a larger surface area at higher latitudes and is less intense.

duces temperature ranges capable of supporting life.

We have said that the Earth keeps a fairly constant temperature throughout the year. Of course, temperatures do vary somewhat with the seasons. This variation is due to several factors. Some of them are suggested in Figures 6.6 and 6.7.

At its summer solstice, the Northern Hemisphere receives intense, direct sunlight. Because the Earth's north pole is tilted toward the sun, summer is also the period when this hemisphere has its longest days. At its winter solstice just the opposite occurs. The Northern Hemisphere is tilted away from the sun and thus receives indirect, less intense rays. In addition, days are shorter during the winter months. These factors determine the rise and fall of winter and summer temperatures. Seasons are not due to the greenhouse effect or to the circulation of air.

The seasons are reversed in the Southern

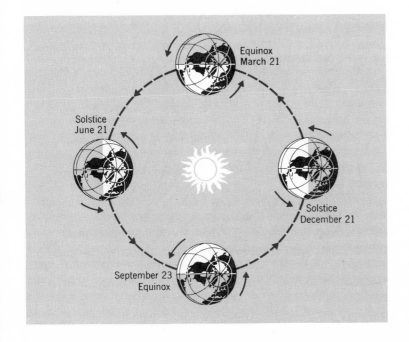

FIGURE 6.7 Earth-sun relationships during the four seasons. Each position represents the relationship of the Earth to the sun at the beginning of a season.

Hemisphere. That is, when the Northern
Hemisphere is having winter, the Southern
Hemisphere is having summer. The winter
conditions that exist in the Northern Hemisphere do not occur until six months later in
the Southern Hemisphere.

Air Masses of the United States

Each air mass has its own characteristics,
depending on where it developed (Figure
6.8). Masses that develop over water (maritime air masses) are moisture laden. If the
parcel originates over land, it is termed a
continental air mass. Continental air masses
are comparatively dry.

The general temperature of an air mass
also depends on its point of origin. Maritime air masses that originate over the cold
seas of the Arctic have lower temperature
than the ones that originate over the warm
seas of the tropics. Temperature is crucial,
since this determines the amount of water the
air mass can hold. Maritime air masses originating in the tropics (referred to as mT
masses) tend to be warmer and moister than
maritime air originating in the Polar regions.
A maritime Polar (mP) air mass tends to be
colder and less moist than an mT parcel.
Meteorologists also designate parcels of air
originating in the northern regions of the
Arctic as *Arctic* masses.

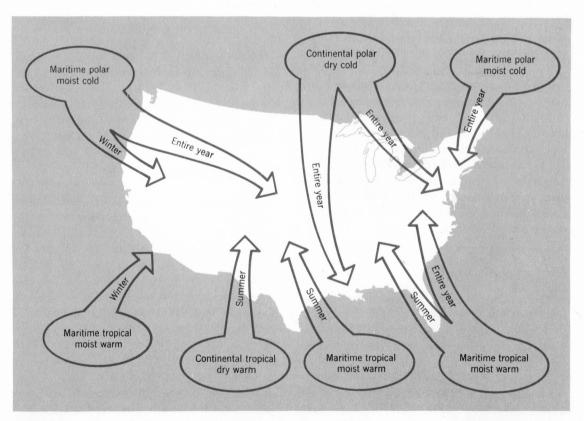

FIGURE 6.8   Major air masses of the United States and their characteristics.

FIGURE 6.9 Fog moving in at Big Sur, California. This fog is a product of the difference between land and water temperatures (A. and S. Wechsler).

Maritime Air Masses

There are two important categories of maritime air masses. The *maritime Polar* (mP) masses originate over high-latitude areas. They are cold and contain some moisture, although not much.

The *maritime Tropical* (mT) air masses originate over the south Atlantic and Pacific Oceans. These are generally warm and quite moist. Maritime Tropical air also originates in the Gulf of Mexico.

Continental Air Masses

The *continental Polar* (cP) air masses originate over large, northern land masses such as Canada, Alaska, and Siberia. These air masses are dry and cold since they pass over few, if any, open expanses of water.

The *continental Tropical* (cT) air masses originate over areas such as northern Mexico. They are most common in the southwestern desert regions of the United States. Having

little opportunity to pass over water, they are hot and dry. However, these air masses have the capability of holding a great deal of moisture and may actually leach moisture from areas over which they pass, thus, further parching an already dry region.

Changes in Air Masses

Air masses are eventually changed by the area into which they move. Maritime Tropical air soon becomes drier and cooler on passing over a cool land surface. When continental Polar air moves over the Great Lakes, it becomes laden with more moisture. The fact that air masses are constantly changing character makes weather forecasting difficult.

Weather Produced by United States Air Masses

When maritime Tropical air masses spawned in the south Atlantic and Caribbean move over the eastern seaboard of the country, they produce the hot, muggy weather

FIGURE 6.10 A squall just prior to a thunderstorm. The wind effects can be seen on the grasses in the bottom of the picture. Dark stratus clouds are moving into the scene from lower right (Max Baur, Photo Researchers).

typical of late July and August. This same Caribbean area spawns the hurricanes that may move toward the eastern and Gulf coasts any time from June through October. On the west coast, maritime Tropical air from the south Pacific affects the southern coast of California in the same manner during the winter. Fog, drizzle, and overcast skies are common here when this air sweeps inland. A smaller maritime Tropical air mass also moves across the Texas-Louisiana Gulf Coast. On contact with the land it produces squalls and rainy weather. A small continental Tropical air mass from Mexico affects the southwest and produces dry, hot conditions.

In winter, cold moist maritime Polar air from the north Atlantic produces the storms that are called Nor'easters in New England whaling stories. This cold wet air affects the entire northeast coast and may reach as far as the southern states during midwinter. The west coast has its counterpart from the North Pacific Ocean. When cold, moist air is forced to climb the mountain ranges of California, Oregon, and Washington, great amounts of rain fall on the western, or windward side. Adiabatic cooling of this air leaves as much as 200 in. of rainfall a year in the western regions of the state of Washington. As this air descends the eastern or leeward slopes, very little rain falls. Desertlike conditions exist in certain areas on the lee side of the mountains.

Canada and the polar regions are the spawning grounds of the very cold continental Polar air mass that sweeps into the Midwest. It can move south, west, and east. But, generally, we associate the cold weather of the northeastern United States with this air mass. The regions over which cP air moves, particularly near the Great Lakes, have severe winters that are in sharp contrast to the warm, dry summers.

Lesser Circulations in the Atmosphere

The character of air and its effects on daily weather are directly related to the topography of the area over which it is passing. Weather conditions are also affected by winds that move from land to water or from the water to land. This is because of the great difference in *specific heat* of these two substances. Specific heat is the amount of heat required to raise the temperature of a quantity of a substance. Given equal quantities, a greater amount of heat is required to raise the temperature of water than the temperature of the land.

Land and Sea Breezes

Land adjacent to a body of water heats up more rapidly than the water, and becomes significantly warmer than the water. Thus, parcels of air over the land become warmer than similar parcels over the water and a low-pressure area develops. This causes air over the water to replace the low-pressure parcel over the land. A convection current is set into motion.

Cooler air from the water moves in to replace the warmer air over the land. The air moving along the water toward the land is a cool sea breeze (Figure 6.12). This is the type of wind that cools a seashore resort during the summer months.

At night, the situation reverses itself and the land cools more rapidly than the water. We now find a low-pressure cell developing over the water. The warm air over the water is pushed upward by cooler, higher pressure air from over the land (Figure 6.13). This land breeze is an exact reversal of the prevailing winds during the day.

The difference in specific heat between land and water produces a modification in climate for areas near the water. Since the

FIGURE 6.11 A snowdrift shortly after a heavy snow. The effects of wind cause drifts which are deeper than the original fall (Lawrence Smith, Photo Researchers).

ocean does not lose its summer heat until after the land has cooled, autumn sea breezes help raise land temperatures slightly above those of inland areas. In summer, when the land is hottest, the cooler sea breezes help keep temperatures below those of inland areas.

Mountain and Valley Breezes

As their names imply, mountain and valley breezes are produced by the effects of topography on air (Figures 6.14 and 6.15).

In mountainous terrain during the evening, air near the peak of a mountain is cooled

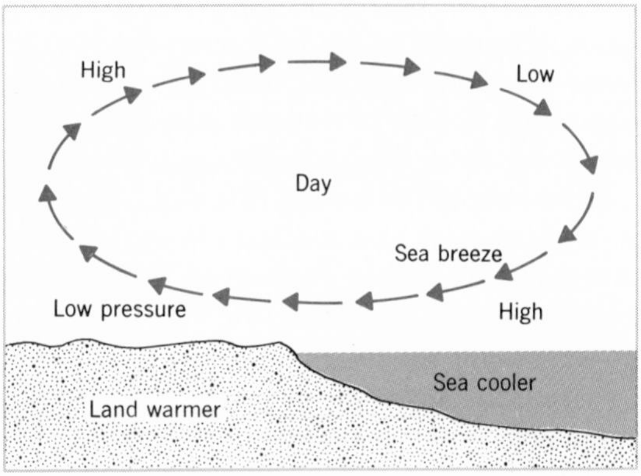

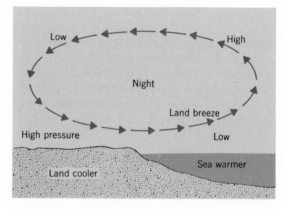

FIGURE 6.12   A sea breeze is formed when the land heats up faster than the adjacent water. Notice the location of the pressure systems and the prevailing air circulation.

FIGURE 6.13   A land breeze is formed when the land cools faster than the adjacent water. It is just the reverse of the sea breeze in Figure 6.12.

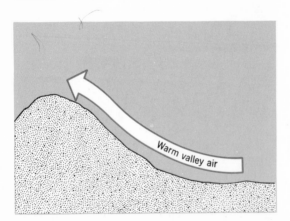

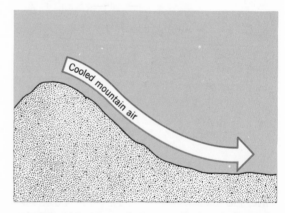

FIGURE 6.14   Valley breeze. During the day, as valley air is heated and expands, it moves upslope toward the mountain peak.

FIGURE 6.15   Mountain breeze. Cooler air from the mountain peak descends the slope readily, cooling the valley below.

and becomes dense and heavy. This air then moves down the side of the mountain as a cooling mountain breeze.

During the day, the warm air over the mountain slope is less dense than the air over the valley. Thus the valley air moves toward and up the slope, displacing the mountain air and warming the slope of the mountain. These breezes are light and local in origin but are so common that they deserve mention.

Monsoon Winds

Another common effect of local topography is the creation of monsoons (Figure 6.16). They are wind systems that cover a large region, but are similar in most respects to the land and sea breezes mentioned earlier.

Monsoons are common in many parts of the world, but are most severe over India and the Indian Ocean. In the winter, relatively dry winds move over the ocean from the land. Their motion occurs because the water is warmer than the land, and a low-pressure system is set up over the ocean. During the late spring and summer, the Indian continental mass is warmed. As the air warms, the pressure on the land decreases, and a low-pressure cell is centered over the land. Then moist ocean air sweeps over the land from the water. When this moist air reaches the Himalayas, it is forced to rise over the mountain slopes and begins to contract and cool. This causes the air to lose its moisture. During the four months of the monsoon season, as much as 400 inches of rain has been known to fall, although the average is closer to 200 inches. The rainfall is primarily the result of the

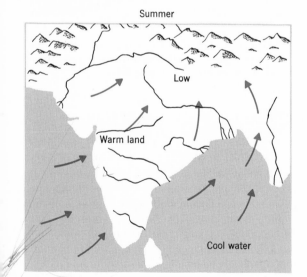
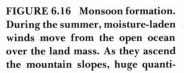
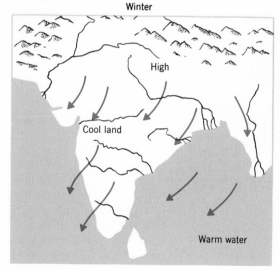

FIGURE 6.16 Monsoon formation. During the summer, moisture-laden winds move from the open ocean over the land mass. As they ascend the mountain slopes, huge quanti- ties of precipitation fall on the land. In winter, the reverse situation prevails and the cooler winds move from land over open water. This is the dry season for the regions of the world that experience monsoon winds. Although they are much larger in scope, monsoon formations are similar to local land and sea breezes.

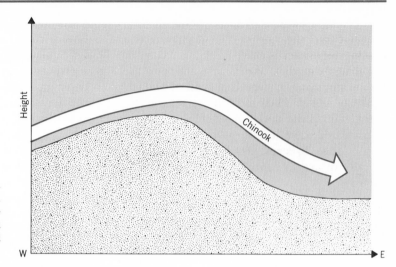

FIGURE 6.17 The formation of a foehn or chinook wind occurs when a high-pressure area forms on the windward side of a mountain and warms as it descends the leeward side.

moist ocean air being forced to rise into the upper atmosphere as it encounters the mountains.

Chinook or Foehn Winds

Another topographical effect on the atmosphere results in the common chinook winds found in the Rocky Mountains of the United States and the Swiss Alps (Figure 6.17). In the Alps these winds are known as *foehn* winds. They occur as wind is forced to rise over a mountain. The air is cooled adiabatically by the rise in altitude, and precipitation of water may occur over the upper slopes. As the wind descends the leeward side, it is warmed. It is warmed more rapidly on the lee side than it was cooled on the windward side. Thus, air temperatures are higher on the leeward side than on the wind-

ward side. Areas on the leeward side of the mountain are suddenly warmed several degrees by the arrival of a chinook wind.

The land and sea breezes, mountain or valley winds, monsoons, chinook, and foehns may influence areas from a few miles to thousands of square miles. Nevertheless, they are very local in origin and effects in comparison with the trade winds or other general circulatory patterns. In all cases, winds are modified or affected by many factors. Topographical features significantly modify the characteristics of air. Winds generally precipitate moisture on the windward side of mountains and are dry when descending the leeward side. In addition, pressure systems and friction with the land surfaces also influence the direction of movement and, therefore, the changes brought about in air masses and in their effects.

CHAPTER 7

The Elements of Weather

AIR AS LOCAL WINDS OR AS BROAD GEO-graphical structures such as air masses, develops its physical characteristics by exchanges of energy with the surface over which the parcel forms. As it moves, these characteristics change and are modified by further exchanges of energy with the regions over which it passes. In turn, these physical conditions of the air determine the weather in an area. For example, an area is referred to as calm, stormy, wet, or dry depending on the physical conditions of the air.

The specific characteristics of air that determine weather are pressure, temperature, wind speed, wind direction, humidity, cloud cover, and precipitation. In this chapter, we shall examine the physical conditions of air, the devices used by meteorologists to measure these conditions, and the significance of the changes that occur in each condition.

Pressure

Air is composed of molecules of various gases which are held to the Earth by the force of gravity. This force gives air its weight. A man pushes on the Earth with a force equivalent to his own weight; a column of air also pushes on the Earth's surface. This force—air pressure—is a measurable quantity. If it were possible to weigh a column of air 1 inch square which reaches from sea level to the top of the atmosphere, its weight would be 14.7 pounds. In the upper reaches of the atmosphere, the height and weight of the column of air are less, and air pressure is proportionately less.

The Barometer

The device used to measure air pressure is the barometer, invented in 1643 by Evangelista Torricelli. Torricelli, a pupil of Galileo, took a glass tube, sealed one end, and filled the tube with mercury. On inverting the open end of the tube into an open container of mercury, the column of mercury dropped in the tube to a level of 760 mm, or 29.92 inches. (See Figure 7.1.) Torricelli suggested that the column of mercury weighed the same as the column of air pushing down on it and that it was, in fact, the weight of the air on the container of mercury that supported the mercury in the tube. Torricelli was the first experimenter to discover that a column of mercury stood at a significantly lower level when the tube was carried to the top of a mountain. This evidence confirmed that the greater the altitude, the lower the air pressure.

At sea level, the pressure of the air supports 760 mm (29.92 in.) of mercury. This is really a measurement of height, and meteorologists prefer to use a system that is a direct measurement of pressure. Today, the unit most commonly used is the bar or millibar (mb), which is 1/1000 bar. (See Table 7.1.) The bar is equal to 1,000,000 dynes per square centimeter. Expressed in these units, the pressure at sea level is 1013.2 mb. This pressure is used as the standard of measure; it is commonly referred to as one atmosphere.

1 atm = 29.92 in., 14.7 P.S.I.
          = 760 mm of mercury = 1013.2 mb

In order that barometric readings at various heights and in various climates can be compared, the readings must be adjusted to a single standard. To correctly predict daily pressure changes, meteorologists must compare the readings taken in several localities. So that all the readings will have the same basis, weather stations above sea level must adjust their barometer readings to sea level before reporting them. Differences in temperature must also be adjusted, because heat causes mercury to expand in the barometer. Temperatures are corrected to a standard temperature of 0°C or 32°F. Barometric adjustments must also be made for gravity. Since the earth is irregularly

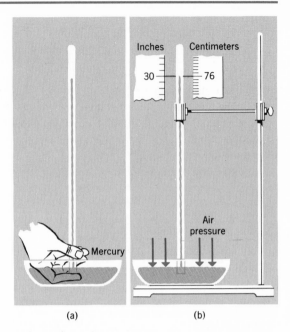

FIGURE 7.1 The production of a mercury barometer. The tube must be evacuated of all air before the mercury is introduced.

Table 7.1 Range of Pressures

MERCURY IN COLUMN INCHES	MILLIBARS
29.0	982.05
29.1	985.44
29.2	988.83
29.3	992.21
29.4	995.60
29.5	998.99
29.6	1002.37
29.7	1005.76
29.8	1009.14
29.9	1012.53
30.0	1015.92
30.1	1019.30
30.2	1022.69
30.3	1026.08
30.4	1029.46
30.5	1032.85

shaped, gravity will affect objects more at the poles than at the equator.

When all these factors have been corrected, readings from different places on the globe can be accurately compared. One can correctly determine the advent of a high- or low-pressure system and the changes occurring in air masses moving into a particular locality.

The mercury barometer has several drawbacks. It is large, difficult to carry about, and fragile. In recent years, a different type of barometer has been devised that lacks these disadvantages. This is the *aneroid barometer*. It is an evacuated metal can which is sensitive to pressure changes. When pressure increases, the can is slightly depressed; when it decreases, the can expands. This motion is carried through a series of levers and gears to a pointer that gives a reading on a dial. A variation on this arrangement uses a needle or lever arm tipped with a pen point and ink reservoir. The pen rests upon a drum turned by a clock mechanism and makes a timed mark on a chart attached to the drum. The chart can be calibrated in days or hours against millibars to give a continuous, permanent record of the pressure of air over a long period of time.

Airplane pilots use another variation of the aneroid barometer called the *altimeter* to find out how high they are flying. This device is marked in height above sea level instead of pressure. When adjusted for the actual pressure reading on the surface, the altimeter gives a direct reading of the altitude of the plane.

Consequences of Air Pressure

Air pressure by itself is not an accurate forecast of weather. A falling barometer usually indicates threatening weather, and a rising barometer usually indicates clearing conditions or more good weather. However, neither of these indications is hard and fast. Low-pressure systems and high-pressure systems have a great deal to do with weather, but many other factors must be considered. Many people who own handsome barometers have found, to their consternation, that the usual face readings of clear, stormy, and fair are frequently inaccurate.

Temperature

Air temperature is an important factor in weather forecasting, since it determines the movement of air. When air is heated, it expands so that it is lighter than the surrounding cooler air and rises above it. Conversely, as cooled air contracts and becomes dense, it sinks below the surrounding warmer air. Contact between a warm air mass and cool air mass changes both of them in ways that affect the local weather significantly.

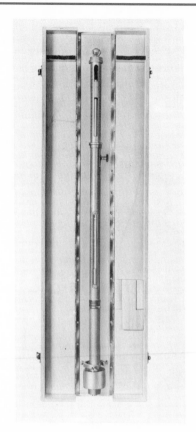

FIGURE 7.2   **Mercury barometer (Kahlsico).**

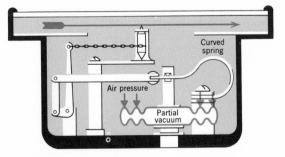

FIGURE 7.3   **The aneroid barometer does not use any mercury. An evacuated can reacts to changes in air pressure. The can, in turn, is attached to a series of levers and springs that move a dial arm. The arm may be tipped with a pen and an ink reservoir to form a recording barometer.**

The Thermometer

The mercury or alcohol thermometer is the instrument most commonly used to record temperatures. However, other thermometers are also used. One of those most frequently encountered is the *bimetallic strip* (Figure 7.8).

This thermometer consists of two different metals bonded together in a strip. When the strip is heated, the metals expand unevenly and, consequently, the strip will curve in the direction of the metal that expands less. The bimetallic strip is set against a properly calibrated dial to record temperature.

The bimetallic strip is ideal for use as a recording thermometer. As in the recording barometer, the strip is tipped by a pen and ink reservoir touching a revolving drum and paper. This makes a permanent recording when the bimetallic strip bends up or down as it is affected by the changes in air temperature.

Temperature Scales

Different scales of measurement are used to determine the temperature. Most of us are familiar with the scale devised by and named after Fahrenheit. In the Fahrenheit system the boiling point of water is 212°F and the freezing point is 32°F. Today the Celsius (centigrade) scale is the most widely accepted system for recording temperature. On this scale, the freezing point of water is 0°C and the boiling point is 100°C. There

FIGURE 7.4 (*above*) Aneroid barometer (Taylor Instrument Company).

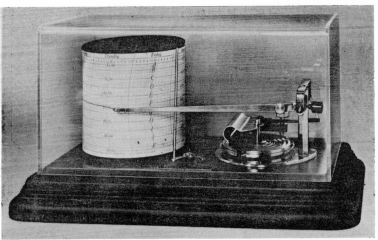

FIGURE 7.5 (*right*) Recording barometer (Taylor Instrument Company).

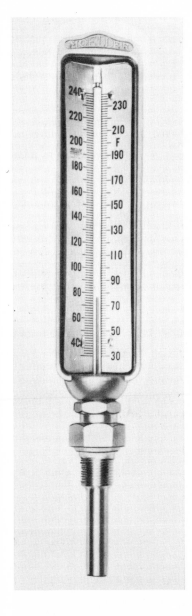

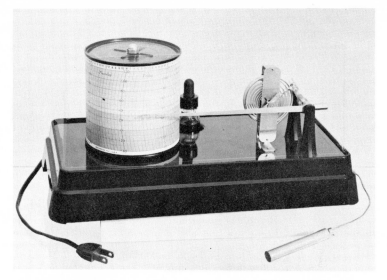

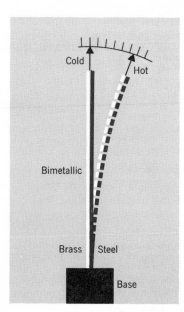

FIGURE 7.6 (*far left*) Mercury thermometer (Moeller Instrument Company).

FIGURE 7.7 (*above*) Recording thermometer (Taylor Instrument Company).

FIGURE 7.8 (*left*) In a bimetallic strip, one metal expands more than the other. The strip that expands less is forced over toward one side by the strip that expands more readily.

are only 100 units between these two standard points instead of 180° as on the Fahrenheit scale. These two scales are shown in Figure 7.9.

In 1848, Lord Kelvin devised a scale that does not employ any negative numbers, a distinct advantage over the other two systems. On the Kelvin scale, sometimes referred to as the absolute scale, the 0° mark is set at that point where there is no heat. This point is known as absolute zero. On the Kelvin scale the freezing point of water is +273°K and the boiling point is +373°K. Notice that the difference between the two

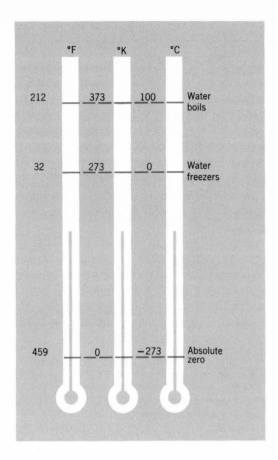

points is 100°, the same as on the Celsius scale. The 0°K point is comparable to −273°C or −459.3°F.

To convert one measurement system to another, a few simple procedures must be followed. They are given in Appendix I.

Significance of Temperature

Temperature readings are extremely important in weather analysis. For one thing, the warmer the air, the more water vapor it can hold, and when warm, moist air is cooled, condensation may produce fog, cloudy weather, and rain or any of the other forms of precipitation.

Thermometers and other weather instruments are usually kept in an instrument shelter, a closed wooden structure with louvered sides that allow for the movement of air through the building. This structure prevents the direct rays of the sun from striking the instruments and affecting the instrument readings.

FIGURE 7.9 (*above*) **Comparison of temperature scales.**

FIGURE 7.10 (*right*) **Instrument shelter (ESSA).**

158

FIGURE 7.11 (*left*) Wind vane and anemometer (Philip Gendreau).

FIGURE 7.12 (*below*) Anemoscope (Kahlsico).

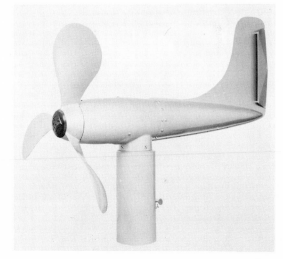

Wind Movements

Since winds move in specific ways around high and low pressure systems, measurements of wind speed and direction give a meteorologist an accurate picture of pressure systems and local conditions that will affect air movement.

The Measurement of Wind Speed

Wind speed is measured by an *anemometer,* a pair of cups attached to the ends of a crossbar with the crossbar attached to a vertical shaft. (See Figure 7.11.) The cups catch the wind and rotate at a speed proportional to that of the wind. The rotating cups in turn cause the vertical shaft to turn. The wind speed is recorded by an electrical counting device located at the base of the vertical shaft. It is usually recorded in knots (nautical miles per hour), which are approximately $1/7$ larger than statute miles.

Wind Direction

Wind direction is measured by a wind vane, which has an arrow that points in the direction from which the wind is coming.

Recently, *anemoscopes* (Figure 7.12) have been used to determine wind direction. At

Temperature of Dry Bulb

	61	62	63	64	65	66	67	68	69	70	71	72	73	74	75	76	77	78	79	80	
41	7	4	2																		41
42	10	8	6	4	2																42
43	14	12	10	7	5	3	2														43
44	18	16	13	11	9	7	5	3	1												44
45	22	20	17	15	12	10	8	6	5	3	1										45
46	27	24	21	18	16	14	12	10	8	6	4	3	1								46
47	31	28	25	22	20	17	15	13	11	9	7	6	4	3	1						47
48	35	32	29	26	24	21	19	16	14	12	10	9	7	5	4	3	1				48
49	40	36	33	30	27	25	22	20	18	15	13	12	10	8	7	5	4	3	1		49
50	44	41	37	34	31	29	26	23	21	19	17	15	13	11	9	8	6	5	4	3	50
51	49	45	42	38	35	32	30	25	24	22	20	18	16	14	12	11	9	8	6	5	51
52	54	50	46	43	39	36	33	31	28	25	23	21	19	17	15	13	12	10	9	7	52
53	58	54	50	47	44	40	37	34	32	29	27	24	22	20	18	16	14	13	11	10	53
54	63	59	55	51	48	44	41	38	35	33	30	28	25	23	21	19	17	16	14	12	54
55	68	64	60	56	52	48	45	42	39	36	33	31	29	26	24	22	20	18	17	15	55
56	73	69	64	60	56	53	49	46	43	40	37	34	32	29	27	25	23	21	19	18	56
57	78	74	69	65	61	57	53	50	47	44	41	38	35	33	30	28	26	24	22	20	57
58	84	79	74	70	66	61	58	54	51	48	45	42	39	36	34	31	29	27	25	23	58
59	89	84	79	74	70	66	62	58	55	51	48	45	42	39	37	34	32	30	28	26	59
60	94	89	84	79	75	71	66	62	59	55	52	49	46	43	40	38	35	33	31	29	60
61	100	94	89	84	80	75	71	67	63	59	56	53	50	47	44	41	39	36	34	32	61
62		100	95	90	85	80	75	71	67	64	60	57	53	50	47	44	42	39	37	35	62
63			100	95	90	85	80	76	72	68	64	61	57	54	51	48	45	43	40	38	63
64				100	95	90	85	80	76	72	68	65	61	58	54	51	48	46	43	41	64
65					100	95	90	85	81	77	72	69	65	61	58	55	52	49	46	44	65
66						100	95	90	85	81	77	73	69	65	62	59	56	53	50	47	66
67							100	95	90	86	81	77	73	69	66	62	59	56	53	50	67
68								100	95	90	86	82	78	74	70	66	63	60	57	54	68
69									100	95	90	86	82	78	74	70	67	63	60	57	69
70										100	95	91	86	82	78	74	71	67	64	61	70
71											100	95	91	86	82	78	74	71	68	64	71
72												100	95	91	86	82	79	75	71	68	72
73													100	95	91	87	83	79	75	72	73
74														100	96	91	87	83	79	75	74
75															100	96	91	87	83	79	75
76																100	96	91	87	83	76
77																	100	96	91	87	77
78																		100	96	91	78
79																			100	96	79
80																				100	80
	61	62	63	64	65	66	67	68	69	70	71	72	73	74	75	76	77	78	79	80	

Temperature of Wet Bulb

FIGURE 7.13 Relative Humidity Chart. When readings are obtained from a wet and dry bulb thermometer, the differences between the temperatures may be related to the individual readings by means of a chart such as this. A direct reading of relative humidity is obtained.

one end, this device has a tail that points toward the wind origin and, at the other end, a propeller that is set spinning by the wind. The propeller acts in the same way as the anemometer. At the base of the anemoscope an electrical device that is calibrated to the propeller movement produces a recording of both wind speed and direction.

Wind vanes and anemoscopes are usually placed at the tops of buildings or in a rather high tower in order to reduce the effects of local conditions and to avoid interference by high structures.

The Significance of Wind Observation

Records of wind readings give the meteorologist an idea of the direction of prevailing winds in a particular locality on a daily, monthly, and yearly basis. Analysis of these records allows the meteorologist to note seasonal differences in winds. For example, greater differences between high and low winter temperatures compared to summer temperatures produces greater wind movement in the winter. This information can be used to make long-range predictions.

The pattern of prevailing winds is another factor used to forecast weather. Forecasts of weather are more accurate when wind patterns are compared with past performances.

Humidity

The percentage of water vapor held by the air compared to the maximum amount which it can hold at that temperature is called the *relative humidity.* This term is undoubtedly familiar to you from the daily weather report.

As the temperature of the air changes, so does the relative humidity. An increase in temperature causes air molecules to move faster and farther apart, so that more water vapor could be accommodated in the additional space. If no additional water vapor enters the parcel, the relative humidity will fall, whereas the *absolute humidity* —the amount of water vapor or moisture content — remains the same. If the temperature is lowered in that same parcel, the relative humidity will rise, since the contraction of the air allows less room for additional water vapor.

The Earth's air temperatures change constantly; thus relative humidity is always changing, even if water vapor does not leave or enter the atmosphere. A chart of relative humidity is given in Figure 7.13.

If air is cooled sufficiently, the relative humidity will reach 100 per cent. The temperature at which this occurs is called the *dew point.* If the air is cooled any further, the excess water vapor may condense as clouds, rain, or dew. If the dew point occurs below freezing, ice crystals, snow, or frost will form.

The dew point can be obtained by placing water in a bright metal can and gradually adding ice to the water. Eventually a mist of condensed vapor forms on the outer surface of the can. When this occurs, the temperature of the water is approximately equal to the dew point of the air for that particular locality. The air in contact with the can has been cooled until the water vapor condenses on it. If all the air were cooled to that point, condensation would occur.

Measurement of Humidity

The humidity of the air is measured by a *hygrometer,* a device using a fiber (sometimes a human hair) that increases or decreases in

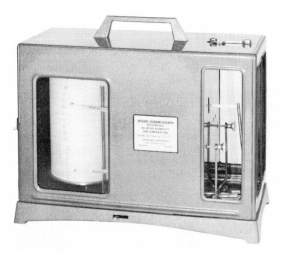

FIGURE 7.14   Recording hygrometer (Friez Instrument Division, The Bendix Corporation).

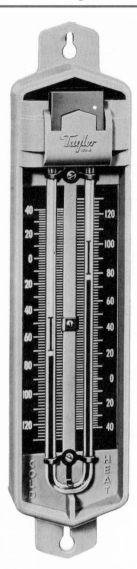

length in response to increasing or decreasing humidity. This fiber is attached to a pointer, and a recording can be made on a revolving drum (Figure 7.14).

A pair of thermometers may also be used to measure humidity. When they are set up with the bulb of one covered by a moistened wick, and one bulb remaining dry, the thermometer with the wet bulb, because of evaporation, produces a lower temperature reading than the dry bulb. (The process is speeded up if the thermometers are placed on a rotating mount and spun in the air.) The dry bulb indicates the room temperature. The difference in temperature between the wet bulb and the dry bulb is proportional to the dryness of the air. The readings obtained from this device can then be used in conjunction with previously calculated tables to arrive at the relative humidity of a locality.

FIGURE 7.15 The minimum-maximum thermometer has a pair of markers that denote the two important points of daily temperature range (Taylor Instrument Company).

FIGURE 7.16 Rain gauge (ESSA).

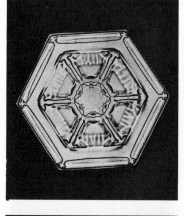

FIGURE 7.17 Various snowflake forms. Although the patterns are dissimilar, they all show the typical hexagonal pattern (R. B. Hoit, Photo Researchers).

Precipitation

When air is cooled sufficiently, condensation (or precipitation) occurs. Which type of precipitation takes place depends on temperature, and on the turbulence and the activity within the air.

Rain, a type of precipitation, is measured by a *rain gauge,* a cylindrical bucket containing a calibrated stick that measures the depth of accumulated precipitation (Figure 7.16).

Types of Precipitation

When water vapor condenses as droplets of moisture around a needle of ice, salt, or other nucleus, rain may be produced. These droplets accumulate more moisture and increase in size and weight. Eventually, they become heavy enough to fall through the atmosphere and reach the Earth's surface as rain.

Snow is produced when water vapor sublimes (changes from a gas to a solid without first going through a liquid state) at temperatures below the freezing point. Snowflakes are always hexagonal, but in other ways are extremely varied. Snow is advantageous because it insulates the ground during winter. It keeps the land warmer than it otherwise would be, and protects organisms within the soil.

FIGURE 7.18 A hailstone. When examined in cross section, hailstones reveal a series of concentric circles representing a series of layers of ice that formed at different levels (American Museum of Natural History).

FIGURE 7.19 Cooled water vapor coats surfaces forming a glaze of ice. Here, a ship has been covered by ice under extremely cold conditions (Woods Hole Oceanographic Institution).

FIGURE 7.20 (*left*) A sleet storm in Times Square, New York City (Jean Paul Jallot, Photo Researchers).

FIGURE 7.21 (*below*) Compacted needles of ice formed along a break in the ice on a lake (Albert Steiner, Photo Researchers).

Hailstones are composed of hard pellets of ice and compacted snow granules laid down in concentric layers. No one is really sure how hail is produced. One theory is that hailstones are formed in cumulonimbus clouds in which there are strong, circulating convection cells of cold air. The up-and-down motions of the hailstones produced by updrafts result in alternate layers of frozen ice and snow as the stones pass from one region of the cloud to another. Another theory holds that hailstones form when frozen droplets pass through several layers of cold, moist air and acquire repeated layers of ice by sublimation in the process.

Other forms of precipitation are sleet, glaze, and snow grains. Sleet is partially frozen rain. Glaze is supercooled rain that freezes on striking cold, solid surfaces. Snow grains are compacted ice needles.

Hygroscopic Nuclei

Salt and other particles that have an affinity for water are abundant in the atmosphere. These particles, termed *hygroscopic nuclei,* enter the atmosphere as spray from the oceans and wind-carried dust. As moisture accumulates in the air, water droplets form around the hygroscopic nuclei and become larger and heavier. When the droplets remain small, they stay suspended in the air as cloud formations.

Clouds

Clouds and fog are composed of fine droplets of water that are still light enough to remain suspended in air. These droplets are thousands of times smaller than raindrops.

Clouds are excellent indicators of impending weather. When the cloud ceiling (the distance between the Earth's surface and a major cloud formation) is low, precipitation is likely. Low visibility (the extent to which clouds, precipitation, and dust obscure horizontal vision) is an indicator of high humidity and possible imminent precipitation.

Cloud Types

Clouds result when warm, moist air has risen and cooled sufficiently for condensation to occur. (See Table 7.2.) Fog is really a surface cloud; it is produced when moist air that is close to the Earth's surface cools.

The different types of cloud are associated with different heights.

1. Low Level Clouds: Ground to 6500 feet.

Stratus clouds are low, sheetlike clouds covering a large part of the sky. They resemble a high fog.

Stratocumulus clouds are patchy masses, or rolls of clouds with a grayish tint. They are indicators of gusty, windy conditions.

Nimbostratus clouds are low hanging and dark gray in color. They are associated with continuous rain or snow.

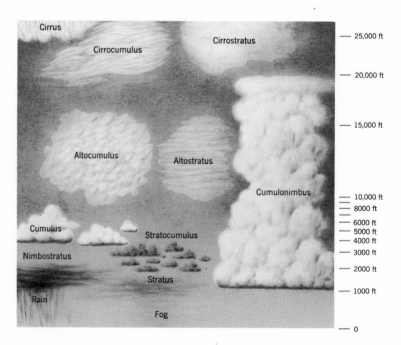

FIGURE 7.22   The families of cloud types.

2. Middle Level Clouds: 6500 to 20,000 feet.
Altostratus clouds are thick. sheetlike formations. They are dark colored and forecast imminent rain.

Altocumulus clouds are flattened, layered, patchy clouds which closely resemble altostratus formations. Although they are not associated with precipitation, altocumulus clouds are often forerunners of storms.

3. High Level Clouds: 20,000 feet upward. These are generally composed of minute ice crystals.

Cirrus clouds are white, feathery, wispy patches. They are generally fair-weather clouds.

Cirrostratus clouds are thin and whitish. The sun and moon can be seen through these clouds with a halo about their edges. The presence of these clouds indicates the possibility of a storm in 24 hours.

Cirrocumulus clouds are small puffs that are usually associated with fair weather. Their ripply appearance has earned them the name "mackerel sky."

4. Vertical Development Clouds. These are single clouds that range from low to high altitudes. They are the direct result of convection currents.

Cumulus clouds are thick, flat-bottomed. high-domed. whitish clouds. They are found in the lower levels of the atmosphere and are associated with fair weather.

Cumulonimbus are heavy, towering clouds. Sometimes called thunderheads, they have dark, flat bases with anvil-shaped tops. Cumulonimbus clouds usually produce thunderstorms accompanied by rain or snow.

Table 7.2 Cloud Types and Associated Conditions

TYPE AND DESCRIPTION OF CLOUD	WEATHER
1. Cirrus Wispy and feathery.	Fair weather
2. Cirrocumulus Mackerel sky.	Fair weather

3. Cirrostratus
  High veil.
  Produces halos.
	Overcast sky

4. Altocumulus Widespread. cotton ball.	Fair weather

5. Altostratus
  Thick to thin overcast;
    high.
  No halos.
	Rain possible

6. Stratocumulus
  Low, cotton ball, wide-
    spread.
  Wavy base of even height.
	Rain possible

7. Stratus
  Uniform cloud layer.
  Like high fog.
	Rain possible

8. Nimbostratus
  Low, dark gray, rainy
    layer.
  Persistent rain maker.
	Usually rains

9. Cumulus
  Fluffy, billowy clouds.
  Flat base, cotton ball top.
	Fair weather

10. Cumulonimbus
  Thunderhead.
  Flat bottom and lofty top.
  Anvil at top.
	Thunderstorms

Fog

Fog is produced in somewhat the same manner as clouds. When heat from the Earth radiates back to the upper atmosphere, the air mass close to the ground cools and condensation occurs. This type of fog is known as *radiation fog*. When this same parcel receives additional heat, usually the next morning, the fog particles evaporate.

The other major type of fog, called *advection fog*, is produced when warm moist air is transported over a cool surface area. When it meets a cooler parcel of air, the cooling of the warm, moist air produces condensation in the form of a fog. These frontal fogs are due to the meeting and intermingling of two parcels of air with different characteristics of temperature and moisture.

A

B

Development of Frontal Systems

The boundary line between two colliding air masses is called a *front*. Air of differing physical characteristics intermingles along the front, and great changes occur. The differences in temperature, density, and moisture content of the two air masses usually result in the formation of fog, clouds, and other forms of visible precipitation.

Cold Fronts

A cold front occurs at the boundary face along which a warm air mass is being displaced by a colder, heavier air mass (Figure 7.24). These fronts have a steep leading edge where the warm air is being pushed sharply upward. They are usually accompanied by thunderhead cloud formations and a long series of thunderstorms. These storms may be very narrow in width but hundreds of miles long. A squall line of storms may precede the front by 100 miles, although the front itself is about 10 to 50 miles in width.

A cold front during the warm months in North America is relatively fast moving; its pressure increases as it moves. It generally moves in from the northwest or west. It is a region of fairly violent activity. When the winds in its area shift from southwest to northwest, as the warm air is pushed upward by the cold, clearing soon follows.

C

D

FIGURE 7.23 Cloud types. (A) Cirrus (ESSA). (B) Cirrocumulus (Philip Gendreau). (C) Cirrostratus (ESSA). (D) Stratocumulus (ESSA). (E) Stratus (Davidson, Monkmeyer Press Photo Service). (F) Nimbostratus (Royal Meteorological Society). (G) Cumulus (Philip Gen-

E

H

F

I

G

J

FIGURE 7.23 (continued)
dreau). (H) Altocumulus (Monkmeyer Press Photo
Service). (I) Cumulonimbus (Frink, Monkmeyer Press
Photo Service). (J) Altostratus (Royal Meteorological
Society).

Warm Fronts

Warm fronts are found along boundary faces where warm air is touching against a colder air mass (Figure 7.25). The lighter warm air rides up on top of the cold air mass along a gently sloping line. These fronts are relatively slow moving and remain over a re-

gion for a longer period than cold fronts. Warm fronts cover a broad area and are associated with drizzle, showers, and thunderstorms. They are preceded by cirrus clouds followed in succession by cirrostratus, altostratus, stratus, and nimbostratus clouds. Pressure falls as the front advances, with a

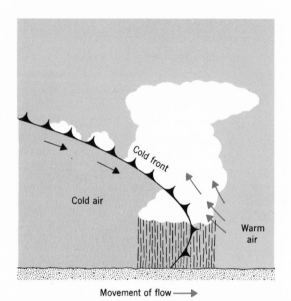

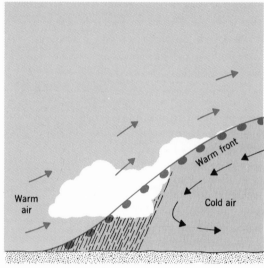

FIGURE 7.24 (*above left*)   A cold front (After Navarra and Strahler).

FIGURE 7.25 (*above right*)   A warm front (After Navarra and Strahler).

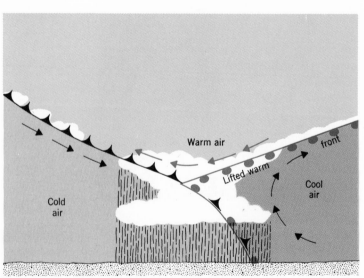

FIGURE 7.26 (*right*)   An occluded front (After Navarra and Strahler).

general accompanying rise in temperature. Winds shift from east to south with the arrival of the warm front, prior to clearing.

Occluded Fronts

An occluded front occurs when a cold front overtakes a warm front or, less frequently, when a warm front overtakes a cold. (See Figure 7.26.) An overtaking cold front traps the warm air mass behind the cold mass in front of it, and pushes the warm mass off the surface of the Earth. The warm air, together with the two cold air masses in contact with one another on the ground, produces an area of cloudiness, precipitation, and localized thunderstorms.

Stationary Fronts

A stationary front results when there are not sufficient winds to move the front one way or another. Then the front remains over an area for an extended period. As long as this front lasts, it may produce day after day of continuous rain, drizzle, and fog.

Storms

A variety of disturbances called storms occur in the atmosphere. These storms, which are produced by topography and surface conditions, may be cyclonic storms, hurricanes, tornadoes, or thunderstorms.

Cyclonic Storms

In the Northern Hemisphere, a cold front is associated with a bulge of cold air advancing from a larger cold air mass. The bulges, or cyclone waves, produce disturbances centering around low-pressure systems. Cyclone waves may have diameters as great as several hundred miles. The cyclonic storms produced by these waves are counterclockwise wind systems converging on the center of the low-pressure system (Figure 7.27). These storms carry moist air, and they last as long as moist air is in good supply. Usually they are not violent. However, local convection currents can produce cyclonic storms and low-pressure systems that are quite severe.

Hurricanes

Hurricanes, or typhoons as they are called in the western Pacific, are tropical cyclones. They are the result of vast swirls of air moving in a counterclockwise direction around a very calm low-pressure center, forming the *eye* (Figure 7.29). Hurricanes begin when

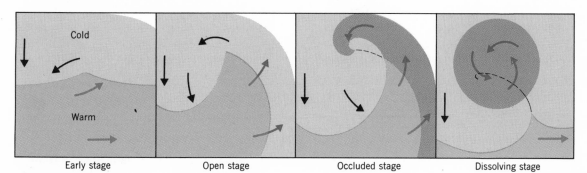

| Early stage | Open stage | Occluded stage | Dissolving stage |

FIGURE 7.27   **Formation of a cyclone bulge.**

FIGURE 7.28 A cyclone funnel moving across a vast expanse of prairie (U.S. Department of Agriculture).

warm, moist air rises to higher altitudes and water condenses, forming moisture-laden clouds that move toward the center of the increasing storm. The air in the eye descends and is heated, increasing the warm, low-pressure area. As the cyclonic movement continues to draw more air up the center column, the storm increases in size and strength. The eye may be as large as 12 miles in diameter, and the winds well above 75 miles per hour, with clouds extending to 18,000 feet.

When hurricanes are no longer fed with warm, moist air, they begin to lose their force and eventually die.

At the first impact, a hurricane produces destructive winds and heavy rainfall. Then there is a short period of calm as the eye passes overhead. As the rest of the storm passes, there is a second impact of winds, this time from the opposite direction. The barometer also begins to rise reflecting the higher pressure in the winds around the eye.

Hurricanes spawn in the Bahamas, the Caribbean Sea to the Gulf of Mexico, the Pacific Ocean west of Mexico, the Philippines near the China Sea, the Bay of Bengal, the Indian Ocean east of Madagascar, and the South Pacific Ocean from Samoa and Fiji to the west coast of Australia (Figure 7.31). The region of greatest production is near the equator among the trade winds. As a result of the trades, hurricanes generally

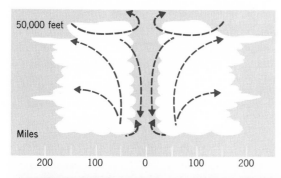

FIGURE 7.29 (*left*) **Vertical cross section of wind circulation within a hurricane.**

FIGURE 7.30 (*below*) **A radar track showing the extent and formation of Hurricane Beulah (ESSA).**

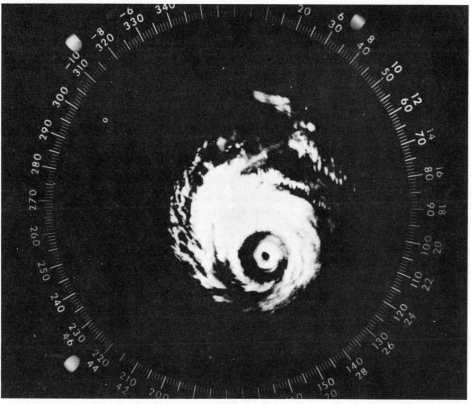

move westward at about 30 miles per hour, although the Coriolis Effect causes storms in the Northern Hemisphere to veer gradually to the right. (See Figure 7.32.) In the Southern Hemisphere, hurricanes move gradually to the left of their intended track, and the wind direction within the storm itself is clockwise rather than counterclockwise, as in the Northern Hemisphere.

Tornadoes

Tornadoes are violently twisting, counterclockwise spiraling winds which may reach velocities of 200 to 300 miles per hour. Although tornadoes are only a few hundred feet across, they are the most destructive storms known. The center, or eye, is a region of extreme low pressure. Buildings, trees, and other large objects are sometimes destroyed by updrafts of rapidly moving wind. When the eye centers over a building, the higher pressure within the building may cause the walls to collapse outward.

Tornadoes are thought to begin when warm, moist air is overrun by cold, dry air. The air begins to whirl about, and forms a funnel-shaped cloud. An updraft is produced by the extreme low pressure within the funnel, and the tornado is born.

A tornado generally moves from southwest to northeast at a rate of 25 to 40 miles per hour. It frequently rises and falls, thus, skipping some spots and striking others. Fortunately, tornadoes are short-lived.

Tornadoes are most common during the month of March in Alabama, Georgia, Mississippi, Kentucky, and Tennessee. From Florida to the Carolinas and in Texas and in Arkansas, April is the month when they gen-

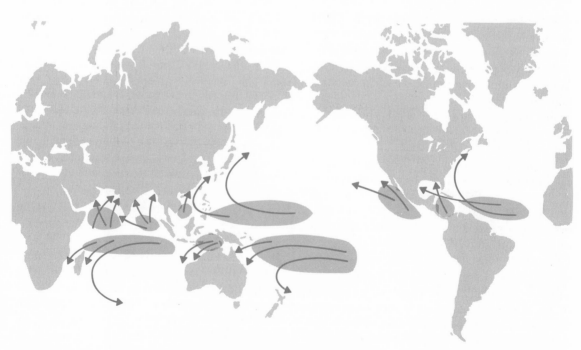

FIGURE 7.31   The principal hurricane spawning grounds of the world. Notice that they are all tropical areas.

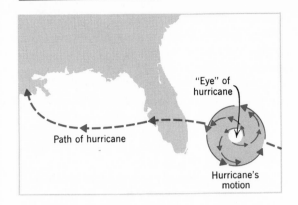

"Eye" of hurricane

Path of hurricane

Hurricane's motion

FIGURE 7.32 A typical hurricane moves along a path that takes it along the east coast of the United States. Occasionally, the path of the hurricane takes it inland in the Gulf coast region.

erally strike. May and June are the tornado months in the Midwest. Although these areas are the most common ones of occurrence, tornadoes have struck nearly every state of the union.

Thunderstorms

Thunderstorms are short-lived, local disturbances produced by strong convection currents. These storms, which are associated with the "thunderhead" or cumulonimbus clouds, produce a great deal of turbulence and heavy precipitation. They most commonly occur during warm weather, when warm air is readily supplied for upward air movement and rapid downdrafts.

Thunderstorms develop in several stages, as shown in Figure 7.34. In the initial stage of updraft, called the *cumulus* stage, the development of the cloud begins. Condensation of moisture is marked by the beginning of showers from the lower levels of the cloud. Particularly strong downdrafts are also observed at the lower levels of the cloud. Updrafts in the higher levels cause the cloud to move upward from 25,000 feet to an altitude of as much as 60,000 feet. This is the stage during which intense storms rage over the area.

The final *dispersal* stage begins as the up-

draft ends, cutting off energy to the convection cell. Precipitation gradually stops, and the cloud eventually disperses and evaporates.

Lightning is common in this type of storm. It occurs when static electrical charges are produced by friction among water, ice particles, and the air. As the particles of water and ice are blown up and down within the convection cell, static charges increase. It is thought that negative charges build up in the lower portions of the cloud and positive charges in the upper portions of the cloud.

When the potential difference between the charges in the cloud is great enough, a discharge passes from one part of the cloud to the other. Lightning may be a discharge between two parts of the same cloud, between two different clouds, or between the Earth and a cloud. A "lightning bolt" is actually composed of a thin leader discharge, followed by a return stroke, and may, in fact, involve several discharges in a split second. The thunder accompanying lightning is the result of air being rapidly heated by the passage of the lightning. The air expands and then rushes back into the partial vacuum, to produce the thunderclap. The process is diagrammed in Figure 7.35.

Lightning occurs in several forms. Forked,

A

B

C

FIGURE 7.33 The development of a tornado. The tornado builds (A) and begins to form a funnel (B). The fully formed funnel advances on the city (C) and builds in intensity before moving off the scene (D). The long, thin, ropelike funnel (E) departs for a new area (Dewey Berquist, Monkmeyer Press Photo Service).

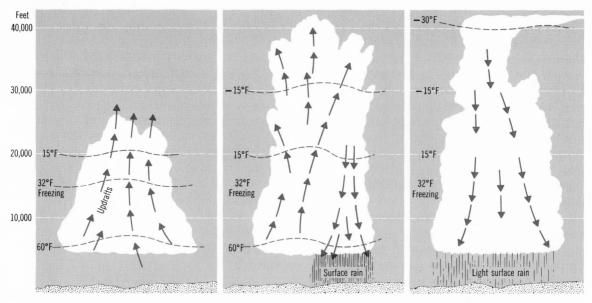

FIGURE 7.34 Three stages in the development of a typical thunderstorm. The first (cumulus) stage is marked by rapid updrafts. In the mature stage, surface winds and rain are prevalent. In the last, or dispersal stage, updrafts cease and the storm breaks up as light rain falls.

zigzag, and streak lightning look exactly as their names imply. Actually, lightning is often a branching streak that is not easily described.

Although the characteristics of air masses are directly related to their areas of origin, these characteristics are drastically changed as the air masses move into new regions of differing topography and characteristics. Observation of the various elements of weather and the contact made by air masses of different characteristics allow for meteorologists to determine the changes in weather that will be brought about.

Records of past occurrences in a region permit the meteorologist to increase the accuracy of his predictions. Although various phenomena such as storms, frontal systems, and pressure changes may be local in origin, their effects may be felt many miles away from their original regions of formation.

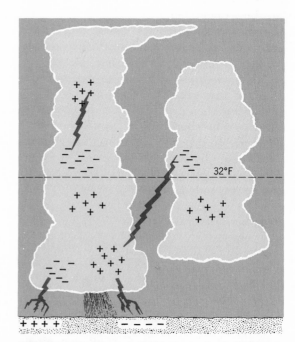

FIGURE 7.35 The typical directions taken by lightning discharges.

178

CHAPTER 8

The Science of Meteorology

EVER SINCE CIVILIZATION BEGAN, MAN HAS been trying to predict the weather. Until the last century, however, he had only everyday observations to guide him. In the last hundred years, new methods of prediction and measurement have been developed. Only recently has meteorology, the study of weather, moved out of the realm of the mysterious and become a strong, scientific field of inquiry. This study is based on record-keeping and the accurate analysis of weather phenomena. It utilizes the exact scientific laws of physics, chemistry, and other related sciences in an effort to extrapolate answers from available information.

The meteorologist is concerned with the changes that take place within specific air masses and their eventual effects on the Earth. Each specific characteristic is measured and examined in order to make predictive analyses. These observations and predictions are what make meteorology the scientific discipline it is today.

Weather Lore

At present, modern weather prediction is fairly accurate within a time span of one week. As predictions take in longer periods of time, for example, 30 days, accuracy decreases.

There are still many survivals from the pre-scientific days of weather prediction in the form of weather lore. Some weather lore is based on sound observation of events, but much of it is extremely undependable, and a few proverbs and sayings have absolutely no basis in fact.

Many of the long-range predictions given in almanacs and other periodicals are based on general information and averages for a given area. Suppose a periodical predicts rain in the northeast for March 3. The person making the prediction can be fairly certain that somewhere in this area precipitation will fall at this time of year. However, to say that he has made an accurate scientific prediction would be somewhat less than the truth!

Over the years, many weather predictions have been derived from observations of animal behavior. Animals, just prior to a storm, may take cover because of their sensitivity to air pressure, wind changes, and wind speed. But predictions beyond immediate observations are doubtful. To date, scientific investigation into animal behavior shows no correlation between such activities as more intensive food storage and growth of longer fur, and an unusually severe winter.

Some weather proverbs are based on scientific fact:

Pink sky in the morning
sailor's warning,
Pink sky at night
sailor's delight

is a proverb that has some basis in fact. The sky color is the result of the refraction of sun-

FIGURE 8.1 (*right*) One activity of rodents preparing for winter is food gathering (Robert H. Wright, National Audubon Society).

FIGURE 8.2 (*below*) Weasels change the color of their coats to blend in with the winter snow. [(A) John H. Gerard, National Audubon Society; and (B) Arthur Ambler, National Audubon Society.]

light as it passes over or below the horizon. In the morning, as the sun rises, the sky color may be the result of high humidity, which is the result of approaching storm clouds. At night, the clouds have already passed overhead and, thus, clearing conditions for the next day are indicated.

A halo about the moon is thought of as being an indicator of impending stormy weather. Although this may occasionally be true, the halo is not a dependable sign. It is the result of cirrus clouds, which indicate an approaching storm if they are followed by lower, thicker clouds. However, the cirrus clouds often simply disappear, along with the expected storm.

The whaler's dread of the Nor'easter is summed up most succinctly in:

> The wind from the Northeast
> Is good for neither man nor beast.

In the northeastern states this is usually a cold, moist wind blowing toward a low-pres-sure system and is a sign of bad weather. However, the same characteristics cannot be attributed to all northeast winds. To a person living on the west coast, a northeastern wind comes from inland and is usually a dry wind. The weather accompanying a storm on the west coast is brought by a northwestern wind from the north Pacific.

The sky has been described in various fashions, with a specific weather condition ascribed to each description. This type of prediction is not very reliable. "Mare's tails" or "mackerel sky," for instance, is supposed to foretell a storm. But these conditions of the sky are only followed by storm if they are accompanied by a falling barometer, eastern winds, and a thick cloud cover. Thus, many other factors must be considered before an accurate prediction can be made.

The most famous chestnut of all is the belief about Groundhog Day, February 2. This is the day the groundhog is supposed to come out of his burrow and look around and,

FIGURE 8.3 A swell at sea nearly swamps the deck of a vessel (Philip Gendreau).

if he sees his shadow, six more weeks of winter weather will follow. There is absolutely no basis for this prediction, nor has any correlation ever appeared between the arrival of spring and the wanderings of the groundhog. Indeed, the fact that the groundhog could forecast calendar dates would be amazing in itself!

Perhaps the best practice is one of, at least, slight skepticism. Although some of the proverbs we hear are partially true, we can be greatly misled by them.

Weather Analysis

Today there is a far-flung network of meteorological stations, and experts are joined in a cooperative effort to understand the weather and its effects on the Earth.

The Development of Meteorological Services

Weather analysis and forecasting requires a widespread compilation of data. Daily records of temperature, humidity, prevailing winds, and other elements are necessary for meteorologists to organize information and synthesize a probable sequence of future events. This type of information gathering and systematic record-keeping was begun in the early nineteenth century by the United States Government Land Office and the military services. After the Civil War, the United States Weather Bureau was established, and it began to collect data on climate and make tentative forecasts as a somewhat official department. Until considerably later, however, most of the burden of keeping what scarce and scattered reports there were rested on the Smithsonian Institution and the United States Army.

The Weather Bureau in the Twentieth Century

Since the beginning of the twentieth century, the air mass and wave cyclone (cyclonic wind flow) theories have become the basis

FIGURE 8.4 The groundhog seeing his shadow is one of the most timeworn and erroneous pieces of weather lore (John H. Gerard, National Audubon Society).

FIGURE 8.5 The United States Coast Guard Cutter *Escanaba* houses meteorologists and equipment for official weather news and forecasts (U.S. Coast Guard).

for weather analysis. As aviation has increased, as business has become more dependent on accurate weather prediction, and as population has concentrated in large urban centers, accurate analyses of weather forecasting have become ever more necessary.

In response to this need, the United States Weather Bureau has been established as a scientific investigation department. Some of its responsibilities are operating weather stations, communicating weather data, compiling data, synthesizing forecasts, disseminating forecasts and storm warnings, developing and testing new equipment, conducting meteorological research, and issuing publications in the fields of meteorology and climatology.

Nearly every country in the world now operates some type of weather bureau. These bureaus cooperate in a worldwide system of information exchange and research on the effects and processes of weather. These weather bureaus also receive information from sources other than land stations. Reports come in from oceangoing vessels deal-

FIGURE 8.6 An electronic package of weather apparatus, the radiosonde, is carried into the upper atmosphere by a balloon. Information is radioed back and the instrument is recovered later (American Museum of Natural History).

FIGURE 8.7 Weather satellite (NASA).

ing with ocean conditions, and the military issues special weather reports. The Army, Navy, and Air Force still compile and disseminate weather data. The Coast Guard regularly makes weather observations, and it also mans the oceangoing weather stations that are specially designed for the weather bureau.

Recent Advances in Data Gathering

Within recent years, sounding rockets have been employed to lift radiosondes into the upper atmosphere (Figure 8.6). Radiosondes are miniaturized instruments which record and send atmospheric data back to Earth via radio transmitter. The information is automatically recorded at the station and correlated with other data. Formerly, helium-filled balloons were used to lift radiosondes. When the balloon reached its limit of altitude, it burst, and the radiosonde package was parachuted back to Earth. Radiosondes have been in use since the 1930's.

The use of rockets in weather forecasting dates back to the late 1940's and early 1950's. At that time, early experiments showed that rockets could obtain photos covering wide areas of the Earth. These photos made it possible to develop a different type of

184

weather map covering a large portion of the Earth. These maps are termed *synoptic,* because they form a synopsis of weather information.

The Weather Satellite

The TIROS weather satellite system, invented in 1960, is an aid to meteorologists in forecasting weather. TIROS takes its name from *Television* and *Infra-Red Observation Satellite.* These satellites carry two types of camera. One wide-angle camera gives a picture covering several hundred miles, depending on the altitude of the camera. The other gives a view somewhat less than 100 miles across.

TIROS satellites take a series of pictures of the Earth's atmosphere and land surfaces. A composite of these photos gives meteorologists a view of a large portion of the Earth's surface, so that cloud patterns and the development of storms can be observed and studied. More TIROS satellites as well as other types of weather satellites are planned

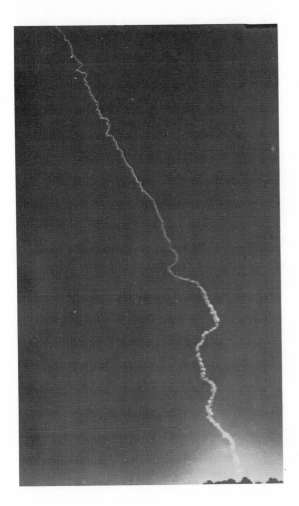

FIGURE 8.8 Wind measurements are made by photographic measurement of the smoke trail generated be a rocket. Wind turbulence is revealed by the fluctuations in the trail (NASA).

in a worldwide system of orbiting meteorological observatories. In 1964, a second family of satellites was inaugurated, NIMBUS. This more sophisticated series of weather satellites carry on the work begun by the TIROS series.

Radar and Storm Center Detection

Of the other types of electronic gear that are useful in forecasting weather, the most important and widely used is radar. Radar sends out radio waves which will be reflected back from a storm center that is in the making. With radar, the observer can detect a storm, determine its distance, measure its size, and evaluate its direction of movement. Severe storms, such as hurricanes, can be observed and tracked at great distances, allowing earlier storm warnings to be given.

Weather Forecasting

Although collecting data is a prominent part of the meteorologist's job, it is not his most important responsibility. The end result of his investigations is the understanding of weather patterns: why weather occurs in different patterns, and what weather developments will take place in the future.

Factors Affecting the Movement of Air Masses

A meteorologist knows that air masses in the United States generally move in an easterly direction. You will remember that the prevailing wind movement in the second cell of worldwide air circulation is from southwest to northeast. Our low-level air generally follows this pattern.

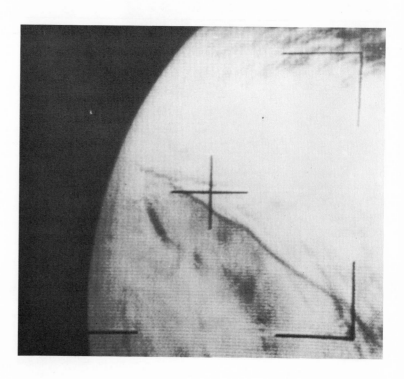

FIGURE 8.9 A Tiros 10 statellite shows the location of the jet stream over the North Atlantic Ocean (ESSA).

FIGURE 8.10 (*above*) Satellite photos can be pieced together in a montage showing the weather over an entire continent. This Nimbus II montage shows the entire United States in July 1966 as relatively cloud free (NASA).

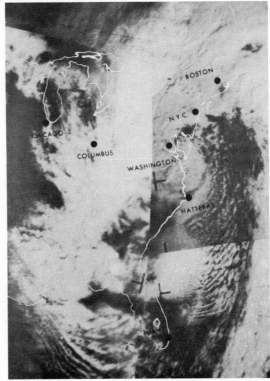

FIGURE 8.11 (*right*) The Environmental Science Services Administration (ESSA) now carries on satellite experimentation programs that are used for weather observation. This new family of satellites takes large area photos. This one of the East Coast was taken from ESSA satellite 8 (ESSA).

FIGURE 8.12 A simplified weather map showing isobars, frontal systems, and highs and lows.

As we have seen, pressure systems are influenced by other pressure systems in the same vicinity. A low-pressure system moves away from a high-pressure area toward an area of still lower pressure. Thus, pressure systems move in fairly predictable patterns. The greater the pressure gradient (difference in pressure of adjacent systems), the faster the movement of the pressure cell.

In mapping winds, the meteorologist must consider not only the pressure gradient, but the Coriolis Effect as well. The pressure gradient causes the initial movement, but the Coriolis Effect causes the turning of the air mass. Buys Ballot, a Dutch meteorologist, stated the relationship between pressure and

air movement nearly a century ago: if the wind is against your back, low pressure will be on your left and high pressure on the right in the Northern Hemisphere. This wind, now known as the *geostrophic wind,* is simultaneously under the influence of the rotation of the Earth and the pressure gradient. Its movement is also influenced by friction with the Earth and with other air masses.

The Synoptic Weather Map

Forecasters use many tools in defining present weather conditions and in predicting future developments. Both upper and lower level atmospheric conditions must be noted and charted. Pressure systems and their configurations must be mapped, and the juncture of air masses must be known. Moreover, other elements such as temperature, cloud types, and precipitation must all be placed in some understandable scheme.

When these elements are measured over a large area of the globe, they are placed on maps so that their spatial relationships can be observed (Figure 8.13). These synoptic weather maps are compilations of all the information on all the weather elements that are known for a particular day and locality.

A typical weather map carries several types of symbols. Each type contains important information. The first symbol is the

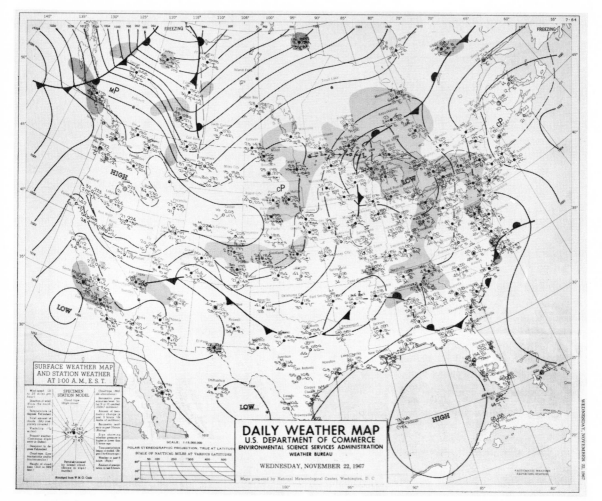

FIGURE 8.13 The synoptic Daily Weather Map is a composite of all the weather information known for a particular date. Pressure systems, frontal systems, and a variety of data at each station model are included (ESSA).

circle, which represents a particular weather station. The circle is filled in or left blank to simulate the appearance of the sky. A black circle indicates a completely overcast sky. A half-filled circle means partial cloudiness; a clear circle indicates clear skies, and so on.

If winds are present, an arrow is attached to the circle coming from the direction of the wind. The tail of the arrow carries a series of featherlike hachure marks that indicate wind speed according to a code known as the Beaufort Scale. (See Table 8.1.) The code, when used, is usually given on the back of the Sunday map (Figure 8.19).

Some fifteen different symbols may be found around the weather-station circles, and numerals indicating temperature, visibility, dew point, amount of precipitation, barometric pressure, and other weather elements may also be included. There are symbols for cloud types, cloud heights, and other general weather conditions. (Figures 8.14, 8.15, and 8.16 show some of these symbols.)

The circles representing weather stations are only a beginning. Other information

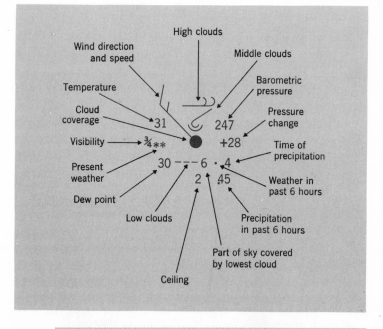

FIGURE 8.14 Weather station models appear on all weather maps. The symbols are placed in fixed positions around the model and reveal all the pertinent information for the area.

FIGURE 8.15 The amount of shading within the weather model circle represents cloud covering.

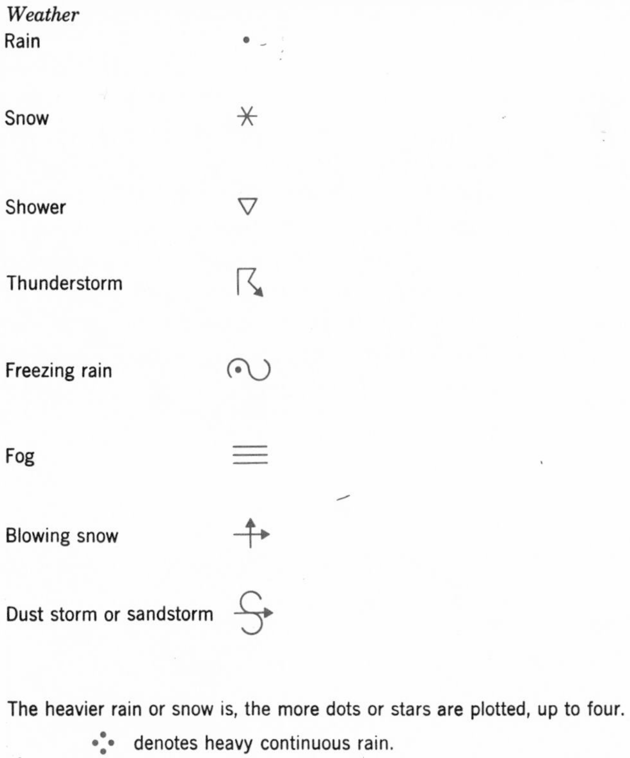

Weather

Rain	• ⌐ .
Snow	✳
Shower	▽
Thunderstorm	▷�
Freezing rain	(• ‿)
Fog	≡
Blowing snow	⇥
Dust storm or sandstorm	⌇→

The heavier rain or snow is, the more dots or stars are plotted, up to four.
∴ denotes heavy continuous rain.

FIGURE 8.16 Some common weather symbols used on station models by the United States Weather Bureau.

must be placed on the weather map if it is to mean anything to the meteorologist, or even to the person reading the daily newspaper. Most daily weather maps include a series of lines called *isobars,* the distribution of which represents pressure patterns. These lines, which represent a pressure difference of four millibars, are placed at a specific distance from one another.

Usually there are closely spaced, irregular isobars around the centers of high- or low-pressure systems, with more widely spaced lines between pressure centers. The pressure gradient conforms with the spacing of isobars. Where the isobars are spaced more closely, the pressure gradient is greatest.

Those spaced more widely apart represent regions where winds are generally light.

Weather maps may carry another type of line representing temperature. These irregular lines joining points of equal temperature are called *isotherms.* The isotherms, when used, show temperature distribution across the nation.

The synoptic weather map also shows the boundary lines between air masses, with symbols denoting the type of front. If cold air contacts warm air to produce a cold front, the line marking the area bears a series of triangles that point in the direction of movement of the front. If the cold air is moving eastward, the triangles point in an eastern

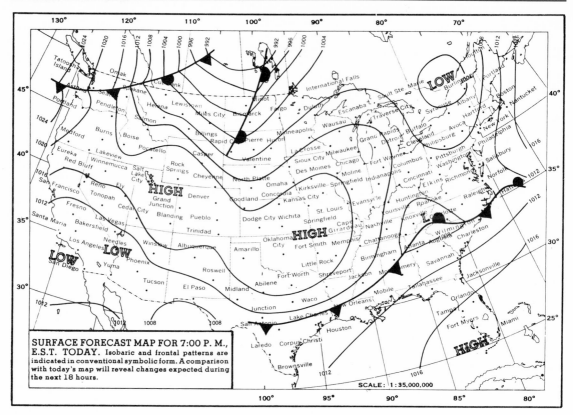

FIGURE 8.17 Specialized weather data are shown on smaller maps such as isobar maps (A) showing lines of equal pressure, and maps (B) showing regions of highest and lowest temperatures (ESSA).

direction. A warm front is denoted by a series of half circles on the line, with the circles pointing to the direction in which the warm front is moving. An occluded front is represented by a line with alternate triangles and half circles pointed in the direction of general movement. Along a stationary front, since there is no movement, the half circles are on one side of the line and the triangles on the other.

Other symbols are also found on the weather map. Shaded areas represent precipitation. Special symbols indicate hurricane conditions or other severe storms with high winds. Upper air conditions, such as prevailing winds, temperatures, and pressure, do not appear in the daily newspaper, although they are reported to the professional meteorologist. Specialized reports, such as marine forecasts, are made at some locations.

One can follow weather conditions quite accurately and can even develop some facility in predicting weather by reading charts of cloud types, observing wind direction, and by knowing the topography of the area, the nearby water bodies, and the annual averages of weather conditions. As we have learned in

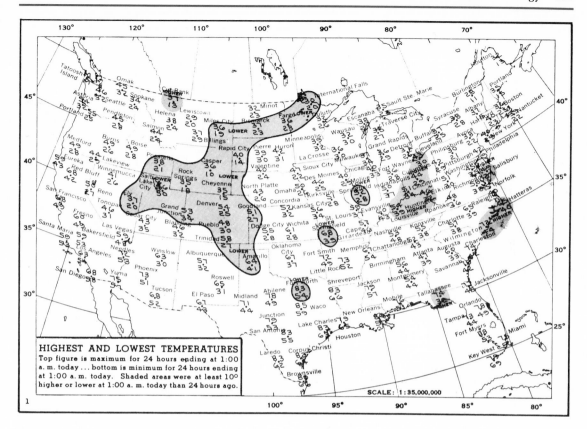

HIGHEST AND LOWEST TEMPERATURES
Top figure is maximum for 24 hours ending at 1:00 a. m. today ... bottom is minimum for 24 hours ending at 1:00 a. m. today. Shaded areas were at least 10° higher or lower at 1:00 a. m. today than 24 hours ago.

SCALE: 1 : 35,000,000

Symbols used in analysis are as follows:

Cold front

Warm front

Occluded front

Stationary front

Frontless trough or
wind shift line

Solid shading indicates areas where precipitation is currently falling.   **FIGURE 8.18**

CLOUD ABBREVIATION	C_L	DESCRIPTION (Abridged From W.M.O Code)	C_M	DESCRIPTION (Abridged From W.M.O. Code)	C_H	DESCRIPTION (Abridged From W.M.O. Code)
St or Fs-Stratus or Fractostratus	1	Cu of fair weather, little vertical development and seemingly flattened	1	Thin As (most of cloud layer semi-transparent)	1	Filaments of Ci, or "mares' tails," scattered and not increasing
Ci-Cirrus	2	Cu of considerable development, generally towering, with or without other Cu or Sc bases all at same level	2	Thick As, greater part sufficiently dense to hide sun (or moon), or Ns	2	Dense Ci in patches or twisted sheaves, usually not increasing, sometimes like remains of Cb; or towers or tufts
Cs-Cirrostratus	3	Cb with tops lacking clear-cut outlines, but distinctly not cirriform or anvil-shaped; with or without Cu, Sc, or St	3	Thin Ac, mostly semi-transparent; cloud elements not changing much and at a single level	3	Dense Ci, often anvil-shaped, derived from or associated with Cb
Cc-Cirrocumulus	4	Sc formed by spreading out of Cu; Cu often present also	4	Thin Ac in patches; cloud elements continually changing and/or occurring at more than one level	4	Ci, often hook-shaped, gradually spreading over the sky and usually thickening as a whole
Ac-Altocumulus						
As-Altostratus	5	Sc not formed by spreading out of Cu	5	Thin Ac in bands or in a layer gradually spreading over sky and usually thickening as a whole	5	Ci and Cs, often in converging bands, or Cs alone; generally overspreading and growing denser; the continuous layer not reaching 45° altitude
Sc-Stratocumulus	6	St or Fs or both, but no Fs of bad weather	6	Ac formed by the spreading out of Cu	6	Ci and Cs, often in converging bands, or Cs alone; generally overspreading and growing denser; the continuous layer exceeding 45° altitude
Ns-Nimbostratus	7	Fs and/or Fc of bad weather (scud)	7	Double-layered Ac, or a thick layer of Ac, not increasing, or Ac with As and/or Ns	7	Veil of Cs covering the entire sky
Cu or Fc-Cumulus or Fractocumulus	8	Cu and Sc (not formed by spreading out of Cu) with bases at different levels	8	Ac in the form of Cu-shaped tufts or Ac with turrets	8	Cs not increasing and not covering entire sky
Cb-Cumulonimbus	9	Cb having a clearly fibrous (cirriform) top, often anvil-shaped, with or without Cu, Sc, St, or scud	9	Ac of a chaotic sky, usually at different levels; patches of dense Ci are usually present also	9	Cc alone or Cc with some Ci or Cs, but the Cc being the main cirriform cloud

WW PRESENT WEATHER (Descriptions Abridged from W. M. O. Code)

	0	1	2	3	4	5	6	7
00	Cloud development NOT observed or NOT observable during past hour	Clouds generally dissolving or becoming less developed during past hour	State of sky on the whole unchanged during past hour	Clouds generally forming or developing during past hour	Visibility reduced by smoke	Haze	Widespread dust in suspension in the air, NOT raised by wind, at time of observation	Dust or sand raised by wind, at time of observation
10	Light fog	Patches of shallow fog at station, NOT deeper than 6 feet on land	More or less continuous shallow fog at station, NOT deeper than 6 feet on land	Lightning visible, no thunder heard	Precipitation within sight, but NOT reaching the ground	Precipitation within sight, reaching the ground, but distant from station	Precipitation within sight, reaching the ground, near to but NOT at station	Thunder heard, but no precipitation at the station
20	Drizzle (NOT falling as showers) during past hour, but NOT at time of observation	Rain (NOT freezing and NOT falling as showers) during past hour, but NOT at time of observation	Snow (NOT falling as showers) during past hour, but NOT at time of observation	Rain and snow (NOT falling as showers) during past hour, but NOT at time of observation	Freezing drizzle or freezing rain (NOT falling as showers) during past hour, but NOT at time of observation	Showers of rain during past hour, but NOT at time of observation	Showers of snow, or of rain and snow, during past hour, but NOT at time of observation	Showers of hail, or of hail and rain, during past hour, but NOT at time of observation
30	Slight or moderate dust storm or sand storm, has decreased during past hour	Slight or moderate dust storm or sand storm, no appreciable change during past hour	Slight or moderate dust storm or sand storm, has increased during past hour	Severe dust storm or sand storm, has decreased during past hour	Severe dust storm or sand storm, no appreciable change during past hour	Severe dust storm or sand storm, has increased during past hour	Slight or moderate drifting snow, generally low	Heavy drifting snow, generally low
40	Fog at distance at time of observation, but NOT at station during past hour	Fog in patches	Fog, sky discernible, has become thinner during past hour	Fog, sky NOT discernible, has become thinner during past hour	Fog, sky discernible, no appreciable change during past hour	Fog, sky NOT discernible, no appreciable change during past hour	Fog, sky discernible, has begun or become thicker during past hour	Fog, sky NOT discernible, has begun or become thicker during past hour
50	Intermittent drizzle (NOT freezing) slight at time of observation	Continuous drizzle (NOT freezing) slight at time of observation	Intermittent drizzle (NOT freezing) moderate at time of observation	Continuous drizzle (NOT freezing), moderate at time of observation	Intermittent drizzle (NOT freezing), thick at time of observation	Continuous drizzle (NOT freezing), thick at time of observation	Slight freezing drizzle	Moderate or thick freezing drizzle
60	Intermittent rain (NOT freezing), slight at time of observation	Continuous rain (NOT freezing), slight at time of observation	Intermittent rain (NOT freezing) moderate at time of obs.	Continuous rain (NOT freezing), moderate at time of observation	Intermittent rain (NOT freezing), heavy at time of observation	Continuous rain (NOT freezing), heavy at time of observation	Slight freezing rain	Moderate or heavy freezing rain
70	Intermittent fall of snowflakes, slight at time of observation	Continuous fall of snowflakes, slight at time of observation	Intermittent fall of snowflakes, moderate at time of observation	Continuous fall of snowflakes, moderate at time of observation	Intermittent fall of snowflakes, heavy at time of observation	Continuous fall of snowflakes, heavy at time of observation	Ice needles (with or without fog)	Granular snow (with or without fog)
80	Slight rain shower(s)	Moderate or heavy rain shower(s)	Violent rain shower(s)	Slight shower(s) of rain and snow mixed	Moderate or heavy shower(s) of rain and snow mixed	Slight snow shower(s)	Moderate or heavy snow shower(s)	Slight shower(s) of soft or small hail with or without rain, or rain and snow mixed
90	Moderate or heavy shower(s) of hail, with or without rain or rain and snow mixed, not associated with thunder	Slight rain at time of observation, thunderstorm during past hour, but NOT at time of observation	Moderate or heavy rain at time of observation; thunderstorm during past hour, but not at time of observation	Slight snow or rain and snow mixed or hail at time of observation, thunderstorm during past hour, but not at time of observation	Moderate or heavy snow, or rain and snow mixed or hail at time of observation, thunderstorm during past hour, but NOT at time of obs.	Slight or moderate thunderstorm without hail, but with rain and/or snow at time of observation	Slight or moderate thunderstorm with hail at time of observation	Heavy thunderstorm, without hail, but with rain and/or snow at time of observation

MILLIBARS → 956 960 964 968 972 976 980 984 988 992 996 1000 1004 1008 1012 1016 1020 1024 1028 1032 1036 1040 1044 1048 1052 1056

INCHES → 28.2 28.3 28.4 28.5 28.6 28.7 28.8 28.9 29.0 29.1 29.2 29.3 29.4 29.5 29.6 29.7 29.8 29.9 30.0 30.1 30.2 30.3 30.4 30.5 30.6 30.7 30.8 30.9 31.0 31.1 31.2

PRESSURE

FIGURE 8.19 From 1955 issue of *Nature*, 175:834.

R_t	TIME OF PRECIPITATION ❹	h	HEIGHT IN FEET (Rounded Off)	HEIGHT IN METERS ❺ (Approximate)	N	SKY COVERAGE ❻ (Total Amount)	N_h	SKY COVERAGE ❼ (Low And/Or Middle Clouds)
0	No Precipitation	0	0 - 149	0 - 49	○	No clouds	0	No clouds
1	Less than 1 hour ago	1	150 - 299	50 - 99	◔	Less than one-tenth or one-tenth	1	Less than one-tenth or one-tenth
2	1 to 2 hours ago	2	300 - 599	100 - 199	◔	Two-tenths or three-tenths	2	Two-tenths or three-tenths
3	2 to 3 hours ago	3	600 - 999	200 - 299	◑	Four-tenths	3	Four-tenths
4	3 to 4 hours ago	4	1,000 - 1,999	300 - 599	◑	Five-tenths	4	Five-tenths
5	4 to 5 hours ago	5	2,000 - 3,499	600 - 999	◕	Six-tenths	5	Six-tenths
6	5 to 6 hours ago	6	3,500 - 4,999	1,000 - 1,499	◕	Seven-tenths or eight-tenths	6	Seven-tenths or eight-tenths
7	6 to 12 hours ago	7	5,000 - 6,499	1,500 - 1,999	◕	Nine-tenths or overcast with openings	7	Nine-tenths or overcast with openings
8	More than 12 hours ago	8	6,500 - 7,999	2,000 - 2,499	●	Completely overcast	8	Completely overcast
9	Unknown	9	At or above 8,000, or no clouds	At or above 2,500, or no clouds	⊗	Sky obscured	9	Sky obscured

8	9	❽	ff	(MILES) (Statute) Per Hour	❾ KNOTS	Code Number	a	BAROMETRIC TENDENCY ❿
⧖ Well developed dust devil(s) within past hour	(⟋→) Dust storm or sand storm within sight of or at station during past hour		◎	Calm	Calm	0	╱	Rising, then falling
∨ Squall(s) within sight during past hour	�devilⅠ Funnel cloud(s) within sight during past hour		—	1 - 2	1 - 2	1	╱	Rising, then steady; or rising, then rising more slowly } Barometer now higher than 3 hours ago
≡] Fog during past hour, but NOT at time of observation	⏃] Thunderstorm (with or without precipitation) during past hour, but NOT at time of obs.		⌐	3 - 8	3 - 7	2	╱	Rising steadily, or unsteadily
↟ Slight or moderate drifting snow, generally high	↟ Heavy drifting snow, generally high		⌐	9 - 14	8 - 12	3	╲	Falling or steady, then rising; or rising, then rising more quickly
⩔ Fog, depositing rime, sky discernible	⩔ Fog, depositing rime, sky NOT discernible		⌐	15 - 20	13 - 17	4	—	Steady, same as 3 hours ago
• Drizzle and rain, slight	• Drizzle and rain, moderate or heavy		⌐	21 - 25	18 - 22	5	╲	Falling, then rising, same or lower than 3 hours ago
✳ Rain or drizzle and snow, slight	✳ Rain or drizzle and snow, moderate or heavy		⌐	26 - 31	23 - 27	6	╲	Falling, then steady; or falling, then falling more slowly } Barometer now lower than 3 hours ago
↔ Isolated starlike snow crystals (with or without fog)	△ Ice pellets (sleet, U.S. definition)		⌐	32 - 37	28 - 32	7	╲	Falling steadily, or unsteadily
⧖ Moderate or heavy shower(s) of soft or small hail with or without rain, or rain and snow mixed	⧖ Slight shower(s) of hail, with or without rain or rain and snow mixed, not associated with thunder		⌐	38 - 43	33 - 37	8	╲	Steady or rising, then falling; or falling, then falling more quickly
			⌐	44 - 49	38 - 42	Code Number	W	PAST WEATHER ⓫
			⌐	50 - 54	43 - 47	0		Clear or few clouds } Not Plotted
			⌐	55 - 60	48 - 52	1		Partly cloudy (scattered) or variable sky
			⌐	61 - 66	53 - 57	2		Cloudy (broken) or overcast
⧖ Thunderstorm combined with dust storm or sand storm at time of obs.	⧖ Heavy thunderstorm with hail at time of observation		⌐	67 - 71	58 - 62	3	⟋	Sandstorm or dust-storm, or drifting or blowing snow
			⌐	72 - 77	63 - 67	4	≡	Fog, or smoke, or thick dust haze
			⌐	78 - 83	68 - 72	5	,	Drizzle
			⌐	84 - 89	73 - 77	6	•	Rain
						7	✳	Snow, or rain and snow mixed, or ice pellets (sleet)
			⌐	119 - 123	103 - 107	8	▽	Shower(s)
						9	⏃	Thunderstorm, with or without precipitation

CELSIUS / FAHRENHEIT — TEMPERATURE ⓭

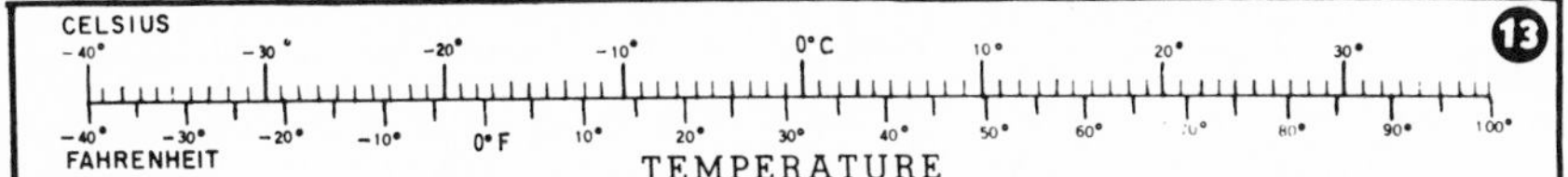

-40° -30° -20° -10° 0°C 10° 20° 30°

-40° -30° -20° -10° 0°F 10° 20° 30° 40° 50° 60° 70° 80° 90° 100°

Table 8.1   Beaufort Wind Scale

SCALE NUMBER	WEATHER DESCRIPTION	MILES PER HOUR	EFFECTS OF THE WIND	BEAUFORT SYMBOL
0	Calm	0–1	Smoke rises straight up.	
1	Light air	2–3	Smoke shows wind direction.	
2	Slight breeze	4–7	Weather vanes turn, flags move rapidly.	
3	Gentle breeze	8–12	Flags blow straight out.	
4	Moderate breeze	13–18	Dust clouds.	
5	Fresh breeze	19–24	Small trees bend.	
6	Strong breeze	25–31	Umbrellas used with difficulty.	
7	High wind	32–38	Walking difficult.	
8	Gale	39–46	Branches break off trees.	
9	Strong gale	47–54	Shingles are torn off roofs.	
10	Whole gale	55–63	Trees topple, telephone wires break.	
11	Storm	64–72	Great damage.	
12	Hurricane	73–82	Disaster.	

these chapters, the rise or fall of the barometer is one of the indicators of expected weather. A sudden sharp drop in the barometer indicates approaching storm conditions. The different types of clouds, along with prevailing wind conditions, are still some of the best weather indicators.

Weather maps from the previous day are published in newspapers. These daily maps are a good guide to expected weather. What types of air masses are prevailing indicate what type of weather to expect. Across the United States, low-pressure systems move west to east at an approximate speed of 20 to 30 miles per hour, or 480 to 720 miles per day; high-pressure systems move at about half that speed. A general knowledge of the major air masses that prevail in various seasons also helps in predicting future weather conditions.

Questions

CHAPTER FIVE

1. What are the various physical and chemical characteristics that differentiate each layer of the atmosphere from the others?
2. Explain what the Van Allen radiation belts are and how they were formed. What would be different on the Earth if these barriers were not present?
3. What is the greenhouse effect? In what way does this effect allow life to exist on the Earth?
4. How is one portion of the atmosphere used in communications? What makes this possible?
5. Define and contrast the wet and dry lapse rates.
6. What are some of man's effects on the composition and characteristics of the atmosphere?

CHAPTER SIX

1. What is the pressure gradient and how does it produce movements of air masses? What characteristic of the Earth affects wind movement? In what way?
2. How do highs and lows affect the movement of air around their respective centers?
3. Describe the sequence of events that produces a convection current.
4. List the important seasonal air masses of the United States. Describe each.
5. Describe and diagram the general atmospheric circulation around the Earth.
6. What is specific heat? How is this physical characteristic related to land and sea breezes?

CHAPTER SEVEN

1. List at least six elements of weather. How is each measured by meteorologists?
2. What are hygroscopic nuclei? What commercial importance do they have?
3. Describe the development of the various frontal systems and outline the weather associated with each.
4. Describe each stage of a thunderstorm.

5. How do cyclonic storms develop? What causes hurricanes to move as they do across the Earth's face?

6. List the four major cloud classes and describe one or two major cloud types in each.

CHAPTER EIGHT

1. What faulty data-collecting techniques are used to produce most weather proverbs?

2. Describe a typical station model and the information it reproduces for the reader.

3. What is the synoptic weather map? How are new technological advances making this map more synoptic than ever?

4. Briefly describe the development of weather data collecting in this country during the last 150 years.

5. What are isobars? Isotherms? What relationships do they reveal?

Bibliography

BOOKS

Day, John A., *Science of Weather,* Addison-Wesley, Reading, Massachusetts, 1968.

Dobson, Gordon, M.D., *Exploring the Atmosphere,* Oxford University Press, New York, 1968.

Flohn, Hermann, *Climate and Weather,* McGraw-Hill, New York, 1968.

Hubert, Lester F., and Lehr, P. E., *Weather Satellites,* Blaisdell Publishers, Massachusetts, 1967.

Rumney, G. R., *Climatology,* Macmillan Company, New York, 1968.

Scorer, R., and Wexler, R., *Cloud Studies in Colour,* Pergamon Press, Elmford, New York, 1968.

World Meteorological Organization, *Air Pollutants, Meteorology, and Plant Injury,* Unipub Inc., New York, 1968.

PERIODICALS

Andrews, J. F., "Circulation and Weather," *Weatherwise,* February, 1970.

Asimov, Isaac, "How Was the Earth's Atmosphere Formed?," *Science Digest,* November, 1966.

Ewing, A., "Earth Air Shakes Like Jelly," *Science,* February, 1966.

Falconer, R. E., "How Weather Is Made," *Conservationists,* December, 1969.

Ludlum, D. M., "Century of American Weather," *Weatherwise,* April, 1970.

Olsen, W. S., "Atmospheric and Hydrospheric Evolution on the Primitive Earth," *Science,* May, 1968.

Rasool, S. I., "Evolution of the Earth's Atmosphere," *Science,* September, 1967.

Terselic, R. J., "The World Weather Watch — A Concept for Improvement," *Weatherwise,* June, 1966.

Thomsen, D. E., "Getting Out from Under — Astronomy Above the Atmosphere," *Science,* August, 1968.

The Hydrosphere

WHEN FUTURE HISTORIANS WRITE ABOUT THE TWEN-tieth century, they may very well call it the Age of Exploration. Our century has experienced an exploration of the Earth, the sky, and outer space that is unparalleled in man's history. Moreover, explorations of the oceans are unlocking the secrets contained in those vast bodies of water which cover more than 70 per cent of the Earth's surface. The ocean basins hold the key to the origins and evolution of our planet and they may also yield clues to its future development. In addition, the oceans are a storehouse of natural resources which may answer many of man's needs for future generations. The life that abounds in the sea and the mineral resources in the ocean floors far surpass those of the land masses.

Only through the development of technology will man be able to explore and to use the ocean basins. And only through control of the ocean basins will man be able to support himself in the future on this planet.

CHAPTER 9

Exploring the Oceans of the World

THE OCEANS HAVE ALWAYS BEEN AREAS of great mystery, since man has been frustrated whenever he attempted to explore the ocean depths. But the oceans hold the key to many of the puzzles that man has encountered in examining the crust of the continents, and recently interest in the study of the oceans has been renewed.

The science of oceanography includes all studies relating to the sea. Oceanographers attempt to integrate and to understand the knowledge gained in the study of subjects such as the physics and chemistry of seawater, submarine geology, and marine biology, as well as meteorology and climatology.

The Beginning of the Science of Oceanography

Modern oceanography began about 120 years ago through the work of Matthew Maury and Edward Forbes. Maury, an American Naval officer, began a systematic study of tides, currents, and channels; Forbes, an English scientist, studied life in the shallows and the deep sea. By utilizing dredging techniques, Forbes pioneered the science of marine biology during the 1840's and the 1850's. He created theories regarding life zones in the ocean, began the analysis of ocean water, and developed geological interpretations of the marine life found on the ocean floor. One of his most famous (and erroneous) theories—that of an *azoic zone* in the ocean—was finally disproved in 1873 during the voyage of the *Challenger*. Forbes believed that no life existed beyond a depth of 300 fathoms (1800 ft). Although he is often remembered best for this incorrect idea of the azoic zone, his theories which combined biological and geological information are credited with developing the science of modern marine biology.

Matthew Maury was one of the first Americans to make a major contribution to oceanography. As a lieutenant in the Navy, he had a great deal of experience at sea prior to being placed in charge of the United States Naval Depot of Charts and Instruments. Once in this office, he studied the winds and currents from the viewpoint of a navigator.

Maury made it a practice to arrange for the captains of sailing ships to send in logs (records) of ship's voyages. Most nations with naval services cooperated to provide Maury with a great wealth of information and a worldwide collection of data followed. Maury assembled and organized this information, and eventually developed navigational charts that were made available to the fleets of all nations.

There were very practical values associated with Maury's charts. His sailing directions shortened the time required to sail from the east coast of the United States to Rio de

FIGURE 9.1 Matthew Maury, an American Naval officer, produced one of the first large-scale studies of ocean currents by using ships' logs. His work led to a vastly improved system of navigational charts (Bettmann Archive).

FIGURE 9.2 Edward Forbes developed early ideas about life in the oceans (Radio Times Hulton Library).

Janeiro. In fact, ships using Maury's charts saved as much as 10 days on their voyages to Rio de Janeiro. His directions cut trips to Australia by 20 days, and they shortened by 30 days the time of travel from the east coast of the United States to California by way of Cape Horn.

The cooperative exchange of information that Maury started is still carried on today, and the modern *Pilot Charts of the Hydrographic Days,* which are used all over the world, are the direct descendants of the charts he began.

Another event of great significance to modern oceanography was an historic voyage which began in 1873. This voyage by the British ship *Challenger,* which lasted more than three years, visited every ocean and

collected thousands of biological specimens from each (Figure 9.3). The scientists on the *Challenger* also collected numerous samples of ocean water, made hundreds of depth soundings, and mapped ocean currents. The data were invaluable in the early days of oceanographic explorations.

Some great work was done by the crew of the *Challenger.* At that time it was thought that Forbes was correct, and that no life could exist at the bottom of the oceans. But the information gathered by the *Challenger's* dredging operations seemed to indicate that pressure and cold are not factors that exclude life. The *Challenger* discovered more than 1500 animal species below 1000 m, and approximately 20 specimens were taken at 6250 m. Furthermore, the dredging showed the bottom to be covered with clay and mud, and not primeval ooze as had been thought.

Early Ideas about the Sea

From the earliest times, man has been concerned with the sea. The ancient Greeks believed that the ocean was an endless stream flowing forever around the edge of the world. They thought that the stream blended with the sky at some place that was a region of whirlpools. If a traveler entered this region, he would be drawn into a muddled world from which there was no return.

The Phoenicians were great sailors, but they were interested in trade rather than exploration. As a result, they tended to keep their ships close to the shores of Europe, Asia, and Africa. For more than 2000 years, from approximately 2700 B.C. to 600 B.C.,

FIGURE 9.3 (*left*) The H.M.S. *Challenger*. The voyage of the *Challenger* marked a high point of 19th century oceanography (The Challenger Report).

FIGURE 9.4 (*below*) A map of the world by Ptolemy. This was the world known to the Greeks (New York Public Library, Picture Collection).

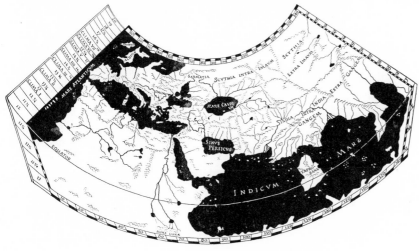

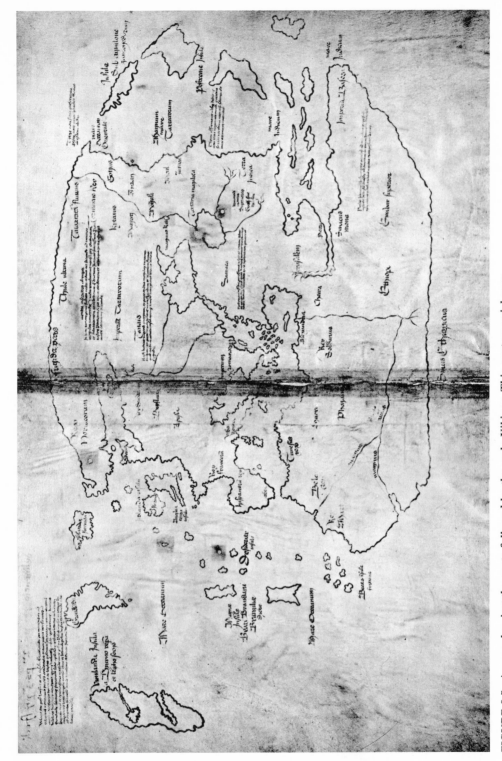

FIGURE 9.5 A map showing the route followed by the early Vikings. This controversial map places the Vikings in the New World hundreds of years before Columbus. (Reproduced by permission of the Yale University Press from *The Vinland Map and the Tartar Relation*, by R. A. Shelton, Thomas E. Marston, and George D. Painter. Copyright 1965 by Yale University.)

they sailed the shores of the Red Sea to Syria, Arabia, India, and China. There is some evidence that they even sailed along the western coast to Europe as far north as Scandinavia. To protect their trade, the Phoenicians kept their comings and goings secret, and unfortunately there is no written record of their journeys.

During the third century A.D., the Vikings sailed from Scandinavia to Spain and through the Mediterranean Sea to Italy. Sometime during the tenth century they established colonies in Greenland, and then moved across the Atlantic Ocean to North America. They were great navigators, since they were able to accomplish all of their feats without a compass or astronomical instruments. They navigated by sighting the sun, moon, and stars.

Even though great journeys were made into the far reaches of the ocean, the old idea of an outer ocean surrounding the Earth persisted. The legend of a dead and stagnant sea filled with dangers was still alive when Columbus ventured into the Atlantic Ocean. Once Columbus had shown the way to the West Indies; however, more great voyages followed. Balboa discovered the Pacific Ocean, and Magellan sailed around the globe. As a result of Magellan's voyage, an idea grew

that there was a great southern continent below the then known lands. This continent, Antarctica, was first sighted by Captain N. Palmer, in 1820, of the *Hero,* one of a fleet of eight seal-fishing vessels from Connecticut.

Another idea, which persisted for centuries, was that a northern passage to Asia existed. Many voyages were made into the Arctic Sea to find the Northwest Passage. It was not until 1897 that one succeeded. This voyage was made by the Swedish ship *Vega,* which at last passed from Gothenburg to the Bering Strait. In 1969, the *Manhattan* made it.

Through centuries of voyaging, ideas about the sea have gradually changed, and increasingly more useful and accurate information has been obtained for navigating the seas.

Ocean Tides

The regular rise and fall of the sea – the ocean tide – can be observed on almost any shore. This phenomenon has been observed by man throughout history. The ancient Greeks and Romans were aware of the tides and their relationship to the passing moon. However, this relationship remained a mystery until 1687, when Sir Isaac Newton was able to give us a reason for the tidal

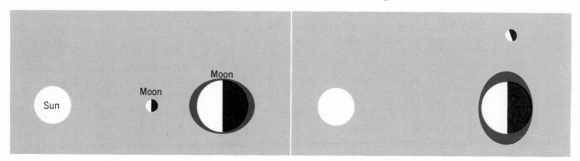

FIGURE 9.6 Spring tides occur when the Earth, moon, and sun are in line. The gravitational effects of the sun are added to those of the moon and create a higher tide than normal. Neap tides occur when the Earth, moon, and sun are at right angles to one another. Then the effect of the sun opposes that of the moon and the high tides are lower than normal.

forces applied to the Earth by the moon 240,000 miles away from the Earth.

As the moon passes over the ocean, a distinct tidal wave develops. A second tidal wave occurs on the side of the Earth opposite the moon. Thus, there are two tidal waves that move continuously throughout the oceans of the world. (See Figure 9.6.)

As the Earth rotates relative to the moon, the water nearest the passing moon is pulled by the moon's strong gravitational force, and a distinct tidal wave develops. This wave is a result of the combination of forces involved in the respective rotations of the Earth and the Moon. The second wave is developed by the fact that the Earth rotates faster than the Moon. Both waves point toward the Moon. The high tides or maximum points in the water level of the ocean occur in a line passing from the center of the moon through the center of the Earth. The low tides occur at points on the Earth that are at right angles to the high tides.

The tides do not always rise and fall to the same degree. They vary at different seasons and at different times of the month. It is here that the sun enters into the production of the tides. When the sun and moon both act together, the highest tides occur. These are known as *spring tides*. When the moon and the sun are opposed, that is, exert their force at right angles to each other, *neap tides* are produced.

The rise and fall of the tide is also affected by the shoreline, the ocean bottom, and the prevailing wind. In the Bay of Fundy in Nova Scotia, the difference between high and low tide is as much as 60 feet. The rush of waters in such situations is referred to as the *tidal bore*.

The Waters of the Oceans

The ocean waters cover about 71 per cent of the total surface of the Earth; only about 29 per cent of the Earth's surface is land. Most of the exposed land is in the Northern Hemisphere. In fact, there is twice as much in the Northern Hemisphere as in the Southern.

Table 9.1 gives the percentages of the surface covered by water and land in each 10° latitude band. For example, between 80° and 90°N latitude, 93.6 per cent of the Earth's surface is covered by water; 6.4 per cent is covered by land. However, in the

Table 9.1   Distribution of Water and Land in Different Latitude Bands

SOUTHERN HEMISPHERE			NORTHERN HEMISPHERE		
LATITUDE, °S	PER CENT OF WATER	PER CENT OF LAND	LATITUDE, °N	PER CENT OF WATER	PER CENT OF LAND
90–80	0	100	90–80	93.6	6.4
80–70	24.6	75.4	80–70	71.3	28.7
70–60	89.6	10.4	70–60	30.0	70.0
60–50	99.2	0.8	60–50	99.2	0.8
50–40	97.0	3.0	50–40	47.5	52.5
40–30	88.8	11.2	40–30	57.3	42.7
30–20	76.9	23.1	30–20	62.4	37.6
20–10	78.0	22.0	20–10	73.6	26.4
10–0	76.4	23.6	10–0	77.1	22.9

FIGURE 9.7 Parrsboro, Nova Scotia, at high (a) and low tide (b). The water drains completely at low tide. Notice the markings on the wharf representing the high tide waterline (National Film Board of Canada).

Southern Hemisphere for this same latitude band, that is, 80° to 90°S, 100 per cent of the surface is covered by land.

When we compute the total percentages of water and land in the Northern Hemisphere, we find that 60.7 per cent of the surface of the Earth in the Northern Hemisphere is covered by water and 39.3 per cent is covered by land. In the Southern Hemisphere, 80.9 per cent is covered by water and only 19.1 per cent by land. Thus, water predominates as the surface cover in the Southern Hemisphere.

The Cycle of Changes

Nearly all of the Earth's water comes from the oceans. Evaporation from the ocean surface converts water into water vapor. How-ever, water vapor is also recycled by evaporation of fresh water and by plant transpiration. Once in the atmosphere, water vapor is transported over the Earth by winds and condenses to form clouds. It is from the clouds that all forms of precipitation fall to the Earth's surface. Much of the precipitation runs over the land and works its way back to the sea. The total amount of water that falls on the Earth's surface as precipitation is equal to the total amount of water that evaporates from the Earth's surface.

Sea level tends to remain nearly unchanged from year to year. However, water is occasionally removed from the hydrologic cycle in various ways. Glaciers "tie up" significant amounts of water for long periods of time. At present, glaciers cover approximately 3 per cent of the Earth's surface. If

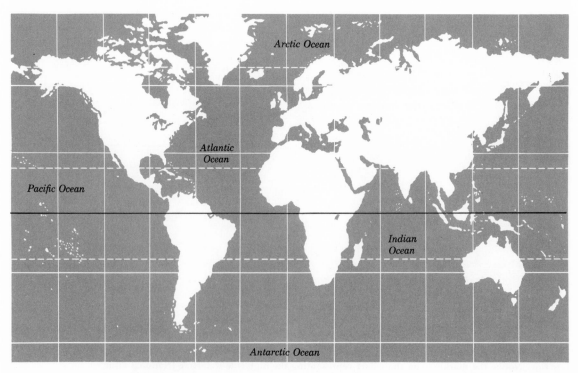

FIGURE 9.8   The oceans of the world.

FIGURE 9.9 Ice floes near the North Pole. The iceberg has broken off the mother glacier and is starting its journey to the North Atlantic Ocean (U.S. Coast Guard).

they should suddenly melt and return their water to the sea, the sea level would rise more than 200 feet. During the ice ages the level of the sea was much lower than at present, since a great deal of the present ocean water was locked up on the land as glaciers.

The Seven Seas

The "seven seas" is a phrase that dates from ancient times. The Phoenicians and Greeks knew of seven large bodies of water, called each one a sea, and thought they were all the bodies of water that the world contained. The seas they knew were the Mediterranean, Red, Atlantic, West African, and East African Seas, the Indian Ocean, and the Persian Gulf.

As the age of exploration developed, it was realized that the seven seas were only a part of the world's oceans, and the phrase dropped into disuse. However, in 1896, Rud-

yard Kipling published a collection of poems which he called *The Seven Seas*. From that point on the phrase became very popular. Although it has little meaning today, we can, if we wish to, divide the Earth's large bodies of water into seven areas that might be generally designated as the seven seas: the Arctic, Antarctic, North Atlantic, South Atlantic, North Pacific, South Pacific, and Indian Oceans.

The Atlantic Ocean (that is, North and South together) gets about half of all the Earth's rain. It gets most of this rain indirectly by means of rivers; most of the larger rivers of the world drain into it. The Pacific Ocean is much larger than the others, and gets the least rainfall. The Arctic and Antarctic Oceans are the least salty. These oceans receive a great deal of glacial ice, which tends to dilute the saltwater. Also, few large rivers flow into either the Arctic or the Antarctic, which might bring dissolved salts from the land.

The Arctic and Antarctic

The Arctic Ocean is almost completely surrounded by land. During the winter, enormous areas of the Arctic are covered with ice floes. The floes drift with the prevailing winds and currents, moving generally from the northern coast of Siberia to the northeastern coast of Greenland, crossing the North Pole on the way. The Arctic Ocean is the most isolated of all the oceans. At times, the water surrounding the North Pole is considered to be part of the Atlantic Ocean and is termed the North Polar Sea.

The Antarctic Ocean completely surrounds a continental area—Antarctica. Ice accumulates on Antarctica until huge icebergs detach themselves from the land and move into the sea. Some may reach heights of more than 200 feet and may extend more than 1500 feet below the surface of the water. Icebergs have been observed that were more than 30 miles long.

The Atlantic, Indian, and Pacific Oceans join in a wide belt of water around Antarctica. Winds, currents, and weather in this area are arranged in a circumpolar pattern, as are the physical and hydrological conditions. Scientists have applied the terms Southern Ocean, Antarctic Ocean, and South Polar Sea to this great expanse of water. The term Antarctic Ocean, when used, refers to the water south of Africa, South America, New Zealand, and Australia.

The Atlantic

The Atlantic Ocean, which receives its name from the legendary island of Atlantis, extends from the Antarctic Ocean northward to the top of the North Polar Sea. In the Southern Hemisphere, the Atlantic is divided from the Pacific by a line running from Cape Horn to the South Shetland Islands, along 70°W longitude. In the Northern Hemisphere the Bering Strait separates the two oceans. The Bering Strait at this point is 58 km wide, with a maximum depth of 55 m.

One interesting area of the North Atlantic is so unlike any other area of that ocean that it is given a separate name, the Sargasso Sea (Figure 9.10). On the west this sea stretches from north of Bermuda to the Virgin Islands and eastward into mid-ocean. Large areas of the Sargasso Sea are covered by seaweed. In fact, the name of the area comes from the weed that accumulates—sargassum. The Sargasso Sea was first sighted by Christopher Columbus. Located between two swift currents, it is a relatively calm area of water, in which the only current is a lazy circular eddy. The surface weed hardly seems to move, and new growth occurs each year in

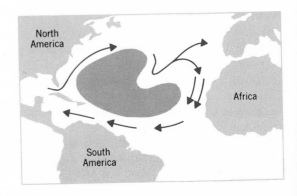

FIGURE 9.10 Contrary to popular belief, the Sargasso Sea, although it is an area rich in seaweed, is not densely overgrown and does not trap ships within it.

the warm, sluggish water. There are many strange stories about the Sargasso Sea, most having to do with ships caught in the weeds. Although it is true that, over the centuries, many ships have been found floating in this area, there is not enough seaweed to entangle even the smallest ship. The Sargasso Sea has no land boundaries at all.

The Caspian Sea is really not a sea at all, but an inland salt lake. The Mediterranean, however, is actually a part of the Atlantic Ocean, as is the Gulf of Mexico and the Caribbean. These large seas have depths which equal the depths of the ocean and are arms of the ocean proper which are located between land masses. The term mediterranean is used to refer to seas between land areas. Thus the Gulf of Mexico and the Caribbean Sea are sometimes called "the American Mediterranean." The Baltic Sea, Hudson Bay, and the Persian Gulf are also mediterraneans.

The Atlantic also includes three *marginal* seas, or deep seas located on the ocean borders near the continents. They are the Gulf of St. Lawrence, the North Sea, and the English Channel. Other marginal seas of the world are the Gulf of California and the Bering Sea.

The Pacific

The cool, sunlit waters of the Pacific Ocean seemed very mild to Magellan after sailing through the straits at the tip of the South Atlantic Ocean around the Cape of Good Hope. He had encountered many storms, so that the mildness of these waters was very striking. In fact, their peaceful quality caused him to give the ocean the name Pacific. Actually, there are great storms that develop in this ocean, but Magellan rarely encountered any of them.

The rim of the North Pacific Ocean is a region of frequent earthquakes, centering near the Bay of Tokyo. It has been esti-

mated that there is an average of four shocks a day in the Tokyo area. However, serious earthquakes occur only on the average of once in six years. The rim of the North Pacific suffers great earthquake damage. Earthquakes are also frequent around the rim of the South Pacific Ocean. There is a large earthquake center in the deep ocean trench off Chile and Peru.

The border between the Pacific and Indian Oceans follows a line through the Malay Peninsula, Sumatra, Timor, Cape Londonberry, Tasmania, and then the meridian 147°E to Antarctica.

The Indian Ocean

The Indian Ocean is famous for the regular monsoons that blow from the East Indies toward Africa in the winter and move in the opposite direction during the summer. When the monsoons change, huge hurricanes often develop in the Indian Ocean, especially in the northern part.

The Indian Ocean was the route used by early traders as they traveled through the Red Sea on their way from Europe to the trade centers of the East. The strong mon-

FIGURE 9.11 **The Sargassum weed, which grows in the Sargasso Sea, east of Bermuda (William M. Stephens).**

FIGURE 9.12 (A) Magellan's ship *Victoria* and (B) the route it followed around the world. Magellan never completed the voyage; he died in the Philippine Islands. [(A) New York Public Library, Picture Collection, (B) Portolan Atlas, Batista Agnese, NYPL Print Division.]

A

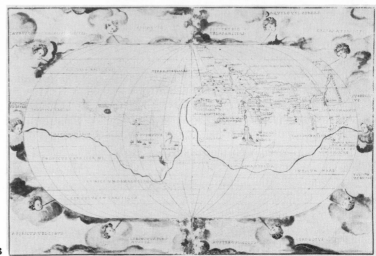

B

soon winds helped the sailing ships complete their voyages between Greece and the Arabian peninsula by way of the Arabian Sea. Unfortunately, this ocean is the last to be studied by modern oceanographers, and much information has yet to be collected regarding its topography and the nature of its currents. The Red Sea is a mediterranean of the Indian Ocean.

The Shoreline

Shorelines are formed by the action of waves and water on coastlines. In general, they tend to become rather straight and regular, whatever the coast was like originally.

214

Classification of Shorelines

Shorelines are often divided into two types: shorelines of *submergence,* which are formed of drowned bays, inlets, and river mouths, such as New York Bay, Delaware Bay, and San Francisco Bay; and shorelines of *emergence,* which are composed of youthful strata that have emerged from the sea in relatively straight lines. Submerged coastlines can be found on the Gulf of Mexico and the East coast of the United States from New Jersey to Florida. Figure 9.13 illustrates the formation of these shores.

Since the coastlines of submergence tend to contain many bays and other irregularities, they are ideal for port cities. Nevertheless, they will eventually be straight and regular. The waves along a submerged coast tend to break closer to shore, so that erosion is severe. Cliffs develop, the surface is highly irregular, deposition of spits (narrow ridges of sediment) and bars are common, and wave-cut terraces are found.

Coastlines of emergence are much less common. They tend to be regular and gently sloping. Waves break further out and beaches are more frequently formed. These coast-

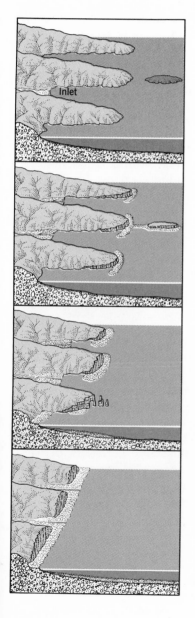
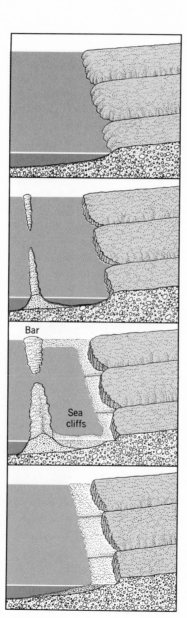

FIGURE 9.13 Coastlines of submergence are characterized by many bays, inlets, and drowned river mouths. Erosion eventually wears away the irregularities and forms a straightened shoreline.

FIGURE 9.14 Alluvium left behind in a dry riverbed. This material, washed from the continents by rivers, finds its way to the coastlines and into the ocean bottom (National Film Board of Canada).

lines result from crustal movements that thrust portions of the continental shelves above sea level, or from a falling sea level.

Coastlines also form in other ways. They sometimes develop from *alluvial deposits* from streams and rivers, and may also be the result of glacial moraine deposits or pits gouged out by the glaciers. The latter is the source of the Scandinavian *fjords.* Shorelines may also be formed by lava flows and by volcanic explosions.

Sea Water Realms

When the sea water is considered in terms of depth, it is usually divided into four domains: the *littoral, sublittoral, continental,* and *abyssal* realms. The littoral realm describes the area between high and low tides; the sublittoral realm to about 100 fathoms, and the

FIGURE 9.15 A fjord is a large depression gouged out of a coastline by glacial activity. This fjord is located in Norway (Norwegian National Travel Office).

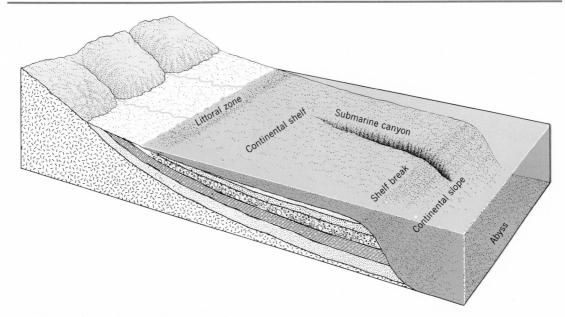

FIGURE 9.16 The edge of the continents slopes seaward and continues beneath the waterline. Although a wide variety of structures is found on the continental margins, the littoral zone generally descends to a gently sloping continental shelf which then drops off sharply at the shelf break, forming the continental slope. This last portion of the continent extends into the sea floor proper—the abyss.

continental to a depth of 500 fathoms. The abyssal realm includes all depths greater than 500 fathoms. The relation of the ocean bottom to the continental margin is shown in Figure 9.16.

The Littoral Realm

The littoral realm has a vast population of seaweed, sea slugs, sponges, mussels, worms, snails, and fish. This is the area in which corals develop, forming reefs and atolls.

The Sublittoral Realm

Many of the animals that inhabit the littoral realm are also found in the sublittoral realm. The bottom area is usually soft and composed of sand, mud, or even a soft clay. Stones and rocks are also found protruding from the bottom material. Shallow-water areas are sometimes present. They provide good foundations to which animals may attach themselves. Sponges, many crustaceans, and miniature forests of sea fans can often be found in the bottom region. Many different groups of the starfish family can also be found in the sublittoral realm. Prawns, lobsters, mollusks, scallops, and the octopus, too, are inhabitants of the sublittoral realm.

Continental Realm

The animal life of the continental realm is the subject of much study. The animals taken in the dredges are often of great interest and beauty. Corals, sea pens, lamp shells, and starfish can be found. Sea cucumbers also inhabit this area, as do the deep-sea sponges. In this rocky, mud-covered region, light becomes less intense and the murky gloom of the abyss begins.

The Abyss

Many special conditions prevail in the abyssal realm and, as a result, the inhabitants of this region show very special features. Conditions are more uniform throughout this region than in any other part of the ocean or on the surface of the Earth. There is absolute darkness in the abyss. The temperature is believed to remain unchanging at slightly above 0°C. The pressure in this region can amount to more than three tons per square inch. The bottom consists of bare rock and ooze, and there is no vegetation. Many of the crustaceans in this region have long legs that

FIGURE 9.17   An oceanic angler fish. With its large mouth and angler bait it can consume fish larger than itself (American Museum of Natural History).

FIGURE 9.18   This copepod crustacean abounds in the planktonic population of the ocean (N. E. Beck, Jr., National Audubon Society).

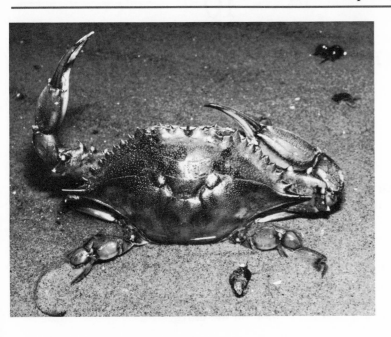

FIGURE 9.19 The blue crab, a larger crustacean of the type commonly found near coasts (Walter Dawn, National Audubon Society).

help them locomote through the ooze, and some sea spiders have legs more than two feet long. Animals with calcareous skeletons are quite rare and mollusks are particularly uncommon. Some deep-sea animals have no eyes at all, although others have eyes at the ends of long stalks. These eyes are probably capable of detecting any fluorescent or phosphorescent light. Many of the fish in this region have large jaws and powerful teeth.

CHAPTER 10

Studying the Ocean Basins

SINCE THE OCEAN BASINS ARE SIMPLY DE-pressions in the Earth that contain great masses of water, it is important that the oceanographer know something about the Earth. The Earth is generally considered to be a sphere, but students of geometry would describe the Earth as an oblate spheroid, for its two axes are not the same length. The axis of rotation is the shorter.

Defining the Location of a Point on the Earth

To define the position of a point on the Earth, scientists use latitude, longitude, elevation, and depth. Latitude and longitude are expressed by angular coordinates; elevation and depth are usually given as vertical distances.

Latitude

Latitude is angular distance from the equator, and is measured in degrees, minutes, and seconds, north and south of the equator. On a true sphere, the linear distance between points would be the same everywhere on the surface. However, since the Earth is not a true sphere, the distance represented by latitude is not the same everywhere. In fact, the distance increases as you move from the equator to the poles. One degree of latitude on the equator is equal to 110,567.2 m. At the geographic poles, 1° of latitude is equal to 111,699.3 m.

The northernmost geographic point of the Earth is the point closest to the pole star — Polaris. Polaris is 90° above the equator. When you sight Polaris, you are measuring how many degrees Polaris is above your position on the Earth.

Longitude

Longitude is the angular distance east or west from a standard point on the Earth's surface. The standard point is the Royal Observatory at Greenwich, England. The angular distance from Greenwich to any point on the Earth's surface is measured from 0° to 180°, east or west.

In developing the idea of longitude, various imaginary planes are run from the North Pole to the South Pole. Because these planes run from the North to the South Pole, they pass through the axis of the rotation of the Earth. The lines which result from the intersection of the Earth's surface by these planes are referred to as *meridians*. When a scientist determines longitude, he is actually measuring the angle between meridian planes. You can compute local longitude by using the following formula: local longitude = 15 times Greenwich time minus local solar time. (Solar time is approximately the same as standard time.)

Ideal Sea Level

Land masses are elevations above the spheroidal geometrical figure of the Earth. Sea bottoms are depressions in this same geometrical figure. Measurements of elevations and depressions are referred to an ideal *sea level*.

Of course, near shore the actual sea level varies with every high tide and low tide. In the open ocean, however, sea level varies by no more than two meters. Thus, soundings taken in the open ocean can be referred to the actual sea surface with minimal error. In fact, the inaccuracies in the equipment will be greater than the difference between the ideal and actual sea levels.

In coastal areas, ocean depths must be ascertained quite accurately, since shoals present hazards to navigation. In these areas, soundings are made with respect to a single reference point, which is usually indicated on the chart. Along the Atlantic coast of the United States (average low tide), the reference point is mean low water. In most sections of the world, low water is used as the reference point.

Subdivisions of the Oceans

There are several systems for naming parts of the ocean. When natural ocean boundaries do not exist, the International Hydrographic Bureau has selected boundaries arbitrarily.

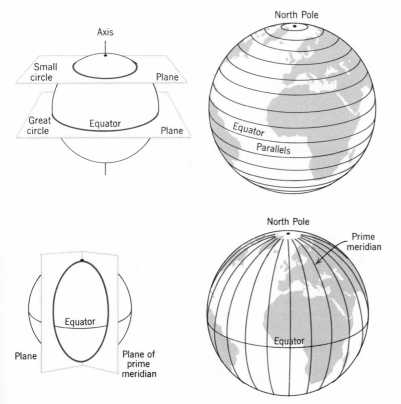

FIGURE 10.1 Latitude lines are imaginary parallel lines drawn equidistantly around the globe. They yield one's location north or south of the equator (After Navarra and Strahler).

FIGURE 10.2 Longitude lines are great circles drawn north to south through the poles. They yield one's location east and west of the prime meridian drawn through Greenwich, England (After Navarra and Strahler).

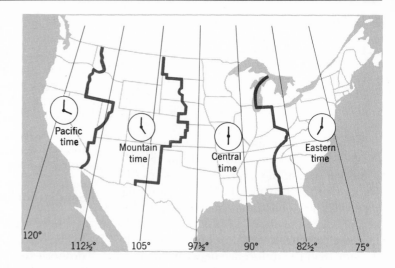

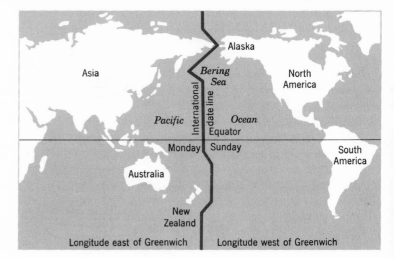

FIGURE 10.3 (A) Time zones of Continental U.S.A. (B) The international date line is located at 180° longitude opposite Greenwich.

Straight or curved lines are drawn on maps to define the particular areas. The Pacific and Indian Oceans, for example, are separated by the 150°E meridian. The Indian Ocean is separated from the Atlantic by a north-south line drawn from the tip of Africa to Antarctica along 20°E. This is the most frequently used system for separating one ocean from another.

Each ocean, however, has distinctive features that permit the separation of the ocean basin into smaller basins or divisions. One method by which this is done is to follow the bottom ridges (mountain areas) that descend to depths of 4000 m, as a basis for defining various parts of the oceans. These ridges form smaller basins deeper than 4000 m (Figure 10.4). The Pacific Ocean, for example, has a number of these basins. In the west, the largest is the Philippine Basin, which contains the Mindanao Trench, a deep that is approximately 10,500 m down. Other basins are also located within the western Pacific Ocean—the Caroline, Solomon, Coral, New Hebrides, and East Australia Basins. The central Pacific holds the North Pacific, Mariana, Central Pacific, and South Pacific Basins. The Aleutian Trench, with a depth of 7680 m, is found in the North Pacific

Basin. In the southeastern Pacific, oceanographers have designated three other basins. The Indian and Atlantic Oceans also contain various basin areas.

A third system for designating areas of the oceans is based on the distribution of living organisms, properties of the sea, and the nature of ocean currents. Unlike the basins system, which is based on the topography of the ocean bottom, this system is directed toward defining natural regions of the oceans. De-

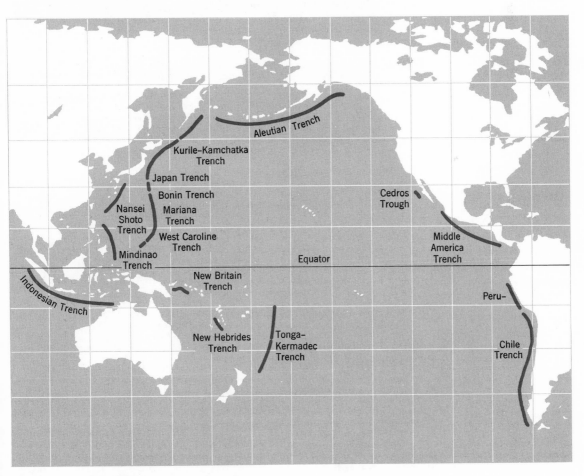

FIGURE 10.4 The Pacific Ocean trench system.

FIGURE 10.5 (*right*) Echo sounding utilizes sound waves from an explosive device or transmitter. The sound waves are reflected and echoed back by the varying layers of sediment on the sea floor. The measurement of the speed of return allows for the calculation of the depth and thickness of the sediments.

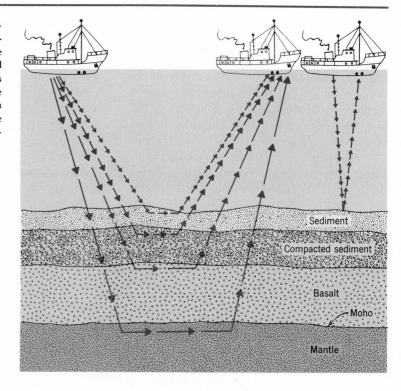

FIGURE 10.6 (*below*)

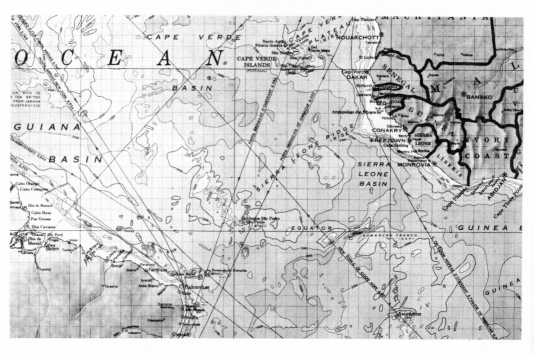

veloping this system is much more difficult than setting up either of the other two systems that have been discussed. Undoubtedly, as man comes to depend on the oceans more for food, and the biological relationships between marine organisms becomes better known, this system will become more widely used.

Topography of the Ocean Bottom

Echo sounding equipment, which was introduced during the 1920's, allows highly accurate surveys to be made in deep water (Figure 10.5). Deep water charts issued by the English-speaking countries give depths in fathoms, whereas meters are the preferred

FIGURE 10.7 (*above*) Turbidity currents recreated in a laboratory tank. The dense current flows along the bottom gouging canyons out of the sediment on the continental margins and sea floor (California Institute of Technology).

FIGURE 10.8 (*left*) A sandfall off Baja California (Scripps Institute of Oceanography).

unit of measure for designating depth. The following conversions may be helpful to interpret depth readings:

1 fathom is equal to 6 feet
1 fathom is equal to 1.8288 meters
5.5 fathoms equals 10 meters

Numerical values alone do not give any real idea of topography. To form any graphic representation, contours must be drawn. The accuracy of these contours depends, of course, on the number of soundings made, and the accuracy with which the position of the soundings is determined.

Charting the Ocean Floor

Maps of the ocean floor are called *bathymetric* charts (Figure 10.6). They are pub-lished by the United States Hydrographic Office and by the International Hydrographic Bureau in Monaco.

Bathymetric charts include contour lines that show the depth of the ocean floor. When contour lines represent equal depths, they are referred to as *isobaths*. Usually the isobaths are crowded together at some places, and are far apart at others. Regions of crowding are regions where the bottom slopes quite steeply; where the floor is flat, the isobaths are farther apart.

Oceanic Sediments

The sediments covering the ocean floor are quite varied. They are of great scientific interest since they yield much information about the past history of the ocean bottom.

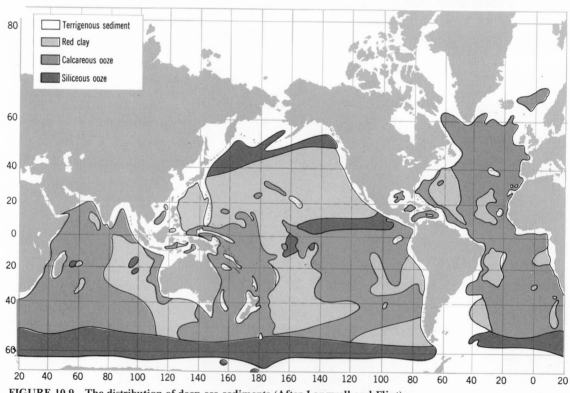

FIGURE 10.9   The distribution of deep-sea sediments (After Longwell and Flint).

A

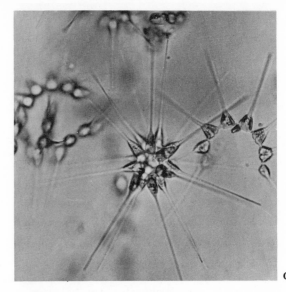

C

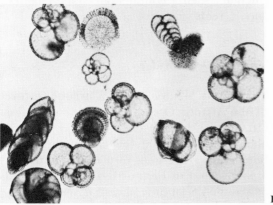

B

**FIGURE 10.10   Remains of plank-
tonic plants and animals give rise to
the various organic oozes on the sea
floor. Globigerina (A) and forami-
nifera (B) are members of the zooplankton, while diatoms (C) are
phytoplankton. [(A) and (B) (American Museum of Natural History;
(C) Walter Dawn, National Audubon
Society).]**

In some of the deep portions of the ocean,
they are 450 to 900 m thick.

Many sediments are carried from the continental masses by rivers emptying into the
ocean (Table 10.1). These are termed *terrigenous sediments*. Sediments that originate in
the sea are *oceanic*.

Many oceanic sediments are organic—they
are the remains of marine plants and animals. Called *pelagic sediments,* these sediments are *calcareous oozes,* the most widely
distributed, consisting of calcium carbonate;
siliceous oozes, the silica remains of planktonic
plants; *brown clays,* the insoluble materials
from the general plankton population.

The oozes originating from plankton are
produced by a variety of organisms. Plankton
are microorganisms found in great abundance in the sea. The plant forms, such as
diatoms, that make up the siliceous oozes are
one-celled, photosynthetic organisms known
as phytoplankton. Other phytoplankton yield
calcareous oozes. The zooplankton, or animal
plankton, yield some of the silica oozes but

A   B

FIGURE 10.11   (A) Calcareous oozes, (B) Siliceous oozes (Ron Church).

are mainly responsible for the calcareous oozes. Larger organisms and many larvae also form part of the general plankton population.

The various forms of plankton are associated with specific water temperatures. Thus, the presence of the various oozes indicates the past water temperatures and climates of the oceans. The calcareous oozes are particularly abundant in warmer waters, and the siliceous oozes predominate in colder regions. In addition, the sediments are thicker in the relatively inactive Atlantic Ocean than in the Pacific where uplifting was more recent.

Submarine Topography

There are elevations on the ocean floor—ridges, rises, seamounts, and sills—and depressions—trenches, troughs, basins, and deeps. These elevations and depressions result from deformations in the Earth's crust. Other features of submarine topography result from erosion, deposition, and biological activity. These features include the con-

tinental shelf, continental slope, banks, shoals, reefs, canyons, and valleys.

The Mid-Ocean Ridge

The Mid-Atlantic Ridge, a striking feature of the Atlantic Ocean, is about 3000 meters (10,000 ft) below sea level, and extends about 6000 to 9000 feet above the ocean floor. This ridge goes in a north-south line from Iceland in the north to Bouvet Island at about 55°S latitude. There is also a north-south ridge in the Indian Ocean, which goes south of India all the way to Antarctica.

The ridges in the Pacific Ocean are not as conspicuous as the ones that we have already discussed. They are also shorter. One ridge runs from the Antarctic to Japan. Another ridge called the East Pacific Rise extends from Antarctica, in the longitude of New Zealand, to the Gulf of California. It is about 4000 meters high.

These ridges, the Mid-Atlantic, Mid-Indian, and the East and West Pacific Rises are all part of one structure which is the largest geological feature in the world. The Mid-Ocean Ridge is more than 40,000 miles long and connects the basins of the Pacific, Atlantic, and Indian Oceans.

The most intensively studied portion of the ridge lies in the Atlantic Ocean. It has been found that the ridge crest is bisected by a rift valley or graben, which is nearly 13,000 feet deep in some places. The overall effect is one of a double mountain chain with the rift valley in between the ridges. In addition, earthquake activity has produced gigantic fracture zones which run across the ridge, giving it a zigzag appearance.

Other Topography of the Ocean Floor

As one examines the ocean floor, the first features encountered are the *continental margins*. They begin with a *continental shelf*, which gently slopes seaward, to widths of 400 miles in some places. The shelf drops off more sharply at a region known as the *shelf break* into a region called the *continental slope*. These slopes mark the last portion of

Table 10.1 *Classification of Kinds of Sediment on the Sea Floor*

1. Terrigenous sediment
   Found mainly on the continental shelves, continental slopes, and abyssal plains. Mud, sand, and gravel, vary greatly from place to place.
2. Brown clay (also called pelagic clay)
   Confined to the deep-sea floor, mostly at depths greater than 13,000 feet. Contains less than 30 per cent calcium carbonate. Chief constituents are clay minerals, quartz, and micas. The clay is red or brown as a result of gradual oxidation during the very slow process of deposition.
3. Calcareous ooze
   Contains more than 30 per cent calcium carbonate, most of it consisting of shells and skeletons. Confined to regions in which surface waters are warm. The resulting shells accumulate on the bottom more rapidly than the inorganic clay. Because it contains much carbon dioxide, deep-sea water dissolves calcium carbonate. As the shells drift down, they are gradually dissolved.
4. Siliceous ooze
   Contains a large percentage of skeletons built of silica. Occurs where organisms with calcareous shells are few in the surface waters.

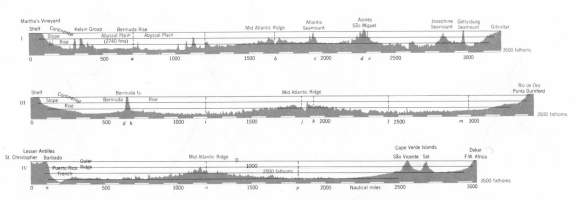

FIGURE 10.12 **Profiles of the Atlantic Ocean floor exaggerated in the vertical 40 times (From B. C. Heezen, Special Paper 65, Geol. Soc. Am.).**

the continents as they descend into the sea floor.

The shelves and slopes are marked in many parts of the world by submarine canyons. These canyons are cut thousands of feet deep by dense, sediment-laden water flowing as strong currents along the bottom. One of them, the Hudson Submarine Canyon, cuts into the shelf and slope at the mouth of the Hudson River in New York City.

In some places, the continental slope extends in ramplike deposits called *continental rises.* In other places, the slopes descend steeply into trenches such as the Puerto Rico trench, which is more than 30,000 feet deep. Although there are only two major trenches in the Atlantic, the Pacific Ocean contains more than a dozen.

The ocean floor itself contains flat *abyssal plains,* which are featureless, sediment-cov-

ered plains. Small *abyssal hills* dot the landscape on the floor. All the ocean floors also contain gently rising arches known as *oceanic rises.* The island of Bermuda is the top of a pedestal of rock that extends upward from the Bermuda Rise, which is about 600 miles long and 300 miles wide.

Particularly prominent in the Pacific Ocean basin are the *island arcs.* These are island groups near the continental borders that are areas of active volcanoes and earthquake activity. They are seaward-curved island chains that seem to be regions of active mountain building.

The Pacific is also marked by east-west fracture zones that have formed steep cliff faces. Mountainous masses known as *seamounts* are common, as are flat seamounts known as *guyots*—named after a Swiss geologist, Arnold Guyot.

The Pacific and Indian Oceans also con-

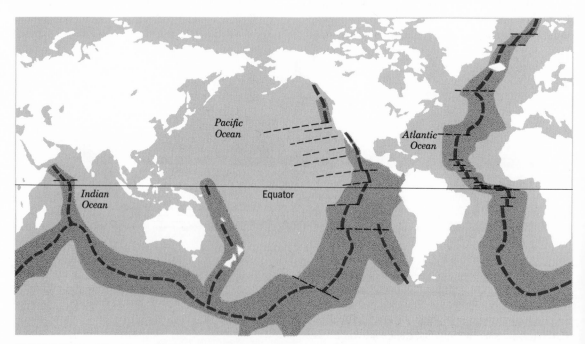

FIGURE 10.13   The ridge system of the entire ocean basin (After B. C. Heezen, 1962).

tain numerous *coral reefs*. These reefs are composed of calcium deposits that form a framework for corals. In warm equatorial waters, often around volcanic islands, these animals build up layers of calcareous deposits known as *fringing reefs*. If the island subsides and the corals continue to grow, they deposit new material near the surface, and a *barrier reef* develops with a lagoon between the reef and the original island. When the island submerges and is lost beneath the water and calcium deposits, the reef is known as an *atoll;* it consists of a lagoon surrounded by the coral reef.

Thus, the ocean floor is as varied as the surface of the Earth. Most notable among its features are the abyssal plains, submarine canyons, mountain formations, continental margins, and the ridge system.

Origin of Ocean Basins

Two theories about the origin of ocean basins are currently being debated by scientists.

The first theory is based upon the fact that the continental masses are composed of lighter rock than the ocean basins. There would be, of course, greater attraction of gravity on the heavier rocks underneath the

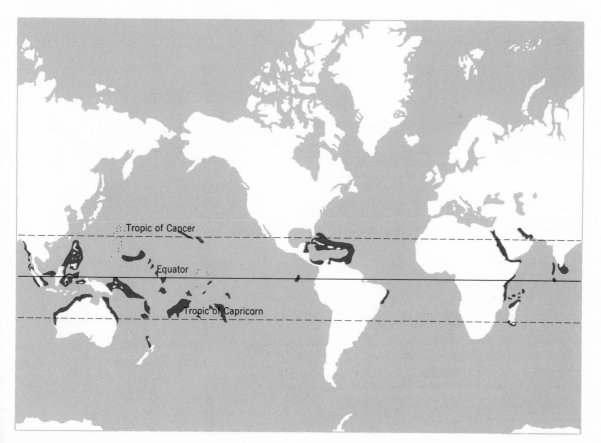

FIGURE 10.14   Coral reef regions of the world.

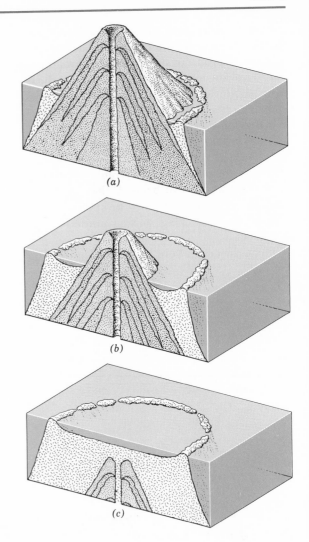

FIGURE 10.15 Formation of an atoll. A fringing reef is attached to the land. A barrier reef lies offshore and traps a lagoon of seawater between it and the island. An atoll is formed by a closed, circular reef around an absent land mass.

ocean, so that they would sink deeper into the underlying magma. This subsidence would crowd the continents upward so that they would float higher than the ocean basins.

The second theory is based on the previously discussed theory of convection current cells circulating beneath the crust. Convection involves the rising of hot material and the sinking of colder material. Doughlike masses of semiplastic material rise from below the crust. Where the edges of two convection cells touch and crowd material together, continents would be formed. The ocean basin would form at the center of the circulating cell of material. Of the two theories, this last is probably the most widely accepted today.

CHAPTER 11

Properties of Ocean Water

OCEAN WATER DIFFERS FROM PURE WATER primarily in the amount of dissolved salt that is present. The great quantities of salt found in the ocean result from the weathering of rocks on land. The dissolved minerals are carried into the sea by streams and rivers.

Evaporation from the surface of the sea removes only water molecules, leaving the salt behind. Thus, the salt content of the oceans has tended to increase. No one knows just how rapidly the salt content of the oceans has increased in the past; it may be stabilizing.

Composition of Seawater

Great quantities of mineral matter (salts) are found in solution in the ocean. More than 32 elements are found, including minute amounts of sulfur and gold.

The first information about the composition of seawater was gathered in 1884 by W. Dittmar. He analyzed 77 water samples that were collected from around the world during the voyage of the *Challenger*. Of the common salts in seawater, table salt (sodium chloride) is the most abundant, accounting for 77.7 per cent. Magnesium chloride makes up about 10.8 per cent of the common salts, and epsom salt (magnesium sulfate) about 4.7 per cent. Gypsum (calcium sulfate) comprises 3.6 per cent, potassium sulfate about 2.4 per cent, calcium carbonate (the chemical of common chalk and limestone) 0.3 per cent, and

magnesium bromide about 0.2 per cent. (See Table 11.1.)

In addition to the dissolved salts, seawater also contains dissolved gases. Among these are oxygen, nitrogen, and carbon dioxide, all of which are important to support sea life.

Salinity

Salinity is the amount of various salts dissolved in the ocean water. It is measured by chemical analysis of water samples that are collected from the surface and depths of the ocean.

The easiest way of determining salinity is by analyzing the chlorine content. Chlorine makes up a part of many salt molecules and accounts for about 55 per cent of the dissolved material in seawater. Scientists report salinity in terms of the concentration of

Table 11.1

SALT	PER CENT OF COMMON SALTS IN SEAWATER
Sodium chloride (table salt)	77.7
Magnesium chloride	10.8
Magnesium sulfate (epsom salts)	4.7
Calcium sulfate (gypsum)	3.6
Potassium sulfate	2.4
Calcium carbonate (chalk. limestone)	0.3
Magnesium bromide	0.2

salt, or mass of salt, in a unit mass of seawater. This is usually expressed in parts per thousand, or "per mille," written as ‰.

The average salinity of ocean water is about 35 per mille. This means there are about 35 lb of salt in 1000 lb (roughly 120 gal) of seawater.

Evaporation and Precipitation

Evaporation and precipitation affects the salinity on the ocean surface. When there is high evaporation from the surface, the salt concentration increases. When high evaporation is accompanied by low precipitation, that is, very little rain or snow falling or being carried into the sea, the salinity increases at a rapid rate. Low evaporation reduces salinity, and heavy precipitation dilutes the seawater by the addition of fresh water.

In the equatorial region, which has the heaviest rainfall on Earth, the salinity is low. However, in the subtropical latitudes (25°N and S), a minimum of rainfall and high rates of evaporation are found. This area of ocean has the highest salinity. Between the sub-

FIGURE 11.1 (*right*) **Range of salinity, and evaporation minus precipitation, in various latitude ranges (After Weyl, 1970).**

FIGURE 11.2 (*below*) **The Great Salt Lake is an example of an enclosed basin in which evaporation is producing an increased salinity of the water. The region is the remains of Lake Bonneville (Spence Air Photos).**

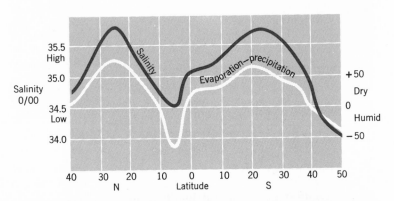

FIGURE 11.3 The high salt content of the Dead Sea has given rise to a barren, lifeless land in the surrounding regions (Sabine Weiss, Rapho Guillumette).

tropical areas and the poles, salinity tends to decrease. This is the result of the increasing rainfall and the decreasing evaporation.

In enclosed basins, such as the Great Salt Lake and the Dead Sea, the water tends to have a high salinity since, usually, evaporation exceeds precipitation in these areas. The salinity in the Dead Sea is greater than 200 mille. The Red Sea and Mediterranean Sea also have a high salinity because they receive relatively little fresh water from rivers. The salinity of these seas run around 40 per mille.

The Density of Ocean Water

The density of pure water (weight per unit volume) is approximately 1 g per cubic cen-

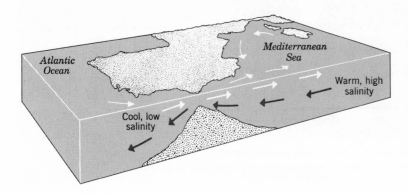

FIGURE 11.4 The Mediterranean experiences low rainfall and rapid evaporation. Thus, it has a higher salinity and is denser than the waters of the Atlantic Ocean. This creates a warm, high-salinity subsurface current that flows past the Straits of Gibraltar to the Atlantic while higher, less saline, cool Atlantic Ocean water flows in over the subsurface current.

timeter. The density of ocean water increases as the salinity increases and is about 2.5 per cent greater than that of freshwater (Figure 11.4).

When pure water is cooled from a temperature of 20°C to a temperature of 4°C, the water contracts and becomes denser. However, where water is cooled below 4°C it expands and the density decreases. When pure water is cooled below 0°C it freezes and expands; the density of pure, solid water, that is, ice, is only about 92 per cent of that

FIGURE 11.5 Estuaries, such as the mouth of the Hudson River in New York State, occur where a river meets the sea. The denser saltwater forms a wedge under the freshwater at the region of mixing (Aero Service).

of liquid water. This is the reason why an ice cube in a glass of water remains about 8/9 below the surface of the water.

The increased density of ocean water produces two effects. First, it causes floating objects to have greater buoyancy than in freshwater. The reason for this effect was stated about 225 B.C. by Archimedes (287– 212 B.C.): the buoyant force that supports a submerged object is equal to the weight of the water displaced by the object. The effect is well known to those who have swum in both fresh- and saltwater. The second effect of the increased density of seawater occurs in *estuaries,* or areas where a river meets the sea. The saltwater tends to sink below and form a wedge underneath the lighter freshwater. The shallow layer of freshwater gradually mixes with the dense seawater as it moves seaward.

Temperature in the Sea

As the salinity increases, the freezing point decreases. Water in estuaries, which usually has a salinity of 25 per mille, will have a freezing point of minus 1.3°C. Ocean water of average salinity, 35 per mille, has a freezing point of about minus 1.7°C. The lowest water temperature found in the oceans is minus 1.7°C or 28°F.

Since ocean water is warmed by the heat of the sun, the surface temperature is related to the intensity of solar radiation. At the equator, the surface temperature is about 80°F; the temperature gradually decreases moving north or south from the equator. In higher latitudes, the average surface temperature is about 30°F. (See Fig. 11.6.)

Neither seasonal changes nor the difference between day and night has much effect on the surface temperature of the deep oceans. The lack of variation results from the movement of seawater which evenly distributes the heat added to the ocean within great quan-

tities of water. Scientists say that the ocean has great *heat capacity,* that is, it can lose or gain great quantities of heat before there is a temperature change. Another way of stating this is that the ocean reacts sluggishly to changes in receipt of solar energy. As a consequence, the ocean acts as a heat regulator for the atmosphere.

The temperature of the surface water in shallow areas of the ocean, that is, along the continental shelf, is influenced more by seasonal changes and weather conditions than that of the deep oceans.

Temperature and Depth

Except in polar waters, the temperature of ocean water decreases with depth. High temperature, even in the region of the equator, is confined to a shallow surface layer. At the equator in the mid-Pacific, where the surface temperature is 82°F, the temperature 600 feet below may be 52°F. In polar regions, on the other hand, the surface temperature may be 30°F, but at 3000 feet the temperature will probably be 33°F. This is quite fortunate! If the temperature decreased with depth in the polar area, the bottom area would freeze and life in the depths could not survive whereas, when water freezes on the surface, life continues to exist in the water below it.

The Thermocline

At moderate depths there is usually a layer of water somewhat less than 1000 feet thick in which the change of temperature is at a maximum. Such a layer of water is referred to as the *thermocline* (Figure 11.7). Below the thermocline the water temperature decreases slowly (except in the polar regions). In fact, the temperature below a depth of 5000 feet is fairly constant all over the world; it is

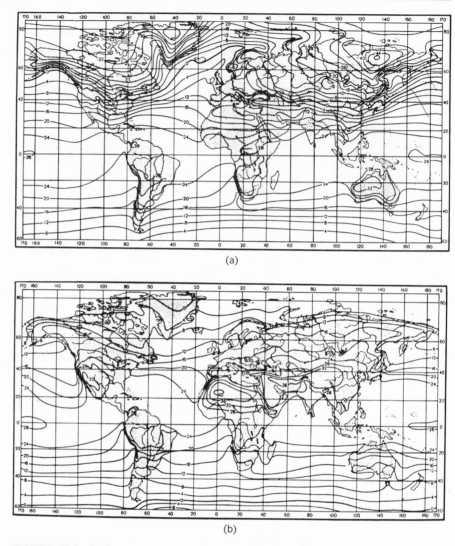

FIGURE 11.6   Surface temperatures of the oceans, in August (a) and in February (b).

around 1° to 3°C. The water below 5000 feet is referred to as *isothermal deep water.*

As indicated, the surface temperature of the sea varies from the equator to either pole. This change is produced by the variation of solar radiation received in each latitude. However, scientists have also found that the sea temperature varies from east to west. This is especially true near the continents as water temperatures are affected by the adjacent land masses and the shallow water areas near the continents.

Gases in the Sea

A number of gases are held in solution in the ocean water. Nitrogen, oxygen, and carbon dioxide are present in large quantities. Small amounts of ammonia, argon, helium, hydrogen, and neon have also been reported.

Oxygen

Oxygen is a most important gas in the ocean. Just as on land, life in the sea is de-pendent on its presence. The mass quantity of oxygen in ocean water is insignificant compared to the quantity found in the atmosphere.

The ocean obtains oxygen from marine plants and from the atmosphere. Oxygen in the air goes into solution through contact with surface water. Thus, water that is exposed to air tends to be rich in oxygen.

The amount of oxygen in seawater is determined by the *Winkler method,* an extremely accurate although rather complicated process. The oxygen content is usually greatest near the surface. However, the fact that there is life in the deepest regions of the sea indicates the presence of some oxygen there as well. This can only occur if there is a general circulation and exchange of water between the different layers of the ocean. Scientists have found that in many places, there are layers in the ocean that contain a minimum of oxygen. An oxygen minimum layer exists within the thermocline; the amount of oxygen in the water then increases to a constant rate below this layer. The amount of oxygen is related to the number of organisms that utilize it.

Carbon Dioxide

Although there is much less oxygen in the sea than in the atmosphere, the reverse is true for carbon dioxide. In fact, there is about 60 times more carbon dioxide in the sea than in the atmosphere.

Obviously, the oceans have a high capacity to dissolve carbon dioxide, and they operate as a regulator for carbon dioxide in the atmosphere. It is also true, however, that the rate at which carbon dioxide dissolves in seawater is regulated by the atmospheric concentration. When the carbon dioxide content of the atmosphere increases, the rate at

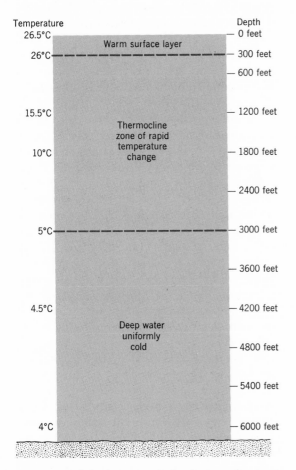

FIGURE 11.7 Ocean waters change in temperature quite gradually except in the regions of the thermocline.

which the gas dissolves in ocean water also increases.

Carbon dioxide reacts readily with seawater and is more soluble than oxygen and nitrogen. Some of the carbon dioxide in the sea is present as free CO_2, but the greater part is present as carbonates and bicarbonates. Carbon dioxide that has combined with water forms carbonic acid, H_2CO_3; bicarbonates and carbonates are derived from H_2CO_3.

Other Gases

Hydrogen sulfide usually develops in water that has an oxygen deficiency, and it may be present in high concentrations in water that experiences little overturning. Generally, in such stagnant water, dead organic material under attack by bacteria uses up the oxygen through the process of decay.

Methane is another gas that may develop as decomposition of organic material takes place in oxygen-deficient stagnant water.

Nitrogen compounds are also produced as products of decomposition of organic material. The qualities vary widely from place to place according to the seasonal changes and the abundance of life forms.

Color of Sea Water

The color of seawater varies from a deep blue to an intense green and is occasionally even brown or brownish-red. The open ocean is usually a beautiful blue color, at least in the middle and low latitudes. Coastal waters are often green, and are sometimes brown or brownish-red.

Scientists use a specially prepared color scale called the Forel scale as the standard in determining color of the water. A special disk called a Secchi disk is used to observe the color.

Causes of Color in Ocean Water

Many theories have been proposed to explain the different colors of the sea. The most acceptable theory for the blue color of the sea relates it to the scattering of light by water molecules or tiny suspended dust particles. These molecules and particles are small compared to the wavelength of the scattered light, and they scatter the blue component of sunlight more than the other wavelengths. According to this theory, the blue of the seawater is comparable to the blue of the sky.

Where large particles (organic or inorganic) are present in the ocean, the water does not have a blue color. Thus, blue is taken as an indication of the absence of small animal and plant life. In this sense, the blue color of the ocean is referred to as its *desert color*.

Scientists do not explain the green color of seawater as a result of scattering light. In developing theories about green water, they refer to a so-called "yellow substance" produced by the metabolic activities of plant and animal life. The combination of the yellow substance and the natural blue leads to a scale of greens. Some scientists also suggest that fluorescent materials contribute to green coloration of the sea. However, those materials are not very significant in producing a green sea when compared to the yellow substance.

When there are great quantities of large particles in the sea, the color of the sea may be determined by this dissolved, or suspended, matter. At such times, the sea is said to be *discolored*. This often happens when masses of material are carried into the sea by heavy rainfall or by rivers. Discoloration can also result from large populations of algae. The color of the Red Sea is produced in this way.

FIGURE 11.8 The United States Coast Guard Cutter *Eastwind* clearing a channel in pack ice in the Ross Sea, Antarctica (U.S. Navy).

Red Tide

An unusual type of discoloration is produced in the coastal waters of Peru, California, and many other places in the world with the advent of the so-called "Red Tide." This is an invasion of microorganisms that are carried by upwelling currents. The organisms produce a deep red color and secrete a poisonous compound which destroys millions of fish. Although the natives have attributed this occurrence to supernatural reasons, we now know it to be a consequence of complex meteorological changes and their effects on ocean currents.

Ice in the Sea

The most familiar ice in the sea is the iceberg. Icebergs are freshwater in the frozen state, found floating in the sea. They are produced when the edge of a glacier breaks off and falls into the ocean. But there is another kind of ice in the sea; this is frozen seawater. Called *sea ice* or *pack ice,* frozen seawater is the most common kind of ice in the ocean.

Sea ice forms when seawater is subjected to cooling at the surface until the surface water reaches the freezing point. The first ice crystals that develop contain little salt and are elongated in shape. As the freezing continues, the crystals form a lattice in which a certain amount of saltwater or even saltier brine is included. The salinity of the ice depends on the rate of freezing: the faster the freezing process, the greater the amount of brine included.

Sea ice usually has many small cavities containing seawater or brine. As the ice ages, it tends to lose the brine. Some ice islands in the

FIGURE 11.9 A water sampling bottle that recovered water samples below the ice in the Weddell Sea, Antarctica. The study included examination of the micro-organisms taken at various depths during operation Deep Freeze in 1968 (U.S. Coast Guard).

Arctic Sea have lasted for many months. Teams of scientists from the United States and Russia have set up camps on these islands and have done a great deal of research.

Sea ice acts as an insulator: it prevents the underlying water from being cooled to the freezing point. It very rarely exceeds a thickness of 10 feet.

There is a great deal of sea ice around the Antarctic continent in the form of huge drifting ice fields. Much is relatively flat and is kept in motion by winds and currents. The ocean in this region is exposed to an exchange with warm seas. Thus, during the summer some of the sea ice melts, leaving parts of the Antarctic coast ice-free.

The Formation of Icebergs

The great continental glaciers in Greenland and Antarctica are the two principal sources of icebergs. The process of ice breaking off the glacier, called *calving,* is accompanied by tremendous noise. Almost everyone has seen films of glaciers calving. The icebergs produced by the Greenland glacier have been photographed more ex-

tensively than any others in the world. The Greenland glacier sends more than 10,000 icebergs into the sea each year. These icebergs tend to drift southward into the Atlantic shipping lanes, and present a great hazard to oceanic shipping. Therefore, the International Ice Patrol keeps a watchful eye on their progress. The danger is at a maximum during the spring and summer, since the icebergs break away from the glacier primarily during the spring and are then carried by ocean currents to the warmer latitudes.

The Greenland icebergs may have a length of 2000 feet, and their height above water can exceed 250 feet. They may last as long as two years before they melt.

Antarctic icebergs originate mainly from *shelf ice.* Shelf ice is a direct continuation of the ice cap covering the Antarctic continent, which extends onto the water surrounding the continent. It breaks off as enormous slabs to produce the icebergs.

The Antarctic icebergs are tremendous. They approach a height above water of 300 feet, which indicates a thickness of 2700 feet, since most of the iceberg is below the surface of the water. Their length can exceed 62 miles. These large drifting masses of ice can easily be mistaken for islands. The melting of these huge icebergs takes a long time, and they survive for as long as 10 years. During this time they can drift and move great distances. They have been found in the Indian Ocean and even the Pacific.

Studying the Movement of Seawater and Ocean Currents

The sea might be pictured as a flat, restless surface that is composed of countless great rivers, flowing in various directions. These ocean rivers or ocean currents, as they are commonly called, have been observed since man first ventured onto the sea. Men long ago recognized that these ocean rivers are thousands of times as large as any of the great rivers floating on the continents.

Ocean Currents

Oceanographers divide ocean currents into three groups: large-scale ocean currents, wind drift currents, and tide and internal wave currents.

Large-scale ocean currents include the Gulf Stream, the Kuroshio, which extends from Formosa past the coast of Japan, and the equatorial currents, among others. All these currents transport great amounts of water. They are related to the distribution of density in the sea. A map of the large-scale ocean currents appears in Fig. 11.13.

Wind drift currents are caused directly by the stress that the wind exerts on the surface of the sea. Wind can cause water to be transported over large areas. This is especially true when it blows consistently from one direction. There also can be changes in sea level because of the direct effect of wind. These changes are easily observed in the Baltic Sea, a shallow and enclosed area.

Tides and internal wave currents are produced by the tides and internal waves that run first in one direction and then in the opposite direction. Internal waves are produced in water physically separated into layers.

Scientists studying ocean currents find that all three types can be present in the same place at the same time. Therefore, it is somewhat difficult to identify each type and to study each independently.

Measuring Ocean Currents

A variety of techniques have been developed to measure the direction, temperature, and speed of ocean currents. The speed of an ocean current is usually measured in *knots,* or nautical miles per hour. One nautical mile is equal to 1.15 land miles. It is actually defined as equal to 1 minute of latitude measured along a meridian. A ship moving at 60 knots will cover 60 nautical miles in an hour. If it is moving north or south, it will also cover a distance of 1° of latitude.

One of the simplest techniques for measuring ocean currents is to observe the motion of a floating object. The motion of a drift bottle, weighted with sand so that it floats just below the water surface, indicates speed as well as direction.

Another device used to measure ocean currents is the *current meter,* which measures the current speed by means of a propeller. The meter must be held stationary to the earth. It is usually supported on a tripod fixed to the ocean floor, but it may also be supported or suspended from an anchored ship.

The *deep drogue* technique is also used. A drogue is any object with a large drag, such as a parachute. Drogues are used primarily for the measurement of deep-water currents. If a parachute is employed, it is weighted so that it will be suspended at a particular depth. It is then attached by a wire to a surface float. As the deep-water current carries the drogue along, the motion of the float is observed. In this way a deep-water current can be measured.

In the early days of oceanography, ships' logbooks were a valuable source of information on the ocean currents. In fact, Matthew Maury derived the first major study of the currents from the logs of seagoing vessels. An officer of a ship records the observed movement of his ship. He must know the compass heading of the vessel and also the speed at which the ship is traveling. The speed he knows is the speed in still water. The actual motion of the ship will not correspond to that speed at that compass heading, because the ocean current will affect the movement of the ship. Therefore, the actual motion of the ship is a result of the ocean current's impact on the ship. The actual motion of the ship is determined by observation of land or by taking navigational fixes on the stars. At the end of the voyage, the ship's officer can calculate how the actual course of the ship has been affected by the ocean current. In this way, the average ocean current during the voyage can be determined.

The electromagnetic method for determining ocean currents is probably the most widely used today. It is based on the fact that an electrical conductor cuts across the force of a magnetic field. Seawater easily conducts electricity because of its mineral content. The earth has a permanent magnetic field. As an ocean current moves, it cuts across the magnetic field, and a measurable electrical effect is produced. To determine the current speed two floating electrodes are towed behind a ship, and the electrical potential induced between these electrodes is recorded. Then the speed can be computed.

Another device used to measure currents at or below the surface of the sea is the Swallow float. It is a cylindrical tube that

FIGURE 11.10 A current meter being recovered. The buoy is located above the meter. The meter records current speed by means of a propeller atop the meter (U.S. Coast Guard).

carries a sonar transmitter and can be weighted to float at any depth. The velocity of the current is calculated from the signals that the float sends.

The Gulf Stream

Southeast from New York or Boston, the Atlantic Ocean changes from a greenish color to a clear and sparkling blue. During the wintertime the air becomes a great deal warmer over the bluish water. The change in the color and temperature of the sea is an indication that one of the great ocean rivers has been entered. This is the Gulf Stream (Figure 11.11).

This ocean current has its point of origin in the Gulf of Mexico. From its origin, the water flows steadily northeastward, toward the Arctic Ocean, moving at a velocity of approximately 2.5 knots. The Gulf Stream swings toward the east, mingling with and finally losing itself in the waters off the northern coast of Europe.

The water of the Gulf Stream is much warmer than the surrounding water, and has an important effect on the climate of the countries that it passes. The average surface temperature of the Gulf Stream is approximately 83°F, and the bottom temperature is about 45°F. More than 436 trillion tons of this heated water flows northward daily, through the Florida straits. Even larger amounts move through the Bahama Channel into the Atlantic Ocean. When a cold wind blows across the Gulf Stream, vapor often condenses above its surface. This is caused by the warm air rising above the Gulf Stream and mixing with the cold wind. The Gulf Stream is sometimes referred to as the *Gulf Stream system,* for it includes not only the northward and eastward flow, which begins at the straits of Florida, but also various currents found in the eastern North Atlantic. The system is generally divided into three portions: the *Florida Current,* the *Gulf Stream,* and the *North Atlantic Current.* The Florida Current is that part of the Gulf Stream system that moves from the Straits of Florida to a point off Cape Hatteras. The portion from Cape Hatteras to a region east of the Grand Banks at about a latitude of 45°N is referred to as the Gulf Stream. The term North Atlantic Current designates the easterly and northern currents of the North Atlantic east of the Grand Banks.

There are a number of reasons why the Gulf Stream moves northward from the Florida Straits. One reason is the action of the trade winds. In the region of the Florida Straits, these winds blow almost continuously from the east. They push the surface water ahead of them, carrying it westward until it finds its way through the West Indian Islands into the Gulf of Mexico. The Gulf is closed by solid land on all but the eastern side. Therefore, the water cannot move farther westward than the land. As

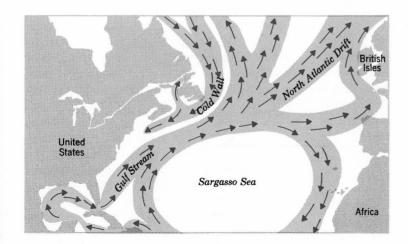

FIGURE 11.11 The Gulf Stream.

the trade winds continue their pressure from the east, the only outlet available is the Florida Straits.

The prime reason for the movement of the Gulf Stream, and other large-scale ocean currents, is not the action of the surface winds, but the difference in temperature between the cold water of the Arctic regions and the warm water near the equator. Heated water expands and rises and cold water contracts and sinks. Thus, the warm, less dense water in the tropics flows north and south to replace the colder and heavier water from the Arctic and Antarctic regions. Since the Earth revolves on its axis from west to east, the water slips backward as the Earth turns, and moves slowly westward.

It is very difficult to see that the water in the Gulf of Mexico is higher than that on the outside, but such is indeed the case. An imaginary visit to the little islands that form the eastern boundary of the Caribbean Sea will permit us to understand this. These islands, known as the *Lesser Antilles,* are only a few miles apart. On their western or leeward side (opposite the wind) the surf beats on the beaches in great white crested rollers, and in the narrow channels between the islands the water is rotating toward the west. On the Atlantic or windward side (direction of the wind), the tide regularly rises and falls more than 5 feet. On the side that faces the Gulf of Mexico, however, there is scarcely more than a few inches difference between high and low water. Thus, when the tide is low on the windward side of the islands, it is several feet lower than that on the leeward side. The level of the Gulf of Mexico remains about 3 feet above the ocean. To sum up, three factors produce the Gulf Stream: the density differences of the water, the winds, and the tides.

The expanse of the Gulf Stream has been known for centuries; the first explorer to make a record of it was Christopher Columbus. The first chart of the Gulf Stream was published by Benjamin Franklin.

While Benjamin Franklin was serving as Postmaster General of the colonies, he noticed how much longer it took for the royal mail packets to travel across the ocean than it did for ships to sail along the coast of the colonies. The American sailors were taking advantage of the Gulf Stream without realizing its extent. Franklin called on them for information about the currents they encountered, and published his chart on this basis. Although crude by modern standards, the chart was excellent for its day, and gave a rather accurate idea of the speed and direction of the Gulf Stream.

Other Large-Scale Ocean Currents

Figure 11.13 depicts the worldwide system of ocean currents. The outstanding feature is the great whorls that represent systems of currents. These whorls center around the subtropical latitudes of the Atlantic and Pacific Oceans. As you examine the illustration, you will notice that the currents seem to rotate in a clockwise direction in the Northern Hemisphere, and counterclockwise in the Southern Hemisphere. Just as the Gulf Stream is driven by the trade winds, each of these surface currents is driven by a wind system.

The Gulf Stream, a poleward warm current, is accompanied on the eastern side of the North Atlantic Ocean by a current of water which moves toward the equator. Called the Canary Current, it is a cool mass of water that flows past the Canary Islands off the coast of Africa. The southern portion of this great whorl of moving water is called the *North Equatorial Current.* It moves westward across the Atlantic Ocean, and joins the Gulf Stream. It dominates the system of currents in the North Atlantic. On the north side of this great whorl of water is the *North Atlantic*

Current, which generally flows eastward. The Sargasso Sea is located in the west central portion of this complex of currents.

Similar systems of currents are also present in the North Pacific Ocean. The *Kuroshio Current,* a warm mass of water also known as the Japan Current, flows poleward on the western side of the Pacific. The *California Current,* found on the eastern side of the Pacific, moves toward the equator. It is a cold mass of water. The *North Pacific Current* moves from the Kuroshio to the California Current on the northern portion of this whorl. On the southern portion of this whorl is the *North Equatorial Current.* This flows westward from the California Current toward Asia.

Large quantities of subarctic water flow

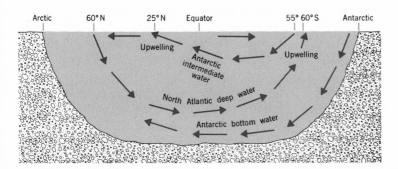

FIGURE 11.12 (*left*) **General circulation of subsurface cold currents in the Atlantic Ocean.**

FIGURE 11.13 (*below*) **Ocean currents of the world. White arrows show cool currents; black arrows show warm currents (After Navarra and Strahler).**

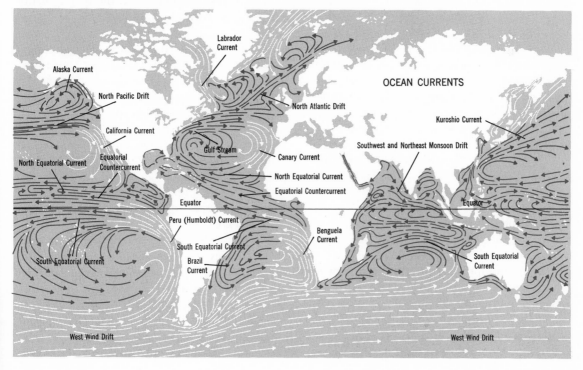

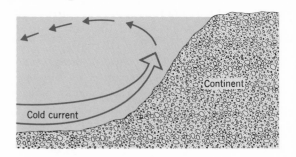

FIGURE 11.14 Upwelling occurs as cold water from below the surface moves upward against the land mass in response to wind stress against the water's surface. The colder water replaces the warm surface water.

into the North Pacific Ocean; these currents are being intensively mapped at present.

The Kuroshio Current is sometimes divided into three portions: the Kuroshio Current, the Kuroshio Extension, and the North Pacific Current. The Kuroshio Current runs northeast from Formosa to Ryukyu, and then runs along the coast of Japan at latitude 35°N. The Kuroshio Extension is the eastern flow of this warm current to longitude 160°E. There the North Pacific Current begins. It moves to the east and sends some branches to the south. The North Pacific Current probably reaches a longitude of 150°W.

At times the *Tsushima Current* is considered as a branch of the Kuroshio system. The Tsushima, a warm mass of water, branches off the left-hand side of the Kuroshio Current. It enters the Japan Sea and follows the western coast of Japan to the north.

The *California Current* has a southward flow between latitude 48°N and 23°N. During spring and early summer, north-northwest winds prevail off the coast of California. These winds give rise to an upwelling of water that begins in March and continues until July (Figure 11.14). In these regions of upwelling, the spring temperatures are lower than the winter temperatures. There are two conspicuous regions of upwelling, located at latitudes 35°N and 41°N, or off Point Conception and Cape Mendocino, respectively.

The upwelling of ocean water in these re-

gions is due primarily to wind stress on the surface waters. Cold water from well below the surface rises upward and replaces the warmer surface water. In the San Francisco area, north winds help to produce the California Current. At the same time, warm surface water is transferred away from the coast. Upwelling occurs as cold bottom water rises to the surface. Thus, along the California coast in the San Francisco area, cold surface water prevails. Plankton are carried to the surface by this process and the fishing is improved because of the additional food supplied to the fish in the region.

In the South Atlantic, the *Brazil Current* is a mass of warm water that flows southward from the equator. The Brazil Current is found on the western side of the Atlantic. On the eastern side of the South Atlantic, a vast amount of cold water travels up along the western coast of Africa. This is the *Benguela Current*, and is the most outstanding current of the South Atlantic Ocean. The denser water of lower temperature is found on the right-hand side of the current, close to the African coast.

As it travels northward, the Benguela Current leaves the coast and becomes the northern portion of the *South Equatorial Current*. The South Equatorial Current flows west across the Atlantic Ocean, generally between the latitudes 0° and 20°S. This current eventually joins the Brazil Current. The *Falkland Current* extends south to about latitude 30°S. The *South Atlantic Current,* a system of shallow currents, flows east on the southern side of this counterclockwise whorl in the South Atlantic Ocean.

The South Pacific Ocean presents an interesting and complicated system of currents. The *Peru Current System* is the only major current of the South Pacific Ocean that has been examined in great detail. It seems to originate in the subantarctic region. As the Peru Current flows northward along the west coast of the South American continent, some of the subantarctic water is deflected west. Just south of the equator it joins the *South Equatorial Current* which moves westward across the Pacific. Complications then arise because of the land masses of Australia and New Zealand. These land masses force the North and South Equatorial Currents southward along the East coast of Africa. From the south Indian Ocean a warm southward moving current called the *Mozambique Current* flows along the east coast of Madagascar and Africa. These currents then form the *Agulhas Current* which moves along the southeast coast of Africa to the Cape of Good Hope.

The stress of the winds plays an important part in driving currents. For example, the westward blowing trade winds maintain the north and other equatorial currents. The prevailing westerlies on the poleward side of the Atlantic and Pacific Currents drive them from west to east. The water of these ocean currents tends to move in the direction of the wind, for it is literally dragged along by friction with the wind. However, the rotation of the Earth also plays an important part. It interacts with the stress of the wind so that the water does not move in the direction of the wind, but instead moves to the right of the wind in the Northern Hemisphere, and to the left of the wind in the Southern Hemisphere. Thus in the Northern Hemisphere a wind from the south sets up a current flowing to the northeast.

The large-scale currents of the world play a tremendously important part in modifying weather and climate. For example, the flow of relatively warm Gulf Stream water around Iceland and the Scandinavian Peninsula keeps the Arctic ports free of ice, even in winter. The flow of this warm water also produces a more moderate air temperature than would be expected in this region. The warm Kuroshio Current plays an important role in modifying temperatures and climate

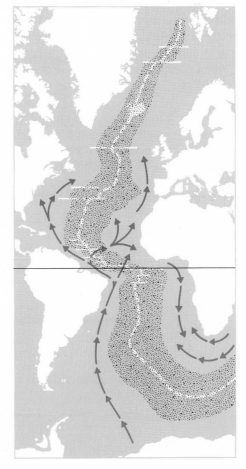

FIGURE 11.15 The deep-sea ocean currents of the Atlantic are affected by topographical features of the floor, such as the ridge system.

of the Aleutian Islands. The cold Labrador and California Currents also have decided effects on the temperature and climate of the regions they pass. As warm, moist air moves in toward the California coast, the California Current cools the lower portion of it. As a result, dense sea fogs occur along the California coast. The famous fog banks which roll into San Francisco are produced by this interaction of warm moist air and the California Current. The Labrador Current produces a similar effect in the North Atlantic.

New York is in the same latitude as Southern Italy and Spain, but it does not have the semitropical climate that this observation would suggest. The reason is that a current of cold water pushes down along this section of the eastern United States. If the Gulf Stream were diverted, the entire east coast of the United States might have the cold, dreary climate of Labrador and Newfoundland. The mild, pleasant climates of British Columbia and Oregon, which are as far north as Newfoundland, result from the warm water of the Kuroshio Current. One thing is certain. Any alterations in the course of these great ocean currents would mean, at the very least, a change in the climate of continents.

Ocean Currents in Deep Water

Scientists have identified ocean currents in deep water as well as near the surface (Figure 11.15). Deep-water currents flow rather slowly, but they play a very important part in transporting water masses from one section of the ocean to another.

At high latitudes, cold heavy water sinks to great depths. This is especially true in the North Atlantic Ocean, where cold water sometimes sinks to depths of two miles. It then travels at this depth below the surface toward the equator. This movement of cold water in vast currents below the surface causes the deep water in the tropics to be very cold. Near the equator, cool water tends to rise, and keeps the water temperature lower than it would otherwise be.

A similar sinking of cold heavy water occurs near Antarctica, in the Weddel Sea. This cold heavy water from the Antarctic region flows northward at great depths.

This sinking of cold water carries great quantities of oxygen down to the ocean depths. The phenomenon is often called the "lungs of the ocean," because the cold water ventilates the deep reaches of the ocean. This sinking of water makes life there possible.

The submarine currents are very complex.

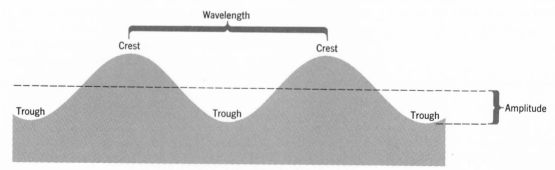

FIGURE 11.16 Wave motion occurs in a series of crests (higher portions) and troughs (lower portions). The amplitude of a wave is measured from the horizontal to the bottom of a trough. The wavelength is the distance from one crest to the same point on another crest.

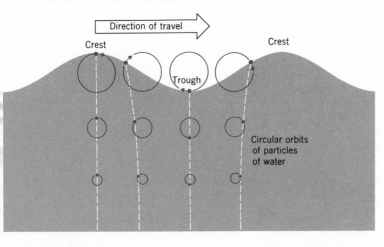

FIGURE 11.17 (*left*) In wave motion the wave energy moves forward while the particles move in small circles and return to their point of origin. The particles are disturbed less with greater depth (After Navarra and Strahler).

FIGURE 11.18 (*below*) As the wave advances toward shore, it becomes steeper and higher as it encounters shallow water. The wave curls over and breaks, and the water flows back near the bottom as undertow (After Longwell, Flint and Sanders).

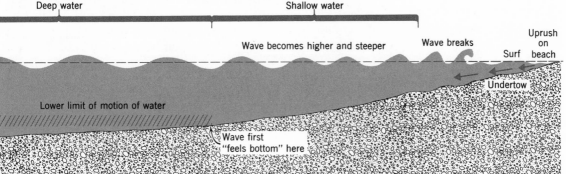

For example, consider the *Pacific Equatorial Undercurrent*. This undercurrent moves eastward at two or three knots, at a depth of a few hundred feet. Above it, on the surface, the South Equatorial Current moves westward.

Waves and Water Motion

Scientists use a variety of terms to describe and to classify waves. The most important are wavelength, period, frequency, speed, and amplitude. (See Figure 11.16.)

The *wavelength* is the distance from any point in the wave to the next repetition of that point. For example, the wavelength can be measured from crest to crest as shown in the illustration.

Scientists define the *period* of a wave as the time it takes for one full wavelength to pass a point. The period of a wave can be clocked by determining the exact moment that the crest of a wave passes a point, and the time when the next crest passes. Then the period is taken as the time interval between the passage of two successive crests.

The *frequency* is the number of waves passing a point in a given unit of time.

The *speed* of the wave is the rate at which the crest of the wave travels. It can be found by taking the wavelength and dividing it by the wave period.

FIGURE 11.19   A plunging breaker (Ron Church).

Amplitude is the height of a wave, but in a special sense. The true height of a wave is defined as the vertical distance from the top of a wave crest to the *top* of a wave trough. When people speak of a 50-foot wave, they usually include the trough in the height. A scientist would say that this wave has an amplitude of 25 feet.

Surface waves in the ocean fall into two distinct groups, according to the depth of the water they pass through. The first group is that of *long* or *shallow-water waves*. These waves occur where the water is relatively shallow compared with the wavelength. The speed of a long wave depends on the depth of the water, so that in a given body of water all the long waves travel with the same speed.

The second group of waves is that of *short*
or *deep-water waves*. Short waves occur when the water is very deep compared with the wavelength. The speed of short waves is regulated only by the wavelength of the wave, so they travel with different speeds. If the wavelength of a deep-water wave is long, it will travel faster than a wave having a shorter wavelength. Thus, deep-water waves can overtake and interfere with one another.

Consider what happens when waves move from the open ocean toward a beach, as shown in Figure 11.18. The deep-water waves in the open area approach the beach, and gradually become shallow-water waves, changing their character completely. Oceanographers describe this change by saying that the wave is "feeling the bottom." As deep-water waves, their speed was dependent only upon their wavelength. However, as they approach the beach, the depth of the water begins to decrease and the speed of the wave also begins to decrease. The gradual slowing down of the wave also causes the wavelength to diminish. Another way of saying this is that shallow water steepens the wave. This steepening of the wave finally causes it to break.

Ocean waves can be tremendously destructive, and many careful studies have been undertaken to measure and to predict their height and power. A pressure-type wave recorder is often used in shallow water. This pressure recorder is placed on the ocean bottom. In a sense, it weighs the mass of water as the wave moves over the recorder. The wave staff is another device used for observing and recording waves in shallow water. It simply records the height to which the water rises on the staff. In deep water the wave staff is attached to a special type of sea anchor so that the rise and fall of the sea surface with respect to the staff may be observed and recorded.

Automatic wave recorders have also been

FIGURE 11.20   Wind blowing across the open water is the primary cause of waves (Monkmeyer Press Photo).

developed for deep water. One recorder, called "the splashnik," measures the vertical acceleration of the sea surface, and radios the information from the floating recording device. Another automatic wave recorder operates on a pressure principle to measure the height of the waves relative to the ship.

Breakers

The breaking of waves along the seashore is a pretty and, at times, spectacular sight. As has already been indicated, the great rollers and breakers are produced by the retarding effect of shallow water. When the water is shallow, the higher portions eventually break and curl over the lower portions. The largest waves produce the most spectacular breakers. Between the waves the water flows back to its normal height.

Out at sea, or in very deep water, even the very largest waves do not break and curl. The only time this occurs in deep water, is when the wave encounters another wave or a large object. At times in deep water the wind "blows off" the top of a wave and causes it to break. In such cases, the wind is traveling faster than the waves, and forces the tops of the waves ahead of the lower portions.

Wind is the most common cause of waves. However, waves are also produced by earthquakes or volcanic eruptions under the sea. The energy that is released underground by the eruption of a shore volcano can also start a wave or a series of waves. Tsunamis are produced in this way. A tsunamis is especially destructive as it approaches a shore line. At sea, such a wave may be only a few feet in height, but its wavelength may be enormous. When such a wave approaches land and the bottom of the wave begins to drag, the mass of water coming from the rear overtakes the dragging forward portion of the wave. A tremendous mass of water begins to accumulate as an onrushing *comber*. Before it eventually breaks, it may be 50 to 100 feet high, and its waters crash down with enough power to destroy anything in their path.

The sea is a dynamic, continuously changing, and evolving structure. The waters of the Earth have sculptured the land and washed a myriad of elements into the ocean basins. These substances help to support a vast array of life forms in the sea. In addition to being the primary source of water for the planet, the sea abounds in minerals and in food supplies.

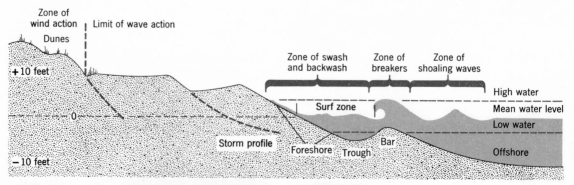

FIGURE 11.21 Sand beaches show a great deal of wave action and mass wasting. The foreshore (the area of wave breakers) may show erosional depositional forms such as troughs and sand bars. Offshore is a region of bottom erosion (After Navarra and Strahler).

The ocean also affects and distributes the Earth's heat. In this way, climates and weather are affected by the bodies of water that surround the land masses on all sides. The rivers of water in the ocean—the currents—carry on the job of equalizing the Earth's supply of heat from space and the atmosphere.

Yet there is much more to be learned about the sea besides its physical properties, chemical properties, and its bottom configurations. In the last chapter, we will examine some of the ways by which man studies the sea and its features. The manner by which we may some day tap the almost limitless resources of the oceans will determine how well man continues to survive on this planet of ours.

CHAPTER 12

Probing the Depths of the Sea

THE LAST 25 YEARS HAVE GIVEN SCIEN-tists an entirely new set of tools for studying the ocean floor. Oceanographers are no longer forced to remain on the surface sending mechanical devices to the bottom to blindly seek out samples of ocean water and sediments. Now scientists can descend to the ocean floor in various types of gear and in submersible craft to make studies at first hand.

United States Oceanographic Agencies and Institutes

The United States Oceanographic Office and the United States Coast and Geodetic Survey operate large fleets of specialized oceanographic research vessels equipped with the latest electronic and mechanical gear. The fleet is used to study ocean-bottom topography, to collect water and sediment samples, to trace ocean currents, to examine biological and geological specimens, and to make general surveys of the oceans. Numerous institutes are also involved in oceanographic research and in training the oceanographers of the future. Among them are the Woods Hole Oceanographic Institution, in Woods Hole, Massachusetts, the Lamont Doherty Observatory of Columbia University in Palisades, New York, and the Scripps Institution of Oceanography of the University of California in La Jolla, California, the first three such institutions.

FIGURE 12.1 The research vessel *Argo* used by Scripps Institution of Oceanograpny (Scripps Institution of Oceanography).

256

Institutes such as these have been responsible for many of the technological advances of the last generation, which have
made data collecting more comprehensive
than ever before.

Water Sampling

Samples of water are taken by means of the
Nansen bottle (Figure 12.4). This is a cylindri-
cal tube that is open at both ends. A series of
Nansen bottles are dropped over the side of a

FIGURE 12.2 (*above*) **A Perry Cub-
mobile. This two-man submarine
dives to depths of 200 to 300 feet
(Woods Hole Oceanographic Institution).**

FIGURE 12.3 (*left*) *Flip* **is a specially designed research vessel. It is
towed to its position and special
tanks are flooded, placing the ship in
a vertical position. When vertical,
Flip has great stability (Scripps Institution of Oceanography).**

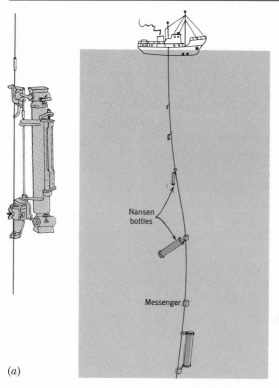

FIGURE 12.4   Nansen bottles are strung out at various depths in order to obtain water samples from different ocean depths. A Nansen bottle set up on a line, ready to be dropped overboard (Scripps Institution of Oceanography).

ship attached to a wire, so that each bottle is suspended at a different depth. When all the bottles have been strung out, each to the required depth, a trip weight is dropped along the wire. The messenger weight, as it is called, trips over each succeeding bottle and causes a valve at each end to close. This traps water from each depth in the respective bottle. Also, a *reversing thermometer* records the temperature of the water at each depth (Figure 12.5).

Measuring Sea Temperatures

A reversing thermometer is a device that records the temperature in the usual manner.

However, on being tripped over with the Nansen bottle, the base of the thermometer breaks. This traps the mercury in the tube, and the temperature at each depth is permanently recorded.

The temperature of surface water is often measured by simply taking samples of water in a bucket and by using an ordinary thermometer. In addition, research vessels use special thermographs and new techniques that measure the infrared radiation emitted by the surface of the sea. Infrared-sensitive devices can also be carried in airplanes to record the infrared radiation emitted by the sea surface.

To measure the temperature of the sea below the surface, a special device called the *bathythermograph* is used (Figure 12.6). This is a device that records changes of temperature with increasing depth, on a smoked plate. The bathythermograph is used in only the upper 150 m of the ocean, because it is in this area that the most pronounced vertical changes in the temperature of seawater occur.

Sound Transmission of Seawater

When sound travels through water, it loses less energy than sound traveling through air. In seawater, the velocity of sound is affected by three variables: temperature, salinity, and pressure. Therefore, when discussing the velocity of sound through seawater, oceanographers must be careful to describe the temperature, salinity, and pressure of the water. The accompanying charts show how sound velocity changes with respect to temperature and to pressure in water that has a salinity of 34.85 per mille. This salinity is used as the standard in the tables of sound velocity in seawater under various conditions that are published by the British Admiralty.

Figure 12.7 shows that in seawater with a standard salinity, at 0°C, sound has a velocity of nearly 1543 meters per second.

Figure 12.8 gives the relationship between pressure and the velocity of sound in seawater. Pressure has a much greater effect on velocity than either temperature or salinity. The curve shown is for seawater with standard salinity (34.85‰) at approximately 0°C, but it would not be very different for other temperatures and other salinities. The effect of pressure is almost independent of temperature and salinity. At the ocean's surface, the velocity of sound is nearly 1430 meters per second. However, at a depth of 8000 meters, the velocity approaches 1600 meters per second.

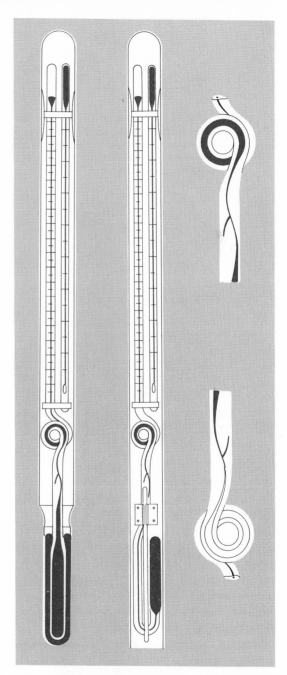

FIGURE 12.5 The reversing thermometer yields a permanent record of temperature when the mercury becomes trapped in the tube.

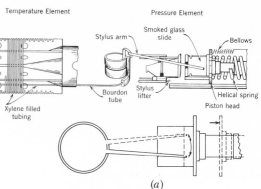

FIGURE 12.6 (A) (*right*) The interior of a bathythermograph. This instrument collects a continuous recording of temperature in the ocean with increasing depth. (B) (*below*) A bathythermograph (ESSA).

Since pressure refers to force per unit area, pressure is directly related to depth, for at greater depths, the force or weight per unit area will increase. The unit of measure used for pressure in chemical oceanography is the *Torr* (from Torricelli, the scientist who invented the barometer). One Torr is the pressure exerted per square centimeter of surface by a column of mercury one millimeter high at a temperature of 0°C. The pressure unit in physical oceanography is the *decibar*— the pressure exerted per square centimeter of surface by one meter of seawater. The Hydrostatic pressure in the sea increases by one decibar for approximately every meter of depth. The depth in meters and the pressure in decibars are, therefore, expressed by the same number: at a depth of 8000 meters the pressure in decibars is approximately 8000. This general rule is quite accurate for determining pressure for most purposes.

Temperature also has a decided effect on the transmission of sound in water. Most depth-finding systems work on a fixed velocity of sound. They are usually calibrated for a velocity of sound somewhere between 1460 and 1500 meters per second. This velocity places the temperature somewhere in a range from 3 to 12°C, and somewhat below 2000 m.

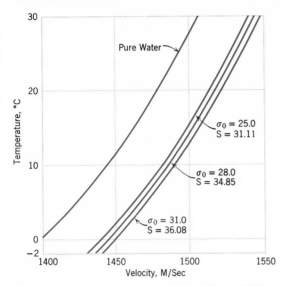

FIGURE 12.7 **Salinity versus velocity of sound (After Sverdrup).**

Measuring Depth in the Sea

During the early eighteenth century, the depth of the oceans was measured with hemp rope. By the mid-eighteenth century, wire was used in place of the rope. A heavy weight was attached to a thin steel wire and dropped overboard. A measuring wheel recorded how much of the wire was released before the weight hit bottom. Wire is still used on modern oceanographic research vessels.

Echo Sounding

The modern technique for measuring depth uses echo sounding equipment, which employs sound to determine depth. A sound signal is sent out from the ship. Sound waves travel out in all directions from the transmitter, and those that hit the ocean floor are bounced back to a receiver. (Refer to Figure 10.5.) The scientists then measure the time interval between the emission of the sound and its return to the ship.

The speed of sound in ocean water is

known: it travels 5000 feet per second. Therefore, if a transmitter sends a "beep" and it returns in one second, the oceanographer knows that the sound wave has traveled a distance of 5000 feet. Since the sound wave must travel both down and back, the depth at that particular point is half the distance the sound wave has traveled, or 2500 feet.

The speed of sound through ocean water is not constant. It changes with temperature —the warmer the water, the faster the sound —and with pressure—the greater the pressure, the faster the sound. The same relationship exists between salinity and the speed of sound. Thus, in tropical waters, at greater depths, or at greater salinities, sound will travel faster than 5000 feet per second.

Depth and Pressure

Another technique for measuring depth depends on the fact that pressure is almost

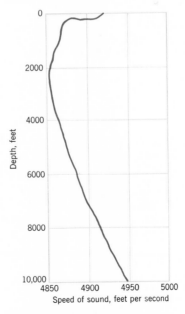

FIGURE 12.8 **Pressure versus velocity of sound (After Sverdrup).**

exactly proportional to depth. The pressure at any given point in the ocean is equal to the weight of the vertical column of water which is above that point, and pressing down on it. The weight of that column per unit area depends mainly on its depth. The pressure of a column of water 33 feet high is approximately equal to one atmosphere, or 14.7 pounds per square inch. Thus pressure in the ocean increases by one atmosphere for every 33 feet of depth.

Scientists also use the *mean,* or *average, depth* which is recorded in feet or meters. (See Table 12.1.) The mean depth in the Atlantic Ocean is 3926 meters (12,355 ft).

The Pacific Ocean is, on the average, the deepest ocean; its mean depth is 4282 meters (14,130 ft). The Indian Ocean has been found to have a mean depth of 3962 meters (13,177 ft).

Bottom Sampling

Samples of the ocean floor are gathered in a variety of ways. Geological specimens may be obtained by means of *dredges* (Figure 12.12). They are made of chain mesh attached to a rigid metal rim. As the dredge is dragged along the bottom, rocks and sediment are trapped in the mesh and can be

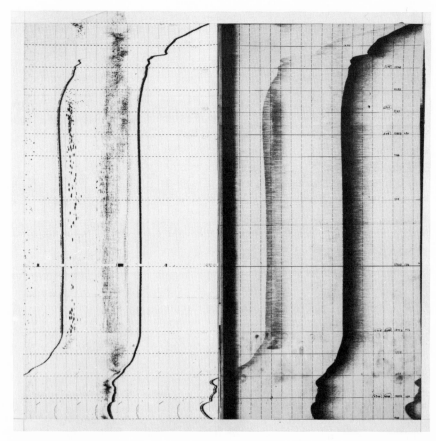

FIGURE 12.9　Echogram charts. They are produced and collected as continuous readings aboard the research vessel (ESSA).

FIGURE 12.10 (*left*) A sonar sounding recording. Varying frequencies produce accurate readings of sea floor depth as well as the substrata underlying the sea floor (ESSA).

FIGURE 12.11 (*below*) A depth recorder profile of the ocean bottom (Woods Hole Oceanographic Institution).

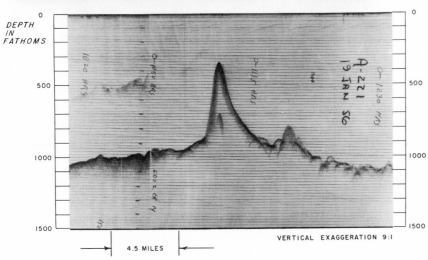

pulled up for examination on the surface. *Grab samplers* are also used (Figure 12.13). They are much like the scoop of a steam shovel. The metal jaws remove samples of the sediment at various depths and return it to the surface for examination.

Each of these samplers disturbs the sediment. In order to collect an undisturbed sample of sediment, *coring tubes* are used. The modern piston corer consists of a freely moving piston inside a metal chamber. The chamber is attached to a heavy weight, which has a series of vanes used for the control of movement.

When the corer is dropped from the ship it falls freely to the bottom. The vanes act like the feathers on an arrow and cause the tube to fall in a straight line. As the coring tube is driven into the sediment by the weight, the piston rides upward in the tube. The tube extracts a portion of the bottom sediment almost exactly as it was laid down by the natural processes acting in the ocean (Figure 12.15). Piston corers have enabled oceanographers

FIGURE 12.12 A chain-mesh dredge used to collect geologic samples of the bottom (Scripps Institution of Oceanography).

FIGURE 12.13 A grab sampler used to collect a rough sample of the sea floor (Woods Hole Oceanographic Institution).

to extract cores in excess of 60 feet in an undisturbed condition. A core this long may represent hundreds of thousands of years of depositional forces at work. The sediment cores are then extruded from the tube and stored in moisture controlled rooms for later examination.

Photography

Ocean bottom photography is another technique available to modern oceanographers. This, too, is a fairly recent method for studying the bottom of the ocean, although it has been used and tested since the turn of the century. However, only in the last 30 years have good deep-water high-speed sequential photographs been made successfully.

Ocean depths, although still measured by means of wire soundings, are more accurately measured today by means of echo sounding. Different radio frequencies will penetrate the sediment to different depths, giving an accurate picture of underlying rock strata.

An International Expedition

Perhaps the best way to understand the practice of oceanography is to consider an oceanographic expedition in some detail. We have chosen the Indian Ocean voyage of the *Anton Bruun*. The following is a version of a report published in May, 1965:

The research vessel *Anton Bruun* arrived in New York February 1 after successfully completing a scheduled two-year expedition in the Indian Ocean. During this time she made nine major scientific cruises, traveling approximately 72,000 miles. The *Anton Bruun*, formerly the presidential yacht *Williamsburg*, was turned over to the National Science Foundation in 1962 for

Table 12.1 *Depths of the Three Major Oceans*

OCEAN	MEAN DEPTH
Atlantic	12,355 feet 3926 meters
Pacific	14,130 feet 4282 meters
Indian	13,177 feet 3962 meters

265

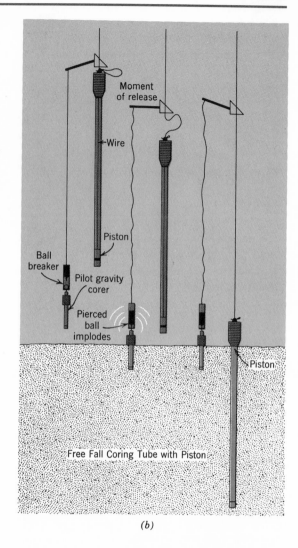

FIGURE 12.14 **(A) A piston corer being lowered into position. The vanes keep the corer on a straight path; the weight drives the core into the sediment (ESSA). (B) The diagram illustrates the operation of the corer (After Shepard, 1963).**

conversion to an oceanographic research vessel for the U.S. Program in Biology as part of the U.S. participation in the International Expedition. This cooperative scientific effort, involving more than 20 countries, was directed by John Ryther and Edward Chin of Woods Hole Oceanographic Institution.

The idea of an International Oceanographic Program covering the entire Indian Ocean was first proposed in 1957. There

were a number of reasons for the selection of the Indian Ocean. First, it is the least known of all the major oceans of the world. No systematic survey of this vast region had ever been undertaken, although numerous expeditions had passed through the waters of the Indian Ocean and made observations as the opportunities occurred.

The Indian Ocean is bounded by Pakistan, India, Ceylon, Burma, Thailand, Sumatra,

Australia, and a variety of countries on the east coast of Africa, including Madagascar. At the time of the expedition, these countries were just developing competency in oceanographic research. They lacked the ability to do more than conduct surveys in the waters immediately adjacent to their coasts.

There is a tremendous protein deficiency in the diets of most of the people located in this section of the world. Many subsist on a diet that is 60 per cent rice. In the interior, this figure may rise to as much as 90 per cent. The Indian Ocean was known to contain an abundant supply of protein.

Estimates of the protein supply which may be obtained from the sea reveal that the oceans are an inestimably valuable source of food. A square mile of good wheat land may produce 600 to 700 tons of wheat per year.

FIGURE 12.16 Ripple marks on a Florida beach. Similar ripples are formed by water moving along the sea floor in the littoral realm (William M. Stephens).

FIGURE 12.15 Cores taken from the sea floor may represent millions of years of Earth history. These cores were recovered intact from the floor (Scripps Institution of Oceanography).

An equal area of the sea might produce 4000 tons of vegetable matter. Better fishing techniques will also produce an increase in sea protein available to these areas of the world. At the present time, about 12 per cent of the world's protein comes from the sea. Yet the Japanese obtain about 74 per cent of their protein from the sea.

No adequate surveys of the food resources

of this ocean had ever been made. The factors that control the distribution and the abundance of life in these ocean waters were unknown at that time.

The oceanographers on the voyage began by considering the factors that probably affected the fertility of the Indian Ocean. One factor was the monsoons, the winds that blow half the year from the northeast and the other half from the southwest. As scientists planned their strategy of the expedition, they decided to investigate the regular wind reversal, and the relationship between the reversal and the surface currents of the ocean. They were quite sure that the reversal had an effect on the physical, chemical, and biological conditions of the Indian Ocean.

Just enough was discovered about the plant and animal life in this area to whet the appetite of scientists. They planned to make a thorough study of the entire ocean. Actually, they did collect information about the plant and animal life of the islands, the intertidal zones, and the shallow waters and reefs. But the deep-sea animal life called the benthon; the free-swimmers, the nekton; and the plankton are still largely unknown and not studied in complete detail. All plankton and other biological data were deposited at the International Biological Center at Cochin, India, which was established under the auspices of the United Nations. The information gathered by this expedition, and it is only one expedition of many carried out in the area

FIGURE 12.17 A bottom photograph. Sea urchins and worm trails can be seen (Woods Hole Oceanographic Institution).

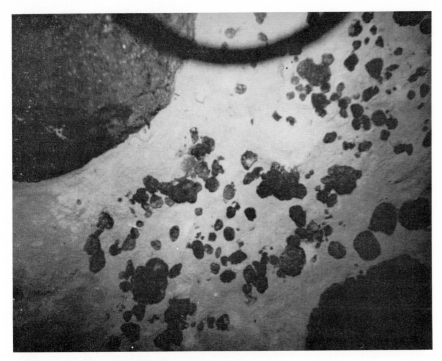

FIGURE 12.18 The continental shelves are rich in manganese nodules. These nodules may become an important commercial source of manganese in the future (U.S. Navy).

during the last decade, will take many years to sort out and examine adequately.

Among the studies carried out by the *Anton Bruun* was the examination of the physical and chemical oceanography of the area. An assessment of the plankton biology in relation to the surface circulation was also made. This included assessing the possibility of commercial fisheries and the resources around the Bay of Bengal on the continental shelf. A mineral supply that is easily obtainable is of utmost importance to any country—but to emerging countries like the ones in the region of the Indian Ocean, it is a matter of life and death. The continental margins represent about 10 per cent of the ocean bottom, yet half the biological population exists on the shelves, and almost all of the commercially valuable elements and minerals. At the present time, we are extracting most of the world's magnesium from the sea. We also ob-

tain sulfur, oil from beneath the sea floor, sand and gravel for construction, and various precious metals on a small scale. The exploitation of the continental shelves off the coasts of these countries would be of great importance in aiding them in raising their standards of living significantly.

Food harvests from the sea remain rather low. Only 20 million tons of edible fish are taken from the sea each year, and the total sea food is less than 30 million tons. Yet it takes about 1000 pounds of plankton organisms to make one pound of fish protein. The process of eating plankton and turning it into fish protein is a very inefficient one. If a way can be found to use plankton directly, and sensible, planned, systematic plankton farming is carried out, the yield from the sea could be increased by a factor of at least several hundred. Also, many of the fish that are taken from the sea are not edible and are

FIGURE 12.19 A tow net recovering plankton samples from seawater. A fine-mesh net is used to collect these organisms (ESSA).

used as fertilizer ingredients. If these were made edible by removing the offensive oils, the food catch from the sea would be increased.

The further advantages of the type of international cooperation represented by the voyage of the *Anton Bruun* are many. Scientists from all over the world have the opportunity to exchange information. The fact that the effort is international makes adequate sums of money available, and places the most modern tools in the hands of scientists who are studying very basic problems. In addition, a great deal of international good will develops from these cooperative undertakings. This kind of cooperation is very important in advancing the frontiers of science.

Submersible Craft

As new and stronger metal alloys have been developed, and more information gathered about the construction of submersible ships, entirely new ways of examining the ocean floors have emerged. Scientists are no longer restricted to the distant decks of research ships, but can now descend to the ocean floor to make firsthand studies. This, the study of the ocean bottom

FIGURE 12.20 (*left*) The bathy-sphere, an early research tool used by William Beebe (American Museum of Natural History).

FIGURE 12.21 (*below*) The *Trieste I* (U.S. Navy).

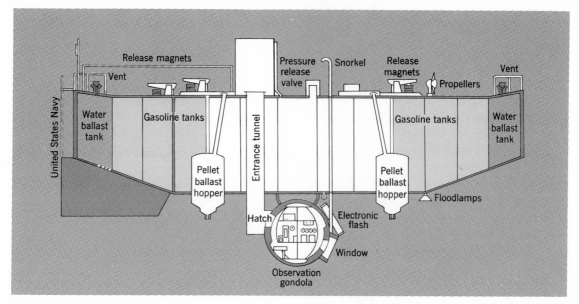

FIGURE 12.22 Cross-sectional diagram of the *Trieste*.

has become a true, firsthand study—*oceanology;* it is no longer a distant survey.

The Trieste

The *Trieste,* a special research vessel termed a *bathyscaph,* takes man into the deepest reaches of the sea. Designed by the famed Swiss physicist Auguste Piccard, the *Trieste* takes its name from the Greek words which mean "deep boat." During January 1960, the *Trieste* was used to probe the Challenger Deep near Guam, which is more than 35,800 feet deep.

As shown in Figure 12.22, the *Trieste* is actually composed of a series of tanks overhanging the sphere that houses the scientists. Approximately 34,000 gal of gasoline are carried in the gasoline buoyancy tanks. As the bathyscaph descends, increasing water pressure compresses the gasoline and ocean water enters the tanks through an opening in the bottom. The fact that the buoyancy

tanks are open to the sea helps to keep the inner and outer pressures equal. This allows a relatively thin hull to be used. Two funnelnecked silos contain ballast in the form of 16 tons of small steel pellets. An electromagnet at the mouth of each silo acts as a stopper. If the electromagnets are switched off, pellets fall from the silos, and the bathyscaph begins to rise to the surface. Propellers near the bow can be used to drive the craft when it is submerged.

The bathyscaph window is a very interesting device for explorations at great depths. Piccard uses a window that is a plexiglass cone. The cone can resist the pressures of eight tons per square inch that are found at depths of more than seven miles. The sea outside the window is lit by the natural luminescence of undersea life. In addition, mercury vapor lamps are used to light the area around the *Trieste.*

A new *Trieste,* the *Trieste II,* was built and launched in 1964.

Bathyscaphs such as the *Trieste* and *Tektite* are valuable instruments in surveying life in the sea. Scientists using the bathyscaph notice that all traces of sunlight vanish at 1500 feet. At depths of 6000 feet, large red shrimplike creatures and tiny white animals are observed.

The *Triestes* make it possible to observe the phosphorescent plankton quite closely. In the Challenger Deep, Jacques Piccard observed luminescent trails at 2200 feet and also at 20,000 feet. At a depth of seven miles, where the pressure is about 1600 pounds to the square inch, Piccard observed a solelike

FIGURE 12.23   The *PX-15*, now called the *Benjamin Franklin*, has no motive power of its own. It is driven by ocean currents and has been used to study the Gulf Stream (Grumman Aircraft).

fish. He reported that the fish swam above the ivory colored ooze at this depth. Piccard's observation proved that vertebrate life exists in the sea's greatest depths. At depths below 32,500 feet, the *Trieste* found shrimp and animals that appeared to be jellyfish. The temperature in the Challenger Deep at 35,800 feet was found to be approximately 38°F. The sediment in the deep appeared as soft as powdered talc, and seemed to be made of the siliceous skeletons of diatoms, which died in the upper reaches of the sea and sank to the bottom.

There seems to be no end to the ingenuity and creativity of Jacques Piccard. In 1969, the *PX-15,* a Piccard-designed submersible, began its examinations of the Gulf Stream and the continental shelf off the east coast of the United States. Built by the Grumman Corporation, the *PX-15* travels silently along the bottom driven only by the ocean currents. It was designed to travel from Florida to Nova Scotia at depths of 300 to 2000 feet. Viewing ports allow observation of marine life; television cameras photograph the surroundings of the craft; and experiments in the acoustics of marine life and the speed of sound are carried on. The *PX-15* also has propulsion tubes that are capable of moving the ship both horizontally and vertically as they swivel about on the hull.

Other Submersibles

Jacques-Yves Cousteau, a French scientist-aquanaut, has also been extremely active in

FIGURE 12.24   The *Deepstar,* **built by Westinghouse. It is equipped with light sources, a claw for obtaining samples, and a collecting basket. It carries three men to 4000 feet (Westinghouse Corporation).**

FIGURE 12.25 Divers in Scuba gear collecting samples from the bottom (Florida Development Commission).

the field of submersible craft design. Cousteau has designed a saucerlike, two-man vessel called the *Denise* for examination of marine life and the shallow-water areas of the continental shelves. Cousteau has also developed, in cooperation with the Westinghouse Corporation, a submersible known as *Deepstar*. This is a three-man vessel that operates under its own power and is capable of performing a variety of jobs at depths of 10,000 feet. Cousteau is one of the chief developers of the well-known scuba gear used by sportsmen and fishermen all over the world.

There are many other designs for submersible craft, ranging from manned craft such as *Trieste, PX-15,* and *Deepstar*, to robot-like craft that have tools and flexible arms attached to their hulls. These craft will roam the bottom doing manual labor in underwater oil fields and mining operations.

The submersible devices presently being used and tested by oceanologists precisely control the pressure to which the operator is exposed, and greatly lessen the danger of the "bends," the ill effects of decompression.

Bends

The "bends," or "caisson disease" as it is often called, is the result of the sudden return to normal surface pressure after a diver has been under water for an extended period of time. As the pressure increases in a deep dive, gases are forced into solution in the blood. Nitrogen gas is not absorbed by the body cells and, as a result, remains in the blood. When the diver surfaces too quickly, the gas may form bubbles or embolisms and block small blood vessels and veins. It is particularly dangerous when those vessels leading to the heart and lungs are blocked.

275

A

B

FIGURE 12.26 The *Sealab* (A) (*left*) and *Tektite* (B) (*right*) experiments conducted by the United States Navy are designed to study how well aquanauts can function under the sea for long periods of time. They are forerunners of the day when man may live under the sea for commercial as well as recreational purposes (U.S. Navy).

The condition causes paralysis and death.

The gas can be forced back into solution by rapid compression once more. Then, the diver must decompress slowly over a period of several hours. Thus, to avoid the condition, a deep diver who has not had the protection of a pressure-controlled environment, must spend hours in a chamber or rise slowly enough so that the gases will be slowly removed from his bloodstream without forming bubbles.

The United States Navy is currently directing a long-range project called the *Sealab* experiments. The *Sealab* is a high pressure tube in which several men are able to live for long periods (10 days or more) at depths of hundreds of feet and pressures of five to six times that of the surface. Ex-astronaut Scott Carpenter has devoted much time to this project. He and the other participants are leading the way to a future that will see men living under water in colonies for scientific examination and experimentation. In addition, men in such colonies will be able to exploit the mineral deposits and the food resources found in the ocean basins.

Questions

CHAPTER NINE

1. Contrast each hemisphere of the Earth in respect to their land-water surfaces. How do the relationships compare from latitude to latitude?
2. What does the term "seven seas" actually describe?
3. Briefly describe the general characteristics of each of the world's major oceans.
4. Which ocean is the center of great geologic activity? Explain the occurrences that take place.
5. Discuss the various sediments found on the ocean floor and their origins.
6. Which theory of Forbes was upset by the *Challenger* expedition? Explain what their findings showed.

CHAPTER TEN

1. Define latitude and longitude. How is each measured?
2. How are the oceans separated? What boundaries are used?
3. Discuss the characteristics of ocean bottom topography. Compare them to features found on the continental surface.
4. What theories have been devised to explain the formation of the ocean basins?

CHAPTER ELEVEN

1. What is the general composition of seawater? Explain salinity.
2. Why is it fortunate that ocean water expands when it freezes? Of what possible significance would this be for ocean life?
3. What is the thermocline? What is marked by its presence?
4. Discuss some of the various factors involved in water color.
5. What causes large-scale and deep-water ocean currents? What is upwelling? In what way do the large-scale surface currents affect man?
6. Relate the period, frequency, speed, and amplitude of a wave. Describe what occurs as a wave approaches the shore.

CHAPTER TWELVE

1. Discuss some of the methods that are used to measure depth in the ocean. How is depth measurement related to the ideas we have developed about ocean floor topography?

2. What advantages have submersible craft like the *Trieste* over past methods of bottom exploration?

3. How are water samples and temperature taken? Explain how ocean currents are measured.

4. How is sound affected by the characteristics of the water it is passing through?

5. In what ways may the survival of mankind on this planet be dependent on the oceans in the near future? What resources will be taken from the ocean in large quantities?

Bibliography

BOOKS

Barnes, Harold, *Oceanography and Marine Biology: A Book of Techniques,* Hofner, New York, 1968.

Behrman, Daniel, *New World of the Oceans: Men and Oceanography,* Little, Brown and Co., Boston, Massachusetts, 1969.

Cotter, Charles H., *Physical Geography of the Oceans,* Am. Elsevier, New York, 1966.

Ericson, David, and Wollin, G., *Ever Changing Sea,* Alfred Knopf, Inc., New York, 1967.

McLellan, Hugh J., *Elements of Physical Oceanography,* Pergamon Publishing Co., Elmsford, New York, 1966.

Pell, Claiborne, and Goodwin, H. L., *Challenge of the Seven Seas,* William Morrow and Co., New York, 1966.

Sears, Mary, Ed., *Progress in Oceanography,* 5 vols., Pergamon Publishing Co., Elmsford, New York, 1968.

Soule, Gardner, *Under the Sea: A Treasury of Great Writings About the Ocean Depth,* Meredith Press, New York, 1968.

Thorne, Jim, *Underwater World: A Survey of Oceanography,* Thomas Crowell and Co., New York, 1969.

Turekian, K., *Oceans,* Prentice-Hall, Englewood Cliffs, New Jersey, 1968.

Weyl, *Oceanography,* John Wiley & Sons, Inc., New York, 1970.

Yasso, W., *Oceanography,* Holt, Rinehart and Winston, Inc., New York, 1965.

PERIODICALS

Asimov, Isaac, "Birth of the Ocean," *Science Digest,* July, 1968.

Baker, D. J., "Models of Oceanic Circulation," *Scientific American,* January, 1970.

Bonatti, E., "Tissue Basalt and Oceanic-Floor Spreading on the East Pacific Rise," *Science,* November, 1969.

Daugherty, J., and Daugherty, M., "How Deep Is the Ocean? " *Science Digest,* August, 1968.

Dott, R. H., "Circum-Pacific Late Cenozoic Structural Rejuvenation — Implication of Sea Floor Spreading," *Science,* November, 1969.

Fleischer, R. L., "Mid-Atlantic Ridge, Age and Spreading Rates," *Science,* 1968.

MacInnis, J. B., "Living Under the Sea," *Scientific American,* March, 1966.

Volt, G., "What Do We Know About the Sea? " *Science Digest,* November, 1969.

The Biosphere

PEOPLE HAVE RECENTLY BECOME MUCH MORE AWARE that technology has changed the environment. Man has become concerned about his relationships to and his effects upon the environment. The question of relationships of living things to their environments is not new. It is a problem that goes back to before the coming of man, when life first evolved on the planet Earth.

It is certain that no living thing, from its origin to the present, is an isolated and independent organism. The relationship between all forms of life and their environments is a critical balance of interrelated cycles which have contributed to the origin of life, the evolution of living things, and the changes in living things. These relationships of living things to their non-living environment and the geological changes that have come about have affected the appearances and disappearances of the many life forms.

Throughout the pages of the Earth's history many factors have caused the Earth to change and to evolve. At the same time, geological changes, temperature changes, and climatic changes have affected the origin and evolution of living things. These continuing natural changes occurred in life and the natural environment before the coming of our present industrial society, and will continue through time. Today, modern technology is affecting this critical balance and we find that we need desperately to understand the role man plays in his relationship to his environment.

Modern man must solve many problems if he is to survive on the planet Earth. Our willingness and power to alter our environment has increased at a much greater rate than our ability to conserve our immediate environment. The changes that have occurred in the past have altered the development and caused the extinction of many life forms. If man is to be the ultimate being he must know how his technology will affect his future. He must know, for example, what effect his changing of the landscape or changing the courses of rivers will have on the social as well as the natural environment. We must learn not to subjugate our environment, but to live with it.

In this unit we will be concerned with the origin and evolution of life, along with the changes that have occurred in the structure of the Earth. We have selected this approach with the hope that knowledge of the past will give us insight into the future.

Photo preceding page by John S. Shelton

CHAPTER 13

The Origin of Life

SINCE LIFE AND THE EARTH ARE SO closely related, a scientist's ability to understand the origin of life depends largely on his interpretation of the evolution of the planet. The understanding of the origin of life can best be achieved by studying the present and past processes of life.

With all of the scientific equipment and techniques available, modern scientists still cannot determine exactly how life began. But it can be assumed that life developed long after the formation of the Earth. With the use of modern scientific equipment, scientists have been able to calculate that the Earth formed four to five billion years ago and that up to two billion years elapsed before the first form of life appeared. That is, they estimate that some form of life existed approximately three billion years ago.

Many hypotheses have been offered to explain the origin of life, but the many mysteries still present about the origin of life are slowly being unraveled. Scientists are developing better understandings of the chemical composition of living things, and, thus, are beginning to develop a more complete picture about the origin of life.

The Substances of Life

According to the modern understanding of life, all living things are made up of protoplasm. Protoplasm is defined as the life substance, or as life itself. It has four basic properties that are not observed in inorganic matter. These properties are the abilities to reproduce, to grow, to respond, and to develop. All living things, whether they belong to the plant or animal kingdom, are made up of protoplasm.

Scientists are capable of chemically breaking down the complex composition of protoplasm into inorganic and organic substances. The inorganic substances (water and minerals) make up about 75 per cent of the protoplasm. The remaining 25 per cent is of organic origin (carbohydrates, fats, proteins, and nucleic acids). Yet the organization of the protoplasmic constituents into a living organism has been beyond the scope of modern laboratory science until recently. In 1967, Arthur Kornberg, a Nobel Prize winning biochemist, constructed a synthetic biological molecule—a molecule capable of reproducing itself. It appears that scientists may now be on the threshold of discovering and of understanding the origin of life. In 1970 Khorama synthesized a gene.

The individuals who study the origin of life must deal with two basic problems. First, they must develop techniques that enable them to analyze the origin of organic compounds. Second, they must be able to explain how life developed from organic compounds. From the scientific point of view, the origin and development of life is a phenomenon that resembles the development of the Earth: scientists are only able to suggest hypotheses that attempt to explain how life developed.

been proposed to answer the question of how life began. One of the most notable hypotheses was that of *abiogenesis* — that life developed spontaneously from a nonliving substance.

Spontaneous Generation

The origin of life has concerned man for many centuries, and many hypotheses have

Abiogenesis

To understand the theory of abiogenesis, it is helpful to compare it to spontaneous

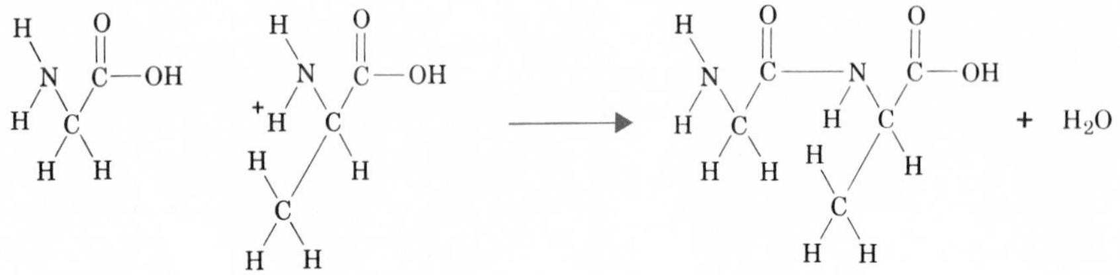

Amino acids add together to make proteins

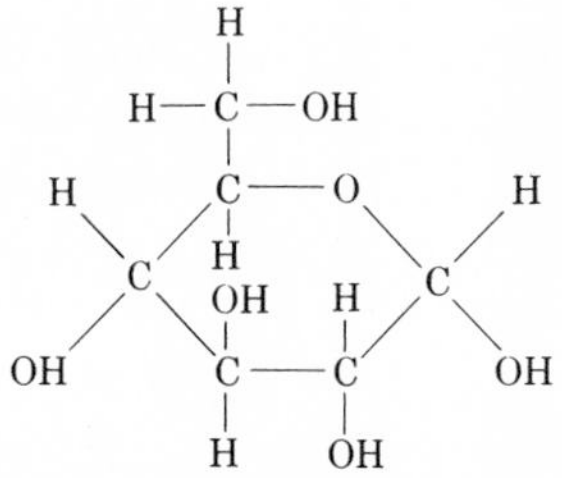

Glucose, a simple carbohydrate

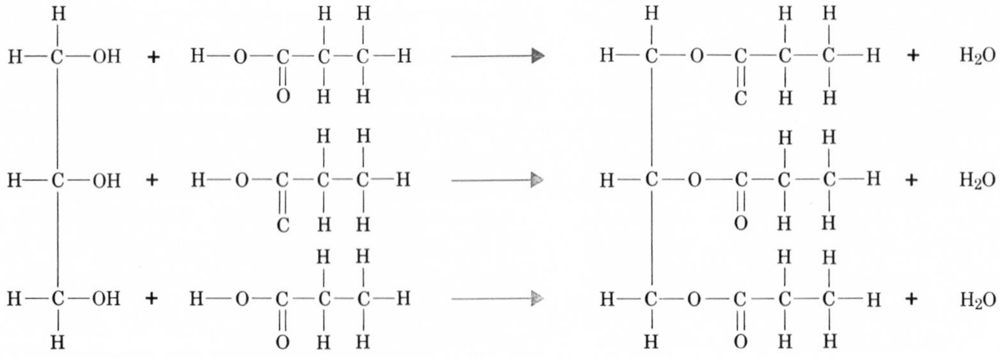

Glycerol and fatty acid ⟶ fat and water

FIGURE 13.1   Diagram of three organic compounds. Protoplasm (life) is made up of organic compounds (fats, carbohydrates, and proteins). These materials are organized into complex organic substances in living organisms.

FIGURE 13.2 (*left*) In December 1967, Dr. Arthur Kornberg successfully manufactured a synthetic DNA molecule which displays the biological activity of the natural molecule in living organisms (UPI).

FIGURE 13.3 (*below*) Legends about spontaneous generation date back to the time of Aristotle more than 2000 years ago. One legend states that lambs were formed from vegetation (Bettmann Archive).

combustion. Spontaneous combustion is defined as the sudden burst of energy from a combustible fuel even though there is no obvious agent available to ignite the fuel. Similarly, the theory of abiogenesis states that nonliving material could have, by itself, produced life or could have itself become alive with no available agent present to produce life.

The theory of abiogenesis was proposed more than 2000 years ago by the famous Greek philosopher Aristotle. He had given much thought to the origin of life, and

finally proposed that life arose spontaneously —from a nonliving substance.

Aristotle proposed that nonliving things contain what he called a "vital substance," which he defined as somewhat like a fertilized egg. This "vital substance" within nonliving matter could, under proper environmental conditions, produce life.

For many years, Aristotle's theory of abiogenesis was completely accepted without any evidence or proof. The theory was still accepted as recently as the sixteenth century. For example, Paracelsus, a sixteenth-century physician, described the spontaneous development of animals from water, air, straw, and wood.

During the seventeenth century, Jean Van Helmont, a Belgian physician, described an "instance" of abiogenesis during which mice were produced spontaneously. He stated that mice could be produced from dirty shirts in contact with sweat. The "vital substance" of this experiment was the human sweat in the dirty shirt.

In 1745 John Needham, an English scientist, attempted to support the theory of abiogenesis. Needham set up an experiment in which he placed heated broth and vegetable juice in sealed bottles. Within a few days, the broth and juice were contaminated with small organisms. Needham believed that he had proved that life could be produced spontaneously from nonliving matter.

The theory of abiogenesis provides us with some insight as to how life could have formed on some other planet. This form of life may have traveled to the planet Earth, as we shall see in our discussion of "Life from Outer Space."

The theory of abiogenesis is not only useful in hypothesizing about life from outer space but can also be a basis for explaining the origin of life within the confines of the Earth's own resources — the conversion of inorganic compounds to organic compounds to life.

Life from Outer Space

One hypothesis about the origin of life on Earth assumes that free-floating bacteria in space were attracted to dust particles or meteorites from different parts of the universe. Scientists proposing this hypothesis assume that the spores of bacteria can withstand the forces of heat, cold, radiation, and friction in outer space, and then reproduce when they reach conditions favorable to life.

This hypothesis attempts to explain how life began on the Earth, but the problem of the origin and development of the bacteria still exists. In addition, it is doubtful that the bacteria could withstand the hostile conditions of space while traveling to the Earth.

Although the hypothesis that life might have come from outer space is open to serious question, it is possible that life has developed on other planets. With the invention of space vehicles, man is now capable of reaching beyond the Earth's atmosphere to determine whether there are any signs of life existing on other planets. If there is life on other planets, this will bring our scientists a step closer to understanding the origin of life and how it has adapted to a variety of environments. This will not be an easy task for it will take many years before man can determine whether life exists on other planets.

Inorganic Compounds Produce Organic Compounds

Some scientists have proposed the theory that life arose from organic (living or once living) compounds that developed from inorganic (nonliving) substances. They hypothesize that a series of forces, over a long period of time, resulted in a molecular reorganization of inorganic molecules. As an end result of this reorganization, the organic molecule evolved. Other forces are then supposed to have acted on the organic compounds in a manner similar to the forces that acted on the inorganic compounds, and life then was developed from these organic compounds.

This hypothesis places the origin of life within the confines of the Earth's own resources. What if this hypothesis is false? Perhaps the reorganization of inorganic and organic molecules occurred outside the Earth. This would mean that life was created somewhere else and made its way to the Earth as has been proposed in the theory about life from outer space.

The Evolution of Organic Compounds

In studying the Earth's history there are two categories of events that are significant. First, there are the physical changes that occurred on the Earth's surface. On the other hand, there are the changes that have occurred in living things from the simple unicellular (one-celled) animals and plants to the multicellular (many-celled) plants and animals of today. Biological changes have been produced by changes in the Earth's physical environment. But in today's modern world, man has found ways of adapting to the physical changes in his environment. He has also been able to produce physical changes within his environment.

At this point, we will not be concerned with the origin of the Earth, but will make an important assumption: that when the Earth first formed it was warmer than it is today. The elements in the atmosphere were combined as compounds of methane, ammonia, carbon dioxide, hydrogen, and water vapor. After these compounds formed, the tempera-

Name	Molecular Formula	Structural Formula	Models	
			Space Filling	Ball-and-Stick
Hydrogen	H_2	$H-H$	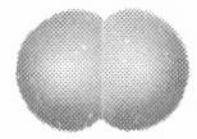	
Water	H_2O	$H-O-H$		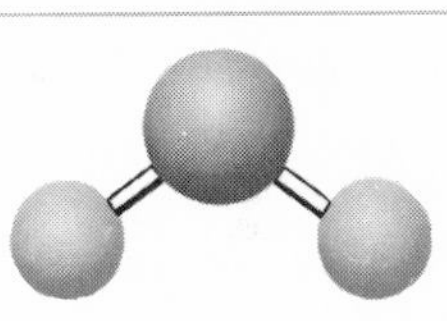
Ammonia	NH_3	$H-N-H$ $\quad\;\;\vert$ $\quad\;\;H$		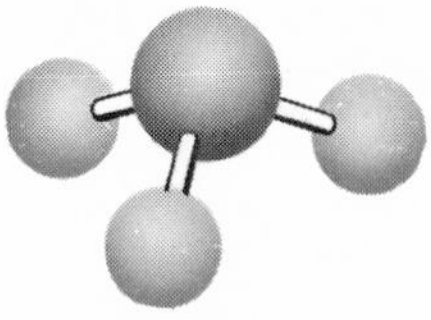
Methane	CH_4	$\quad\;\;H$ $\quad\;\;\vert$ $H-C-H$ $\quad\;\;\vert$ $\quad\;\;H$	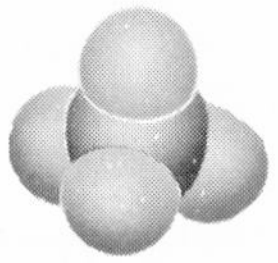	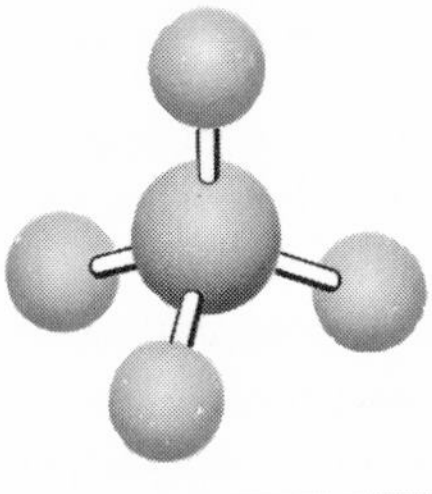

FIGURE 13.4   The sequence of events that might have led to the formation of organic compounds. The gases in the atmosphere were combined by the action of solar energy into simple organic compounds.

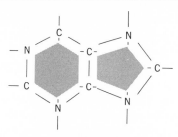

Purine "skeleton"

FIGURE 13.5   The purine molecule is one of the basic components of DNA.

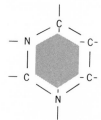

Pyrimidine "skeleton"

FIGURE 13.6   The pyrimidine molecule, with the purine molecule, forms the protein links in DNA.

FIGURE 13.7 (*right*)   A portion of a DNA molecule. The structure of DNA controls inherited characteristics.

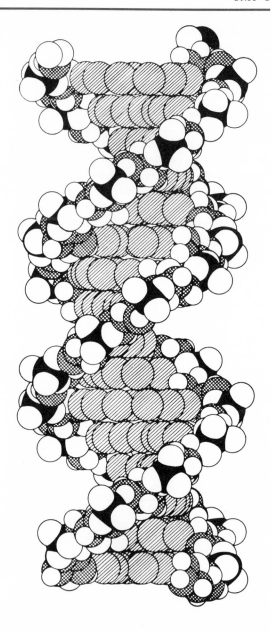

ture of the Earth declined. This cooling process produced a great amount of rain. The rain, which carried atmospheric gases with it, resulted in the formation of oceans, and the gases carried by the rain dissolved in these waters.

Erosion also began as a result of these great rains. Streams and rivers evolved and carried minerals to the oceans. We can further hypothesize that the oceans provided the best environmental conditions for supporting life.

Conditions Necessary to Support Organic Reactions

We know that chemical reactions occur most readily in liquids, and we can assume that the ocean provided the best environmental conditions on Earth for different chemical reactions. The energy necessary for these reactions to take place could have been supplied by the sun.

It is evident that the environment was suitable for many chemical reactions to occur to form many compounds. Of all the compounds formed, five types are biologically important. They are sugars, glycerin, fatty acids, amino acids, and nitrogen bases. The two important varieties of nitrogen bases are *purines* and *pyrimidines* (Figures 13.5 and 13.6). The purine and pyrimidine bases are basic parts of the DNA molecule (deoxyribonucleic acid), which makes up the "thread of life" (the gene).

For those who doubt that organic compounds may be formed in this manner, it should be noted that it has been done under laboratory conditions. In 1953, Dr. Stanley Miller set up (under laboratory conditions) a primitive atmosphere, and he sent an electrical discharge that simulated solar energy through a composition of water, methane, ammonia, and hydrogen. At the end of the experiment he noticed a red fluid. After analyzing this fluid he concluded that it was of organic origin. Further examination revealed the red fluid to be made of amino acids. Urea, another organic compound, has also been synthesized from inorganic elements under laboratory conditions. These

FIGURE 13.8 In 1953, Dr. Stanley L. Miller succeeded in producing complex organic compounds (amino acids) in an apparatus that simulated the Earth's atmosphere of three million years ago (Wide World).

experiments have shown that life could have evolved from inorganic elements.

By assuming that the types of biologically important compounds were formed and further reactions occurred, it is possible that more complex organic compounds such as fats, proteins, and nucleic acids could have been formed. If these organic compounds were produced (especially the nucleic acids that affect protein synthesis), then the origin of living things from inorganic compounds is possible.

Life from Organic Compounds

Recent studies suggest the probability that organic matter appeared under special conditions in which many carbon compounds were present. The carbon atom is the basis for the construction of all organic compounds. The organic compounds called proteins are the basic substances of all life. That such protein-rich substances made up of amino acids were present at an early stage in the development of the Earth is highly probable. This does not necessarily mean that the development of life was acci-

dental, but it was probably in keeping with the physical laws of the universe.

The development of life from organic compounds is one of the most important factors to consider in the development of the Earth. The gap that exists between the complex organic compounds and living protoplasm is great, yet science has gone some way in building a bridge across it.

Coacervates

We have suggested that organic compounds—especially the proteins (amino acids)—formed before life. How, then, did organic structures join together to form complex organic compounds? One answer to this question has been derived from a group of compounds called the *coacervates* (Figure 13.9). Coacervates are groups of proteins or proteinlike substances held together as small droplets surrounded by a liquid. It can be assumed that organic molecules joined together to form small groups that resemble the coacervates.

When protein molecules are dissolved in water, they develop an electrical charge

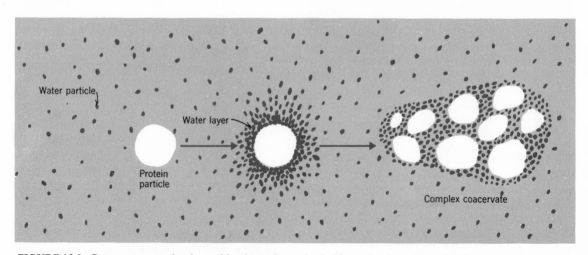

FIGURE 13.9 Coacervates are simple combinations of proteins held together by water molecules.

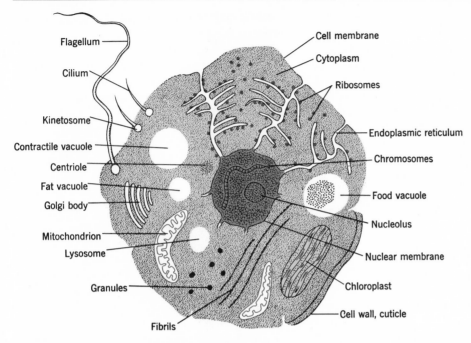

FIGURE 13.10   A generalized cell showing its various components.

of such a kind that they attract molecules of water. The molecules of water surround the protein molecule to form a spherical structure. Similarly, a coacervate is formed when a group of charged protein molecules join together and are surrounded by water to form a sphere. There must have been a high concentration of proteinlike substances with a high molecular weight, like the coacervates, dissolved in the primitive ocean.

Reactions Between Coacervates and Other Organic Matter

It is possible that many of the proteinlike droplets, instead of breaking up, increased in size. This reaction could have been brought about through diffusion, a process by which a substance moves randomly from an area of higher concentration (area of many molecules) to an area of lesser concentration (area of fewer molecules).

As the molecules of organic compounds and coacervates spread out and bumped into each other, small organic molecules could have attached themselves to the coacervates. This would have resulted in the formation of a large complex organic compound.

Some combinations of organic matter and coacervates may have been more stable than others. In fact, it is even possible that, in some combinations, energy was liberated, producing a stable molecule. The reaction between the organic matter and coacervates probably progressed at a very slow rate.

There is often no sharp line of distinction between living and nonliving things. Instead there is a wide spectrum of possibilities. Obviously, the origin of life, or the formation of the organic compounds, did not evolve systematically. The organic compounds and

FIGURE 13.11 Charles Darwin, adventurer, ship's naturalist, and scientist. It was through his efforts that the theory of evolution won acceptance (Bettmann Archive).

the complex chemical reactions involved probably occurred by chance. This is supported by the fact that the first form of life on Earth, as estimated by modern scientific analysis, appeared approximately three billion years after the Earth was formed.

We have proposed several ideas as to how life might have begun. We cannot say which hypothesis is correct. At this time, scientists can only propose theories as to how life might have started. However, with the advances that have been made in science during the past 50 years, and at the rate we are obtaining scientific information, one day we shall probably be able to answer the question of how life began.

Now that we have developed some basic ideas about the origin of life, it is important to remember that not just one form of life

has evolved on earth. The study of living things is a dynamic study. Life has developed, changed, and evolved. Evidence in the form of fossil remains has given scientists insight as to how life has changed from its time of origin.

The major question about evolving life forms is: What has caused these structural changes? In the following pages, we shall be concerned with the factors that have caused living things to change from their time of origin up until the present day. In Chapter 15 we shall discuss the forms that life has taken during its evolution.

The Evolution of Life

Charles Darwin is the scientist most often associated with the Theory of Evolution.

People assume that Darwin was the first man to propose the theory but, actually, several scientists were interested in the concept of evolution long before Darwin. Erasmus Darwin, Charles Darwin's grandfather, in the late 1700's, published a work entitled *Zoonamia*. This publication dealt primarily with the possibility of evolution; it accepted the concept of evolution but did not propose any way to explain the processes of evolution.

In the late 1700's and early 1800's the French scientist Jean Lamarck proposed a theory of how life evolved. His hypothesis was based on the assumption that changes in the environment produced genetic changes that eventually produced physical changes in living species.

Lamarck proposed two theories about the evolution of life. His first theory was called "The Law of Use and Disuse." This law proposed that the more a specific organ is used, the better it develops. The less the organ is used, the smaller it becomes.

His second theory was called "The Law of Inheritance of Acquired Characteristics." This law proposed that animals are capable of passing on to their offspring structural changes in their own body parts. These structural changes became dominant through

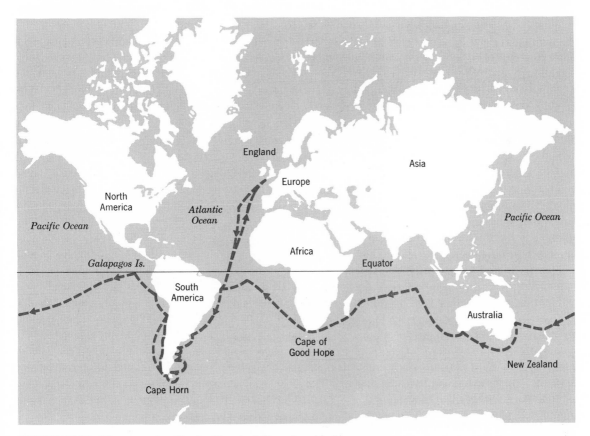

FIGURE 13.12   The route that Charles Darwin followed on his 5-year voyage.

use. For example, Lamarck proposed that the giraffe acquired his long neck because he lived on barren land without grass and had to stretch his neck upward into trees for food. Those characteristics not used did not become dominant and were not passed on to the offspring.

Darwin proposed his Theory of Evolution only after he had become interested in and studied the origin of life. In 1831, he ended a five-year voyage around the world as a ship's naturalist. It was Darwin's responsibility to collect specimens and to record his observations about living things and the Earth. He collected thousands of plant and animal fossils. During his examinations of these fossils he noticed that there were many differences and similarities in the variety of organisms and the places where he found them. Eventually, Darwin became curious about why some organisms and fossils were found in some parts of the world and not in others. He also wondered why the Earth itself had so many different and similar characteristics. The many unanswered questions about the variety of living things and the structural similarities and differences in organisms troubled Darwin. When he returned from his voyage, he set out to answer some of them.

In 1857 Darwin proposed his Theory of Evolution. The theory rejected the idea of a fixed species. Darwin's theory was that living things are not constant from generation

FIGURE 13.13 Autotrophs produce their own food supply by using solar energy and mineral constituents from the soil (Chicago Natural History Museum).

to generation, but change over a long period of time. This change produces a variety of life.

The Autotroph Versus the Heterotroph Hypothesis

When considering the many problems pertaining to the evolution of life, it is necessary to consider the conditions in which life began on Earth. One of the basic requirements of living organisms is the need for nourishment. But it can be assumed that there was no source of food available on the primitive Earth to support life. Two hypotheses have been proposed that satisfy this requirement.

The Autotroph Hypothesis

The first of these theories, called the autotroph (self-feeder) Hypothesis, proposes that the first form of life on Earth was capable of manufacturing its own food. For example, green plants manufacture their own food with the use of cellular constituents and minerals consumed from the soil. This process is called photosynthesis. The autotroph hypothesis assumes that the first organisms were capable of a process like photosynthesis in which they used solar energy to manufacture their own food. Other autotrophs that did not use solar energy could have used energy from lightning, static electricity, or complex chemical re-

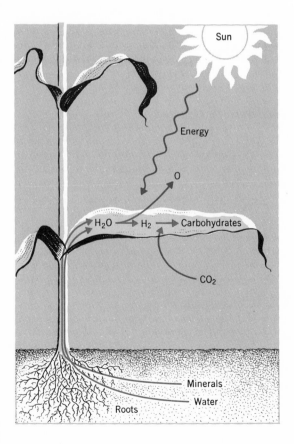

FIGURE 13.14 Photosynthesis utilizes CO_2 gas from the atmosphere, minerals and water from the soil, and solar energy to synthesize organic compounds, such as carbohydrates.

FIGURE 13.15 Heterotrophs must ingest complex organic materials as food (Harry Engels, National Audubon Society).

actions. It is possible, then, to hypothesize that the first form of life was capable of using some source of energy to manufacture its own food.

A serious problem exists with the autotroph hypothesis. In the process of photosynthesis, for example, plants convert light energy (the sun) into chemical energy through a very complicated process (Fig. 13.14). Therefore, the autotroph hypothesis holds that a complex chemical process evolved early in the history of life, so that the first living things would have been rather complex. This is contrary to the Law of Evolution, which states that living things progress from a simple to complex state both chemically and structurally. Therefore, an important aspect to consider in evaluating this hypothesis is that it assumes a complex living thing existed in a very simple environment.

Heterotroph Hypothesis

Unlike the autotroph, the heterotroph is an organism that is not capable of manufacturing its own food. Although it might be able to synthesize a few organic compounds, it must usually rely on its environment to supply the food necessary for its survival. Bacteria and plants such as the fungi are examples of heterotrophs. The heterotroph hypothesis proposes that the first form of life may have developed from nonliving substances.

Heterotroph Hypothesis Versus Abiogenesis

Recalling that the Theory of Abiogenesis and the heterotroph hypothesis both state that living things evolved from nonliving things, then, how do these two hypotheses differ? Abiogenesis proposes that life evolves continuously within a short period of time. The heterotroph hypothesis, on the other hand, proposes that the first form of life to develop was very simple, and that the production of diverse forms of life required a long period of time, probably billions of years.

As previously stated, the heterotroph hypothesis proposes that the first forms of life developed from nonliving substances. It is also important to remember that according to the Heterotroph hypothesis, these organisms were unable to manufacture their own food. Assuming that there were no autotrophs available to manufacture food, what would have been a source of food for the heterotrophs?

It was first proposed by Darwin that the environmental conditions under which the heterotrophs developed were similar to the ones of today. According to this assumption, it would seem, then, that the atmospheric conditions of the Earth could very likely support life. Modern scientists believe Darwin was inaccurate in this assumption because they think that the Earth's atmospheric conditions were different when the Earth first formed. According to Darwin, a heterotroph could have survived if conditions of the Earth were the same as they are today.

Darwin also proposed that life evolved in a small pond. He hypothesized that the pond provided substances such as ammonia and phosphoric salts, heat, light, and electricity, which interacted to form proteins. Under present conditions, these substances would be absorbed by living things. But what if they existed and formed proteins before life developed? If such conditions prevailed, the heterotroph could have survived.

After reviewing both hypotheses, which one seems the most acceptable? From a scientific point of view, both hypotheses provide possibilities and both pose questions as to probability. However, it is important that scientists never overlook any possibility in the quest for the origin of life. Perhaps this question may be answered within your lifetime. Remember, many doubted the possibility of creating life in a test tube, and yet in December 1967, a synthetic biological molecule divided. This magnificent step forward will undoubtedly give scientists some new insight into the origin of life.

It is important to realize that we may never know the truth about the origin of life. The best that scientists can do is to make educated guesses based on scientific experimentation, and to evaluate the information collected from experiments.

CHAPTER 14

The Geological Time Calendar

WHETHER LIFE DEVELOPED OUTSIDE the Earth or on the Earth, it is important to recognize the fact that the Earth formed before life began, and that the Earth has undergone many changes since its formation. In turn, the changing characteristics of the Earth have brought about the development and destruction of many different life forms.

The evolutionary development of the Earth and its associated life forms has always been an area of concern to scientists. Thus far, man's study of the Earth and of life has proved to be extremely complex, because man has only existed on Earth for an infinitesimal portion of the existence of the Earth itself.

Scientists have estimated the age of the Earth to be between four and five billion years. Life has existed for approximately three million years, but man has kept written records for only a little more than 5000 years. What about the life that existed before man? Scientists have had to find ways of interpreting and dating fossils (imprints in rocks) before they could begin to understand the evolution of living things.

What about the evolution of the Earth? Is the Earth the same today as when it was first formed? Geologists have had to search for evidence that is millions of years old, and to interpret the scattered evidence, before they could form any sort of understandable picture.

Why is it important to know about the Earth and how it has changed? On a practical level, knowledge of the development of the Earth has helped modern man to find new ore and petroleum deposits. In addition, this study unfolds the story of the development of life. A discussion of all the geological changes on the Earth is a topic too broad to cover in one chapter. Therefore, the discussion will be limited to the continent of North America. In the next chapter we will discuss how and why certain life forms appeared, and why other life forms disappeared.

Rocks as Pages of History

Geologists learn about the ancient Earth and the organisms that inhabited it by studying layers, or strata, of rock. The history of the Earth is recorded in the life forms that died and left imprints in these layered rocks. These imprints, called fossils, result from the remains of the hard parts of living organisms, such as shells, teeth, and bones. Fossils provide only limited information, but paleontologists—those geologists who specialize in the study of prehistoric life forms—can organize these bits and pieces of information to provide insights into the Earth's past. This evidence of the past is often called the *stratigraphic record*. (See Figure 14.2.)

About 75 per cent of all fossils are found in sedimentary rock. Each layer of the sedimentary rock tells a story about the evolutionary changes that occurred in one period on the Earth, and about the evolutionary development of living things.

The Law of Superposition

The Law of Superposition, on which geological history is based, states that in a formation of undisturbed strata, the oldest layer is on the bottom and each layer in turn above it is younger than the stratum on which it rests. Of course, strata are seldom found undisturbed, except in such dramatic examples as the Grand Canyon.

The Law of Superposition implies that all a geologist has to do is study and analyze fossils found in the different layers of sedimentary rock, and he will discover the entire story of the Earth's evolution in chronological order. Unfortunately, this is not the case. During the course of three billion years these layers of sedimentary rock have been subjected to environmental processes such as erosion, exposure, deformation, and overturning. Therefore, it is a rather difficult job to piece together the history of the Earth and of life in simple chronological order.

The Law of Superposition does not apply to intrusive rock. For example, an igneous rock that intrudes between sedimentary rock is younger than the surrounding sedimentary layers, since an intrusion that cuts across three-million-year old rock must have occurred less than three million years ago. The situation may be further complicated by folding or faulting of the original rock. This

FIGURE 14.1   The Grand Canyon is a good example of the stratigraphic record (John Shelton).

example illustrates only one simple problem in comprehending the evolution of the Earth. Another problem is determining which sedimentary rocks are of land origin and which are of marine origin.

Land Versus Marine Sediments

Sediments deposited on land differ from the ones deposited in the sea in two important ways. One is the color of the rock, and the other is the nature of their fossil remains.

Many land sediments possess a bright red, reddish brown, or brown coloration. Marine sediments are dull gray, green, or black, or a combination of these colors. The difference between the general brightness of land sediments and the dullness of the marine

material results from the oxygen in the air. Oxygen in the air reacts with the iron particles found in the land sediment, producing bright colored limonite and hematite. The marine sediments are not exposed to the air, so that dull hydroxides form instead.

Geologists also use fossils to distinguish between land and marine sediments. Fossils of cephalopods (squid, octopus, chambered nautilus, and cuttlefish), Anthozoa (sea anemone, corals), and echinoderms (sea lily, starfish, brittle star, sea urchin, and sea cucumber) indicate marine sediment. (See Figure 14.3.) Fossils of four-legged animals indicate a terrestrial origin. We return to the discussion of fossils in the next chapter.

Hardened ripple marks and mud cracks also differentiate terrestrial and marine sediment. Mud cracks indicate clay sediments that were exposed to air and sun. Ripple

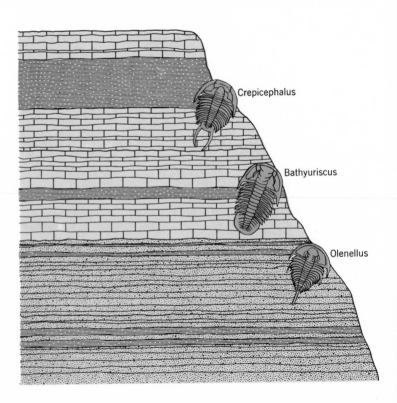

Crepicephalus

Bathyuriscus

Olenellus

FIGURE 14.2 The law of superposition is illustrated by the succession of formations exposed on an escarpment in the Colorado plateau, near Price, Utah.

FIGURE 14.3 The study of fossils found in Burgess Shale reveals that a variety of invertebrates evolved during the mid-Cambrian Period. Among these were the dominant organism—the trilobite—and a variety of coelenterates, brachiopods, sponges, gastropods, arthropods, and corals (American Museum of Natural History).

marks indicate that waves and currents disturbed loose sand on the bottom of the sea.

The Importance of the Geological Calendar

In order to maintain some chronological order in discussing the history of our environment, scientists have divided the development of the Earth into units of time. The geological calendar has been divided into three basic eras, further subdivided into periods and epochs.

Geological Correlation

Geologists have experienced great difficulty in presenting an exact history of the Earth's development because of the many events that have occurred during the Earth's existence. These events include numerous successive erosions and upliftings of the Earth's surface. Thus, the geological calendar has been put together in bits and pieces. In order to do this geologists had to develop a method by which they could determine the ages of different rock strata. This method is called *geological correlation.*

Geological correlation is particularly effective in correlating strata that occur short distances from one another. However, by using the thickness, color, and composition of strata, scientists can often correlate different widely separated strata. For example, if an outcrop composed of gray limestone underlaid by sandstone and overlaid by red conglomerate is similar to an outcrop ten

Geological Calendar

ERA	PERIOD	EPOCH	MILLIONS OF YEARS BEFORE PRESENT	EVENTS
CENOZOIC	Quaternary	Recent Pleistocene	2.5–present	Four glacial periods occurred. Glaciers formed across North America.
	Tertiary	Pliocene Miocene Oligocene Eocene Paleocene	63–2.5	Tremendous volcanic activity in northwest North America. The Columbia Plateau and Cascade Range arose.
MESOZOIC	Cretaceous		135–63	The Rocky Mountains appeared.
	Jurassic		181–135	The Sierra Nevada Mountains formed, from Alaska to California.
	Triassic		230–181	The Appalachian Mountains began to erode.
PALEOZOIC	Permian		280–230	A period of tremendous geological and climatic disturbances. Many deserts in North America.
	Carboniferous	Pennsylvanian Mississippian	345–280	The east coast of the U.S. was a swampy area. Forests grew, died, and became submerged.
	Devonian		405–345	The east coast from Canada to North Carolina rose from the sea.
	Silurian		425–405	Most of the east coast of the U.S. was covered with water. Much volcanic activity occurred from New Jersey to Maine.
	Ordovician		500–425	75 Per cent of North America was covered by a shallow sea. There was volcanic activity, and mountains formed in New England.
	Cambrian		600–500	East and West coastal troughs filled with sediment to form the Appalachian and Cordilleran geosynclines.
PRECAMBRIAN			beginning–600	The Earth's crust formed. Mountains and seas appeared.

miles away, it is very likely that these two outcrops at one time were joined together.

It is sometimes possible to decipher a formation by tracing the surface outcrop to the point where it is found to be continuous with a formation of known stratigraphic position. A classic example of this is the Grand Canyon, where formations can be followed for hundreds of miles.

One of the most important factors in geologic correlation is the identification of fossil types called *index fossils*. Index fossils are organisms that lived for only a short period of time, but became widely distributed and left easily identified traces behind. If no index fossils are available, fossils of long-lived life forms can be used for dating. However, a fossil known to have existed over a shorter span of time makes it easier to date the rock strata in which it is found.

The formations of the Earth (continents, oceans, mountains, valleys, streams) did not form at random; these dynamic structures evolved in chronological order. Their formation can be reviewed in the events of the *geological calendar*.

The Geological Beginning

The study of the Earth's history begins with the Precambrian Era, which is dated from the beginning of time to 600 million years ago. Very little is known about this part of the Earth's history. Much more is known about the Paleozoic, Mesozoic, and Cenozoic Eras. Although these latter eras occurred over a relatively shorter period of time than the Precambrian period, there is more fossil evidence available for geologists to study.

The Precambrian Era (beginning–600)

Geologists have had difficulty trying to learn about the Precambrian Era because of the lack of fossil remains from this time. Most

Precambrian rock has been buried so deeply by overlying younger rocks that it has been metamorphosed (changed) or destroyed. Moreover, most Precambrian rock that is available for study does not contain fossils. This lack of fossil remains does not mean that no life existed from the beginning of time until 600 million years ago. Instead, it is probably because of the fact that animals with shells or hard external structures did not exist to leave imprints in the rocks. Recently, evidence has been uncovered in thick strata (limestone and layers of graphite) that definitely indicates the presence of Precambrian marine life some two *billion* years ago.

Geologists have shown great interest in the Precambrian Era, and in spite of the difficulty in finding fossils and rocks, they have been able to uncover some information about this era. This era is an important one, because some minerals had their origins during it.

Because there is a lack of remains from the Precambrian Era, there has been much speculation about it. The most ancient rocks discovered that date back to this period are metamorphosed sedimentary rocks which contain particles of ancient granite. These rocks have given scientists an indication that the Earth's continents formed more than three and one-half billion years ago, long after erosion and diastrophism had already been working on the Earth.

Even though many late Precambrian strata have been exposed to considerable metamorphism and diastrophism, they still contain unchanged sedimentary and volcanic rocks. This is unexpected, considering the amount of erosion that the Precambrian strata have been subjected to. Yet even with all the metamorphism, diastrophism, and erosion, some history of the past has been revealed.

Scientists have uncovered many varieties of sedimentary rocks in Precambrian strata.

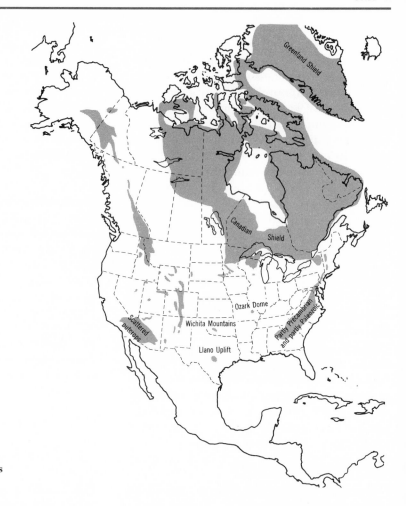

FIGURE 14.4 Precambrian rocks in North America.

They have also found, in late Precambrian rock, deposits of glacial origin. These deposits, which consist of striated rock found on top of grooved rock surfaces, indicate the possibility of at least two glacial periods occurring in the early history of the Earth.

The Precambrian Era in North America

A large area of Precambrian strata has been discovered in North America, as shown in Fig. 14.4. This area extends from eastern Canada to the northeastern parts of the United States. Smaller deposits of Precambrian rock have also been discovered in the area of the Grand Canyon, where the Colorado River has cut and eroded some 6000 to 7000 feet of strata.

Precambrian *outcrops* and *shield areas*, are found on all continents. Shaped when upwarping allows younger rock to be eroded, Precambrian shield areas are exposed in high mountainous regions that have been little affected by flooding and the deposition of marine sedimentation. The Canadian shield area extends from Canada to Greenland, south to Wisconsin and Minnesota, and covers an area of approximately two million square miles. It includes some granitic rock, probably the result of intrusive events that took place over a period of billions of years. Granite is found as intrusives in limited areas that also contain graywacke sandstone, shales, and iron-bearing rocks.

There are small Precambrian shield areas in the United States. These include the Ozark Mountains of Missouri, the Black Hills of South Dakota, the Eastern Appalachian Mountains, the river regions of the Grand Canyon, and parts of the Rockies.

Geologists have also found traces of the Precambrian Era around Lake Superior. This evidence is in the form of older marine iron-bearing sedimentary rocks which became deposits of hematite. The origin of the younger areas is nonmarine and records sedimentation of the late Precambrian Era.

Geologists have found areas in other parts of the United States that also tell something about the Precambrian Era. For example, the colorful appearance of Glacier National Park in Montana is provided by limestone, red shales, and green shale which formed during the late Precambrian Era.

Precambrian rock that is highly deformed and metamorphosed developed during the early Precambrian Era. Undeformed rock and unmetamorphosed rock found in Glacier National Park of Montana and the upper regions of the Grand Canyon formed during the late Precambrian.

The Paleozoic Era

The history of the Paleozoic Era is rather complete, because an abundance of fossils is available.

The Paleozoic Era makes up about 300 to 400 million years of the Earth's history. It is during this era that many different types of invertebrates (animals without backbones) appeared and developed. Invertebrates have produced excellent fossils for interpreting the Earth's history. Therefore, geologists are able to piece together the scattered information, and to form some basic ideas about the development of life and the changes that occurred on Earth during the Paleozoic Era.

North American Paleozoic History

It has been fairly well established that during the Paleozoic Era, North America did not resemble the structure that it is today. (See Figure 14.5.) The southern and central states of North America were submerged in a shallow sea; the only exposed land area was the low-lying central region of North America and the North Canadian Shield. During the Paleozoic Era much reorganization of the Earth took place. Upwarping of the land resulted in the withdrawal of the sea. When the sea withdrew, sediments consisting of thin layers of limestone and shale beds were left behind.

Unlike the interior of North America just described, the outer areas were rather unstable. They provided sites of geosynclines in which seaways were produced. The seaways, in turn, resulted in the accumulation of sediments. The Eastern Geosyncline (Appalachian Geosyncline) extended from Newfoundland to Alabama, then turned west to Arkansas, Oklahoma, and Texas. The Western Geosyncline (Cordilleran Geosyncline) extended north-south toward the Circum-Pacific Zone of geosynclines, volcanic island areas, and mountainous regions. During the Paleozoic Era, the Western Geosyncline (Cordilleran Geosyncline) extended from all the western states to central Utah, Wyoming, and western Montana. A third geosyncline extended from Greenland to the Arctic Islands.

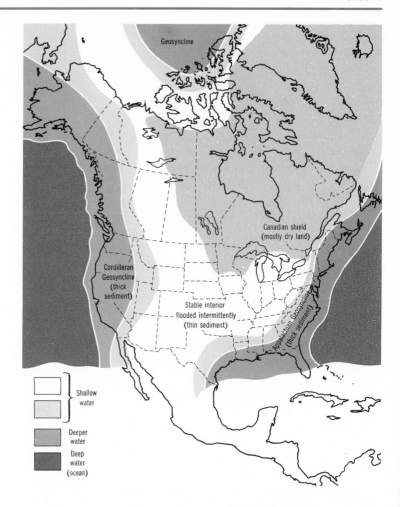

FIGURE 14.5 Diagram of North America during the Paleozoic Era. This diagram shows the Cordilleran geosyncline, stable interior, Canadian Shield, and Appalachian Geosyncline.

Thus, the entire continent of North America was surrounded by geosynclines during the Paleozoic Era.

The Cambrian Period (600–500)

At the beginning of the Cambrian Period the interior of North America was primarily dry land. Because of slow warping, the sea began to flow toward the interior of the continent from the geosyncline seaways. The result was an advancing sea and the deposition of sand on a peneplane of Precambrian rock.

During the late Cambrian the sandstones of the present Ausable Chasms of New York and the Wisconsin Dells formed. Limestone was deposited on sandstone during the late Cambrian Period.

The Ordovician Period (500–425)

During the early Ordovician Period more sediments of limestone were deposited on sandstone. Toward the middle of the Ordovician Period orogenic activity ("orogenic" means "mountain building") developed in

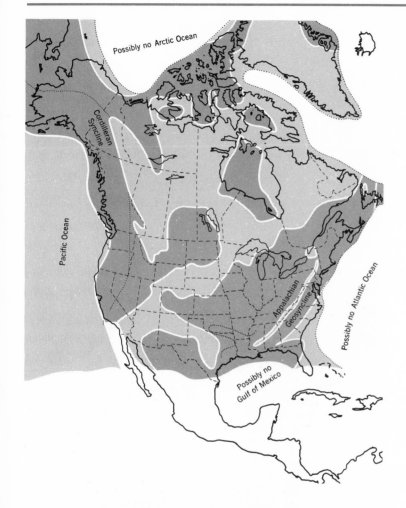

FIGURE 14.6   North America in the Ordovician Period.

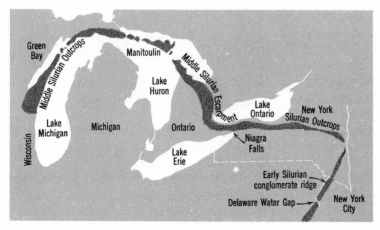

FIGURE 14.7   The major outcrops of Silurian rock found in the north central and northeastern United States. The Silurian rock belt stretches from southern New York, south of Lake Ontario, through Lake Huron, along the east shore of Lake Michigan.

the north Appalachian Geosyncline. As a result of this activity, a body of land began to rise along the eastern side of the Appalachian Geosyncline. (See Figure 14.6.)

During the late Ordovician Period much of the northern part of the Appalachian Geosyncline was affected by orogenic activity known as the Taconian Orogeny. This resulted in the formation of the *Taconian Mountains*. Streams were developed that eroded these lands (Figures 14.7 and 14.8). Mud and sand were carried to form great

deltas that can be seen below Niagara Falls the rock strata and bedrock in Ontario, Canada.

The Silurian Period (425–405)

The Taconian Orogeny of the Silurian Period began with the raising of the north-eastern part of North America. Water began to recede in a northern direction around the northern end of the geosyncline—what we know as the state of New York. Rivers and

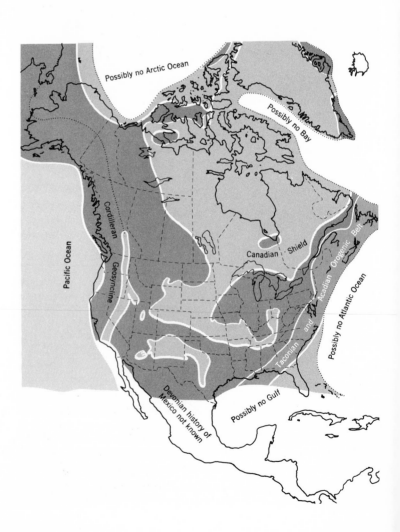

FIGURE 14.8   North America in the Devonian Period.

FIGURE 14.9   The mid-Devonian forest contained treelike plants, ferns, and insects (American Museum of Natural History).

streams formed on the newly emergent land. As they flowed down from the mountains, they carried sand and gravel which was deposited in the sea. Over the years, the sand and gravel hardened to extend the Appalachian Mountains.

During the mid-Silurian Period, limestone and dolomite formed from the state of New York northwest to Lake Huron, then southwest to form Door Peninsula, located between Green Bay, Wisconsin, and Lake Michigan.

It was during the late Silurian Period that the seas drained from what are now the states of Michigan, Ohio, and New York. Basins formed, and the water from these basins began to evaporate. Large deposits of salts formed at the bottoms of the basins. Today these salts serve as an important economic resource in the states of Michigan and New York.

The Devonian Period (405–345)

During the early Devonian Period large quantities of limestone were deposited in the seas of North America. The Appalachian Geosyncline underwent movement. It folded and was intruded by granite, and then uplifted in the *Acadian Orogeny*. (See Fig. 14.8.) Remainders of the Acadian Orogeny can be seen today as deposits of red sandstone, shales, and conglomerates in the Catskill Mountains of New York State.

Orogenic unrest was also present in the Cordilleran Geosyncline. This resulted in a period of folding and uplifting that extended in a northeastern direction to Nevada and north to Idaho.

The Carboniferous Period (345–280)

During the beginning of the Carboniferous Period the Appalachian Mountains be-

309

gan to erode. Deposits of this erosion can be seen in the Pocono Mountains of Pennsylvania. (See Figure 14.10.)

At this time the central part of North America was covered by a warm shallow sea, in which many forms of life developed. In addition, folding occurred in many regions from New England to Texas. The area between what is now Pennsylvania and Kansas was tropical, low lying, and partially submerged under water. The inland sea kept receding and spreading out, producing swampy conditions and the alternation of marine and nonmarine life. This resulted in a buildup of sedimentary layers, which can be observed in the coal beds of Pennsylvania.

The Permian Period (280–230)

The Permian Period was one of great geological activity (Figure 14.11). The sea had completely disappeared from the eastern coast of North America. The Appalachian Geosyncline continued to uplift and eventually gave rise to the Appalachian Orogeny.

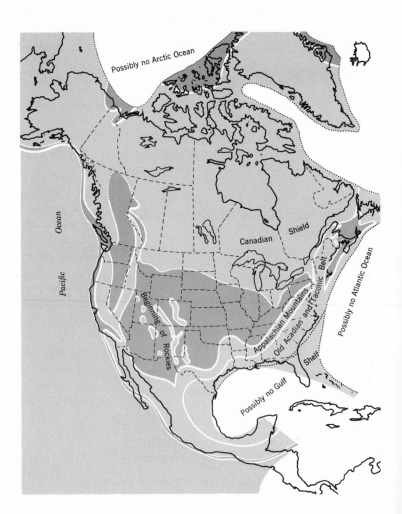

FIGURE 14.10   North America in the Pennsylvanian Period.

The sea was still present in much of the western half of the United States during the Permian Period, but some land rose in what is now Nevada and Idaho. Large reefs developed in southern New Mexico and Texas, which helped localize the circulation of the sea.

The Mesozoic Era

About 150 million years of the Earth's history make up the Mesozoic Era. During this era in history, many of the world's fa-

mous mountain ranges developed, among them the Andes and the Rockies.

Rocks from the Mesozoic Era are more abundant on the west coast than on the east coast of North America. This indicates that erosion first occurred on the east coast and sedimentation occurred on the west coast.

The Triassic Period (230–181)

At the beginning of the Triassic Period (Figure 14.12) the Appalachian Geosyncline

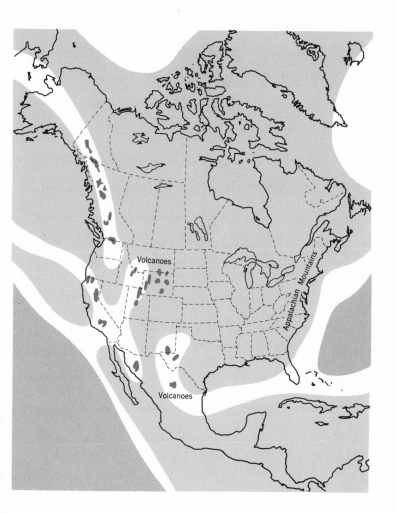

FIGURE 14.11   North America in the Permian Period.

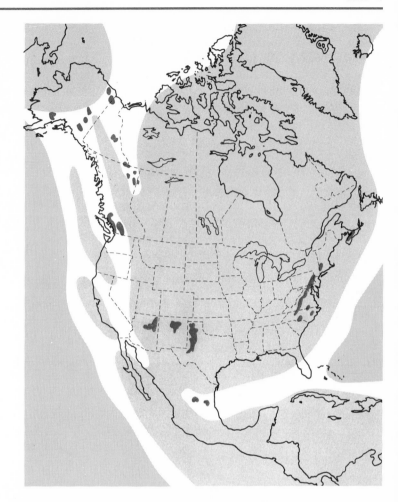

FIGURE 14.12 North America in the Triassic Period.

underwent its developmental process to form the Appalachian Mountains. The east coast was no longer surrounded by geosynclines and most of it was dry land. Some of the south coast was still submerged in water.

Erosion of the Appalachian Mountains continued, with the eventual formation of a peneplane. Faults caused the rocks of the peneplane to tilt and move. Basins formed into which streams carried mud and sand. This mud and sand eventually solidified into hard strata.

Near the end of the Triassic Period, breaks occurred in the Earth's crust and basaltic magmas escaped. The Palisade Cliffs along the Hudson River, which separates New York City and New Jersey, are the remains of a big sill from this period. The Cordilleran Geosyncline was still declining, and what is now the Pacific Ocean reached as far inland as Nevada and Idaho.

Sedimentary sandstone and shale of nonmarine origin can be found in New Mexico, Utah, Arizona, Colorado, and Wyoming.

These rocks lend color and interest to the desert of the western United States.

The Jurassic Period (181–135)

There is very little evidence of the Jurassic Period on the east coast of the United States. It is assumed then, that most Jurassic activity took place on the west coast. Even here, however, geologists find very few outcrops of this period.

Strata of the Jurassic Period that resemble Triassic strata have been found in Nevada,

Oregon, and California. Jurassic sandstone outcrops can be seen in Zion National Park in Utah.

Toward the end of the Jurassic Period there were uplifts on the west coast that began the western mountain formations. The development of these mountains is referred to as the *Nevadan Orogeny*. The uplifts and folds that occurred toward the end of the Jurassic Period can be seen as granitized or granite magma in Yosemite National Park and in other parts of California.

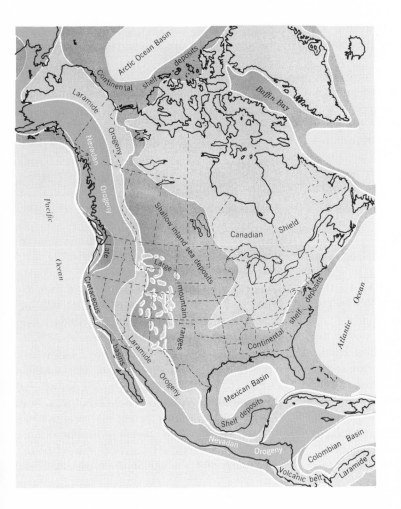

FIGURE 14.13   North America in the late Cretaceous and early Tertiary Periods.

The Cretaceous Period (135–63)

The Cretaceous Period began with tremendous activity. At the very beginning of this period, great mountains developed from the Cordilleran Geosyncline. These mountains then began to erode, and the eroded material was spread out to cover the area we now call Wyoming, Colorado, and Montana. Traces of the Cordilleran Geosyncline can also be seen in small quantities in central Kansas and Nebraska. (See Figure 14.13.)

One can see a line of separation of Cretaceous outcrops in the Mid-Atlantic states. The eastern rocks originated from the Atlantic Ocean (they are marine in form) and show how deposits formed when the Atlantic Ocean spread inland from New York to Mississippi.

The end of the Mesozoic Era witnessed the decline of the Cordilleran Geosyncline, or Laramide Orogeny. The western coast of North America became a mass of huge mountains. To the east of these mountains were plains. At this time, the Pacific Ocean still extended inland. The western parts of California, Oregon, and Washington were submerged.

Cenozoic Era

We are now living within the last pages of the history of the Cenozoic Era. It spans about 60 to 70 million years of the Earth's history. Man has been present for about the last two million years of this era.

Even though the Cenozoic Era represents the shortest era in the history of the Earth, it is the era about which we know the most. The reason for this is that results of events occurring in the Cenozoic Era are comparatively untouched. Most of the Cenozoic rocks have not been buried very deeply nor have they had time to be deformed and metamorphosed. Much of the sediment has not been carried out to sea and, thus, many fossil-bearing formations can be studied rather easily.

The Tertiary Period (63–2.5)

Most of the marine rock of the Cenozoic Era can be found on the west coast of North America. Nonmarine rocks cover the plains and low-lying areas of the western part of the continent.

Geologists have collected material from this era and have been able to reconstruct the shape of the west coast of the continent. It is believed that the Pacific Ocean extended eastward to the mountains that formed from the Nevada Orogeny. On the opposite side of the continent the same story was true; the Atlantic Ocean covered more of the east coast than it does today.

The eastern shoreline during the early part of the Cenozoic Era was similar to that of the Cretaceous Period. It divided the present states of New Jersey, North and South Carolina, and Georgia in half, then turned north to Illinois through the center of Arkansas, and inland about 200 to 300 miles to the present coastlines of Texas and Mexico. The eastern and southern coastal lines of the United States during the Tertiary Period were similar to the present coastlines except that they were much farther inland. Eventually, the waters of the Atlantic Ocean and the Gulf of Mexico receded and the land area of the United States increased. When these waters withdrew, sediments were deposited on the coast of the Atlantic Ocean from the Appalachian Mountains, the mountains of Arkansas and Oklahoma, and the Rocky Mountains. Streams from the Cordilleran Geosyncline carried sedimentation in a western direction toward the west coast.

Folding and raising of the west coast be-

FIGURE 14.14 The extent of Pleistocene glaciers in North America.

gan. In fact, the earthquakes that occur periodically on the west coast today are part of the rising and the folding that are still going on. The most intense earthquake activity is in the state of California. The Baldwin Hills area of California is rising at a rate of about three to four feet every 100 years.

Some of the sediment that was being carried by the streams from the Rocky Mountains was deposited on the plains of Wyo-

ming, Montana, and North and South Dakota. The mountains of the old western geosyncline continued to reduce in size to form a broad peneplane. Then, toward the middle of the period, the Rocky Mountains uplifted. Erosion continued, bringing more deposits of sediment to Nebraska and Texas.

Small mountain ranges rose within the Rocky Mountains, and several uplifts occurred along faults within the Rockies. Between the middle and late Tertiary Period, basaltic magmas erupted from fissures and covered the states of Oregon, Washington, and the southern part of Idaho. Today one can still see the lava flow from the crater at Moon National Monument.

The Quaternary Period (2.5–present)

The Quaternary Period began approximately 2.5 million years ago. The beginning of this period is marked by extrusive igneous activity that resulted in the development of the great volcanoes of the Cascade Mountains. These mountains stretched from northern California to the state of Washington.

It was during this period that the last Ice Age occurred (Figure 14.14). The four different glacial advances that made up the Ice Age produced large areas of erosion and deposits of sediment.

CHAPTER 15

The Evolutionary Calendar

ALL THE THEORIES AND CONCEPTS THAT attempt to explain how life evolved, and how the Earth reached its present form, support one basic idea: that the Earth was formed long before life evolved and that it has supported a wide variety of life forms during the last three billion years.

What were the first life forms? Has the first form of life changed physically since its beginning? What other forms of life have existed? When did these different forms of life appear? Scientists have managed to answer most of these questions by means of studying the fossil records.

Fossils

Fossil imprints usually provide only limited information, resulting as they do from the hard parts of living organisms. But at times complete organisms have been discovered. For example, ancient insects have been discovered preserved in amber. Even the giant mammoth has been preserved and found frozen in the Arctic.

Most fossils are animal remains, since plants do not have hard substances such as teeth and bones that are easily preserved. Therefore, most of the ancient plants have decayed and disappeared, leaving very few records.

Nevertheless, paleontologists have been able to uncover many records of plant life.

For example, they have discovered the structures of ancient tree trunks in petrified wood. Petrified wood results when the original wood is buried and then replaced by deposits of silica from groundwater. Paleontologists have also discovered fossil remains of leaves and wood in coal.

Fossils are not only imprints of plant and animal remains but may also be trails and footprints that animals left in soft material, for instance, mud that was then covered by some form of sediment to preserve the print.

Fossil records are far from complete. Yet the fossils that are present indicate that life has changed drastically from its beginning. Studies of fossils from ancient rocks indicate that life progressed from very simple to complex forms. Fossil remains also indicate that there has been a very broad diversification in life; many creatures that existed during a particular era are no longer present today. Because of fossil diversification, paleontologists have been able to determine the times at which different strata formed, and what living organisms existed then. They have formed a chronological picture of the evolution of life, and have developed the evolutionary calendar as an aid in the study of evolution.

The evolutionary calendar lists some forms of life that are now facing extinction, and are at the end of their evolutionary development. But one thing can be observed from

317

FIGURE 15.1 (*right*) Ancient arthropods have provided excellent
index fossils to help geologists
identify different periods of the
Earth's history. (1 and 2) Cockroaches from Illinois; (3) spider
from England; (4) myriopod, or
"thousand-legs," from Illinois.
These fossils were obtained by
splitting open concretions (American Museum of Natural History).

FIGURE 15.2 (*below*) Skull of an
ancient mammoth (American Museum of Natural History).

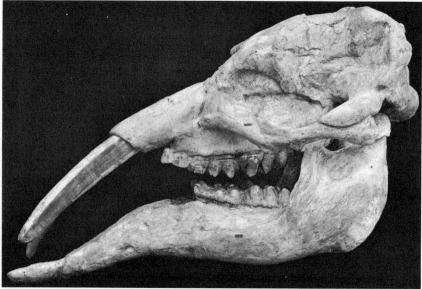

FIGURE 15.3 Limulus trail (Bernard Kummel).

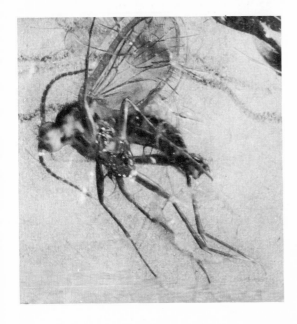

FIGURE 15.4 An insect from the Tertiary preserved in amber. Amber or resin will preserve not only the hard parts of the animal, but the soft body parts as well ×20 (University of Michigan, Museum of Paleontology).

this calendar—that throughout the history of the Earth, there has been much diversification of living organisms. Evolution is a continuing process, and other forms of life will evolve during the history of the Earth.

Geologists have found fossil records indicating that the first life forms evolved in the sea. It is assumed that these first living organisms were plants, probably simple unicellular forms that absorbed the minerals from the sea through the cell wall and carried on photosynthesis in order to obtain food.

Animals are thought to have developed after plants because animals cannot manufacture their own food, and therefore plants serve as a primary source of food. The first form of animal life was probably simple in structure.

The Paleozoic Era

Geologists estimate that approximately one to two billion years elapsed between the formation of the Earth and the development of life. The activity that contributed to the development of early life occurred during the Cambrian Period.

Geologists believe that prior to the Cambrian Period, there was little or no evolution of living things. There is, of course, an exception to this statement, in that life did begin in the Precambrian Period. Although

Evolutionary Calendar

ERA	PERIOD	EPOCH	MILLIONS OF YEARS BEFORE PRESENT	EVENTS
CENOZOIC	Quaternary	Recent Pleistocene	2.5–present	This is known as the Age of Mammals. Man made his first appearance on Earth.
	Tertiary	Pliocene Miocene Oligocene Eocene Paleocene	63–2.5	Bony fish appeared. Sharks grew to lengths of 60 to 80 feet. Mammals began to dominate the Earth.
MESOZOIC	Cretaceous		135–63	Dinosaurs entered a state of decline and eventually became extinct. Modern trees developed.
	Jurassic		181–135	Mammals first appeared. Reptiles remained dominant. This was the period of greatest dinosaur development. Birds also appeared.
	Triassic		230–181	Reptiles became the dominant life form. Dinosaurs appeared. Complex arthropods also appeared.
PALEOZOIC	Permian		280–230	Vertebrates, amphibians, and reptiles progressed structurally. Modern insects appeared. Trilobites continued to decline and became extinct at the end of this period.
	Carboniferous	Mississippian Pennsylvanian	345–280	Insects were the dominant form of life. Reptiles and sharks appeared; fish rapidly multiplied.
	Devonian		405–345	Vertebrates developed, fish acquired lungs and strong fins and developed into amphibians. Treelike plants appeared on land.
	Silurian		425–405	Primitive animals and plants made their appearance on land during a period of dramatic change.
	Ordovician		500–425	Clams, starfish, corals, seaweed, and primitive arthropods were present during this period. Primitive fish and other vertebrates appeared.
	Cambrian		600–500	Trilobites, primitive arthropods, and early mollusks, sponges, and worms were dominant during this period.
PRECAMBRIAN			beginning–600	The first forms of life appeared. These included algae, fungi, and marine animals.

some Precambrian fossils have been found, some geologists believe that they were made from inorganic markings and are not the remains of living structures. Other scientists believe that these fossils indicate that, during the Precambrian Period, primitive plants, such as the blue-green algae, existed (Figure 15.6). Precambrian fossils of primitive animals such as sponges, bacteria, protozoans, and some primitive arthropods have also been found.

Paleozoic Plant Life

Plant life was probably the first form of life to appear in the early oceans. This hypothe-

sis is based on the fact that an aquatic environment is most favorable for supporting primitive life. During the Precambrian Period, marine algae were abundant. However, these plants were very simple. The higher forms did not evolve until the Silurian Period. Most of the major divisions of algae were in existence by the Silurian Period. Blue-green algae (*Cyanophyceae*) were the most numerous.

It was not until the middle of the Silurian Period that land plants evolved. It is believed that prior to this time only lichens inhabited the Earth. The first true land plant was the Psilophyton. Structurally, this plant was about three feet high, and possessed very

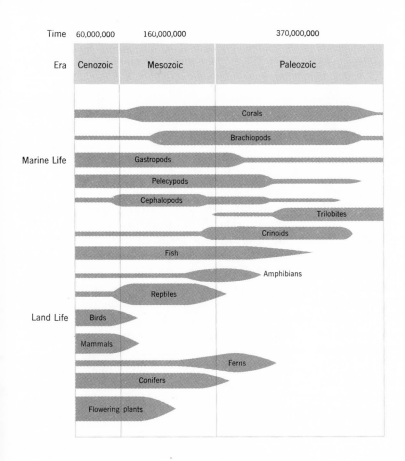

FIGURE 15.5 The evolutionary calendar showing the different forms of life that evolved during the different periods.

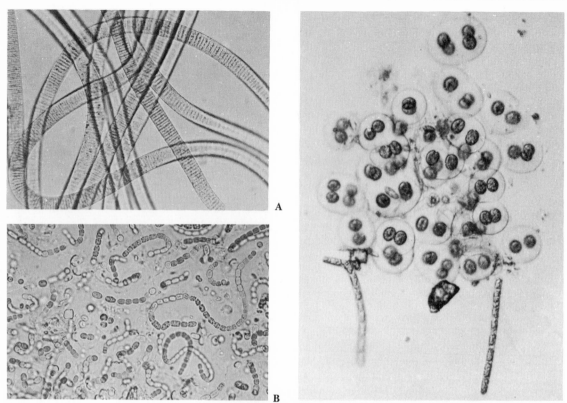

FIGURE 15.6   (A) *Oscillatoria,* (B) *Nostoc,* and (C) *Gloeocapsa* are three varieties of blue-green algae (Hugh Spencer).

shallow rootlike structures, although it lacked true roots, leaves, or branches.

During the middle of the Devonian Period most plants underwent dramatic evolutionary changes. By the end of the Paleozoic Era most plants had changed structurally from the ancestral Psilophyta. Five different groups had emerged (Figures 15.7 to 15.10). The first group, the Lycopsida (scale trees) gave rise to our present-day ground pine (*Lycopodium*) and several other pines. Trees of this group grew to heights of more than 100 feet in the late Paleozoic swamps. The second group of plants were the spore-bearing plants (Sphe-

nopsida). The only existing species of that group today is the *Equisetum* (horsetail). The third group, the ferns, developed during the middle Devonian Period and have continued their development to the present day. The fourth group consists of the seed ferns. The seed ferns structurally resemble the fern that developed during the middle Devonian Period except that these plants produce seeds for reproduction and do not reproduce by spores. The modern cycads and angiosperms (seed-bearing plants) are believed to have evolved from this group. The fifth and final group is a primitive order resembling our present-day conifers. Like the modern conifers, they had needlelike leaves and produced naked seeds (gymnosperms).

Paleozoic Invertebrates

The first form of animal life to evolve in the earliest Cambrian Sea was that of the small invertebrate. Fossil remains have been found of coelenterates, brachiopods, sponges, gastropods, and arthropods. Among the most abundant organisms to appear during this period are the trilobites (Figure 15.12). These extinct members of the phylum Arthropoda have jointed appendages and good sense organs. The trilobites were the dominant organisms during the early' part of the

FIGURE 15.8 The *Lycopsida* are the ancestors of the modern pine (Verna Johnston, National Audubon Society).

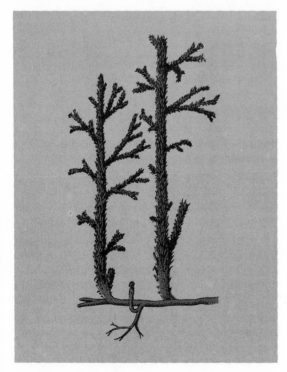

FIGURE 15.7 The Psilophyta were ancestors to most modern plants (After H. N. Andrews, Jr.).

Paleozoic period. Thereafter, members of the group diminished in numbers until they completely disappeared toward the end of the Paleozoic Period. Mid-Cambrian fossils of trilobites have been found in Europe and in eastern North America.

Trilobite fossils are small, but it is believed that some trilobites grew to lengths of one foot. Because the trilobites were so diversified, it is believed that some were swimmers, that others crawled along the bottom of the sea and, finally, that some were capable of

FIGURE 15.9   The horsetail (*Equi-setum*) is the only surviving member of the Sphenopsida (Hugh Spencer).

burrowing into the sea floor. Because of their diversity, and the abundance of their fossil remains, trilobites have become excellent tools in supplying information about Cambrian life.

Further study of fossils in Burgess Shale reveals that many other types of animals also existed during this period. These animals include the other arthropods, the jellyfish, and the annelids (ancestors of modern earthworms).

Throughout the strata of the Cambrian Period in all parts of the world, fossils of the genus *Archaeocyathus* (spongelike organisms) are very abundant. The skeleton of the *Archaeocyathus* (Fig. 15.13) is similar in construction and composition to both the sponges and the corals. As these organisms died, their skeletons accumulated and

formed ocean reefs in the lower Cambrian strata. These reefs can be found throughout the entire world. By the end of the mid-Cambrian Period the *Archaeocyathus* became extinct.

The nautiloids (Fig. 15.14) were some of the largest animals in existence during the Paleozoic Era. They are found throughout the entire would in the strata formed during the Ordovician Period. By the end of the Ordovician Period, the nautiloids had begun a steady decline that continues to the present day. Only one genus of the nautiloids—the *Nautilus*—still survives. There are several possibilities as to why the nautiloids have declined. One is that they could not adapt to ecological changes. Another explanation lies in the rise of predators and in the development of other organisms competing for the same environment. A final possible reason is the rise of the ammonoids after the mid-Paleozoic Period. The ammonoids (Figure 15.15) were a variety of cephalopods (mollusks) that possessed a head surrounded by tentacles and had a straight or coiled shell.

The corals (Figure 15.16) appeared during the end of the Ordovician and the beginning of the Silurian Period, and were relatively active during the latter. It was during the Silurian Period that corals built reefs throughout the world. Paleozoic reefs are not like modern reefs, as the Paleozoic reef

is primarily made up of tabulate (tubelike) corals, and Stromatoporoides (a type of coelenterate that was a free swimming metozoan with a baglike body).

Most tabulate coral reefs were created during the mid-Silurian Period; toward the end of the Paleozoic Era, fewer were formed. Both simple and colonial corals continued to develop, and late in the Paleozoic Era other varieties of corals came to outnumber the tabulates. By the end of the Paleozoic Era the tabulate corals disappeared.

The Stromatoporoides were also active

FIGURE 15.10 During the Devonian Period, most plants underwent dramatic changes. One group to evolve during this period were the ferns. These plants have continued their development up to the present day. The Devonian ferns reproduced their species by spores. The spore reproductive structures appears as little circles located on the undersurface of the leaf (Robert C. Hermes, National Audubon Society).

FIGURE 15.11 Primitive orders resembling our present-day conifers (a modern Slash Pine is shown here) evolved during the Devonian Period. Like the modern conifers they had needle-shaped leaves and produced naked seeds (Robert Lamb, National Audubon Society).

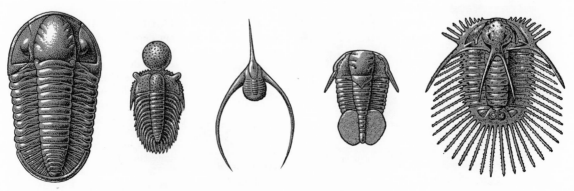

FIGURE 15.12 Diversity of form among the Paleozoic trilobites.

reef-builders during the Silurian, Ordovician, and late Devonian Periods.

Another important group of animals found during the Silurian Period were the brachiopods, marine invertebrates having two unequal shells that are bilaterally symmetrical (Figure 15.17). They resemble the modern clam except that the plane of symmetry of the shells bisects, rather than lies between, the two shells. These small animals were abundant throughout the Paleozoic Era. They first appeared in the Cambrian.

The most abundant brachiopods during the Cambrian Period were the Inarticulata. The Inarticulata have resisted change through millions of years and are still present today.

Geologists have discovered other brachiopods, the Articulata, in the lower Cambrian strata. The Articulata became abundant to-

FIGURE 15.13 In the Cambrian Period these spongelike *Archaeocyathus* organisms gave rise to oceanic reefs throughout the world.

FIGURE 15.14 During the Paleozoic Era the chambered nautiloids were among the largest animals in existence (American Museum of Natural History).

FIGURE 15.15 The most active development of the ammonoid cephalopod took place during the Jurassic Period (Mesozoic Era). Toward the end of the Triassic Period changes in the marine environment resulted in the near extinction of the ammonoid group (American Museum of Natural History).

ward the end of the late Cambrian Period, and underwent most of their development during the Ordovician Period. A great number of articulate shells are found in Ordovician beds.

The environmental conditions that were favorable for the development of the brachiopods were also conducive to the development of the bryozoans (small colonial animals about the size of a pinhead). During the

Silurian and Ordovician Periods, the bryozoans built up masses of reefs, and were also important contributors to Ordovician limestone. Little is known about these animals because special instruments and techniques are required for their study. They were very abundant during the late Paleozoic Era, and some disappeared quickly at the end of this era. Other species abound in today's seas.

The elongated bryozoan was present

FIGURE 15.16 The corals evolved during the Paleozoic Era. During this era, corals built reefs throughout the world. This reconstruction shows rugose corals and associated trilobites that existed during this era (American Museum of Natural History).

FIGURE 15.17 The brachiopods are small soft-bodied marine animals with upper and lower shells that evolved since the Cambrian Period. They were abundant throughout the Paleozoic Era. They have resisted changes through millions of years and are still present today (American Museum of Natural History).

FIGURE 15.18 *Fenestella* from the middle Devonian Period (University of Michigan, Museum of Paleontology).

throughout the Paleozoic Era, but was eventually pushed into the background by the development of the genus *Fenestella* (Figure 15.18). The *Fenestella* was the most successful bryozoan during the late Paleozoic Period.

The echinoderms are marine animals that possess hard calcareous exoskeletons. Many of the early echinoderms were primitive in structure and became extinct during the mid-Paleozoic Period. Among their modern representatives are the sea urchins, sea lilies, brittle stars, and sea stars.

Many of the classes included in the phylum Echinodermata were common fossils. The crinoids (Fig. 15.19) found in the early Pa-

leozoic Era became increasingly abundant throughout the Paleozoic Era. As the small stalks tubular bodies of the crinoids settled to the ocean floor, they formed small plates that were cemented together to create limestone. Much of today's limestone is of crinoid origin. Other echinoderms of importance were the blastoids. Both the crinoids and blastoids have made their marks in evolution as limestone depositors and as index fossils. It was not until the late Mississippian Epoch and early Permian Period that they began to decline.

During the early Paleozoic Era the most numerous animals were the nautiloids. By the end of the Paleozoic Era, however, the nautiloids began to disappear. During the Devonian Period coiled cephalopods (mollusks) appeared. The ammonoids, the first cephalopods, appeared before the end of the Paleozoic Era and were the dominant cephalopods of the late Paleozoic Era.

Snails, clams, and other mollusks were also present during the entire Paleozoic. The mollusks of the late Paleozoic Era were not very different from those of the present.

The arthropods (animals having segmented bodies and appendages and a chitinous exoskeleton) were abundant during the early Paleozoic Period. However, they left few fossil remains because they were poorly adapted for preservation. At death, their exoskeletons degenerated rapidly. Burgess Shale fossils provide some insight into the development of the arthropods.

During the Paleozoic period there evolved groups of animals known as the ostracods and eurypterids. The ostracods were small segmented animals. The eurypterids (Figure 15.20) had spiderlike bodies, and some grew to be ten feet long. Their bodies were long, segmented, and covered with chitin.

FIGURE 15.19 The crinoid was an Echinoderm that originated in the Ordovician Period. It was a flowerlike marine animal that was anchored to the sea floor by a stalk opposite its mouth. During the Mississippian Period they began to decline (American Museum of Natural History).

FIGURE 15.20 During the Paleozoic Period there evolved an arthropod called the eurypterid. The eurypterid had a long, segmented, spiderlike body. It grew to be as much as 10 feet long and is thought to have been a predator (American Museum of Natural History).

The eurypterids were among the largest animals of that period, and it appears from their claws that many were predators.

Insects, a class of arthropods, did not appear until the middle of the Devonian Period. Because they are so brittle, they have left very few fossil remains. The first insect is believed to have been wingless, and thus structurally different from today's insects. The first winged insect appeared during the Pennsylvanian Period. (See Figures 15.21 and 15.22.)

Scientists think that the coal forests supported most of the insect life. This is because there was a great abundance of food in these forests. The lush plant life also provided a source of protection. The insects that existed during this period were giant cockroaches and cricketlike insects, most of which are extinct today.

A great variety of insects had evolved by the end of the Paleozoic Era. Because of their ability to reproduce in great numbers, they are still the dominant life form today. They have undergone few structural changes since the Paleozoic Era and are one of the most successful forms of life.

The Paleozoic Vertebrates

Fossil remains indicate that vertebrate life first evolved during the mid-Ordovician Period. The oldest complete vertebrate fossils appear in the Silurian beds of England and northern Europe. The first vertebrate discovered belonged to the class Agnatha — primitive animals that resembled the present-day lamprey (Figure 15.23). During the Devonian Period better swimmers and feeders evolved, and the Agnatha began to disappear.

Other early vertebrates that inhabited the Earth during the mid-Paleozoic Period were the ostracoderms. It is believed that they spent most of their time swimming on the bottom of the sea as scavengers searching for food. They were probably not efficient swimmers when compared to our modern fish.

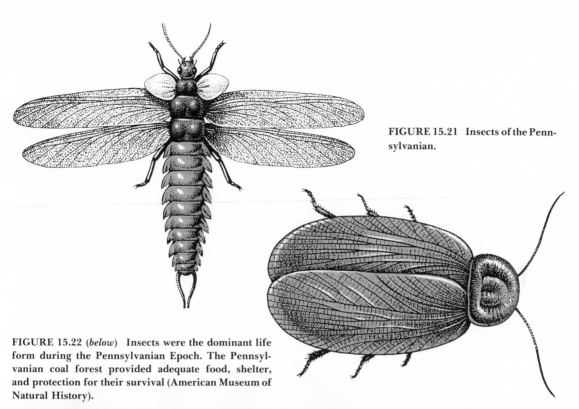

FIGURE 15.21   Insects of the Pennsylvanian.

FIGURE 15.22 (*below*)   Insects were the dominant life form during the Pennsylvanian Epoch. The Pennsylvanian coal forest provided adequate food, shelter, and protection for their survival (American Museum of Natural History).

One of the most important factors to consider in studying the evolution of the vertebrates is the development of teeth and jaws. The progress of this development is clearly seen in the group Ostracoderm. The early members of this group had gills located on the side of the head supported by cartilaginous gill arches. Scientists believe that the first two pairs of cartilaginous gill arches eventually developed into the upper and lower jaws, and the third cartilaginous gill arch developed into the teeth.

The first organisms equipped with lower jaws were a group of fish called the placoderms (Figure 15.24), which evolved during the late Silurian and Devonian Periods and became extinct by the close of the Paleozoic Era. It is interesting to note that they are the only vertebrates without descendants. But it has also been observed that, in physical appearance, the placoderms resembled today's shark.

The shark evolved during the Devonian Period, and has continued its evolutionary development up to the present. The shark's cartilaginous skeleton was previously thought to be primitive, but it is now generally accepted that the cartilage is a secondary development. The bony skeletons of the ostracoderms and placoderms were actually the primitive structure.

One shark discovered to have been domi-

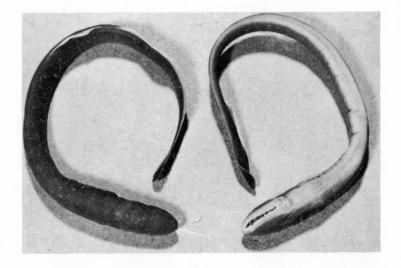

FIGURE 15.23 Vertebrate life first appeared during the mid-Ordovician Period. The first vertebrate belonged to the class Agnatha, primitive animals that resembled the present-day lamprey (U.S. Department of the Interior, Fish and Wildlife Service).

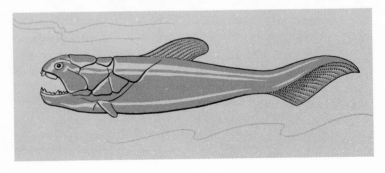

FIGURE 15.24 The placoderm was heavily armored in the region of its head, with jointed, bony plates. These fish grew to lengths of 30 feet.

FIGURE 15.25 The Arthrodire, a middle Devonian bony plated fish (American Museum of Natural History).

nant during the Devonian Period is the *Cladoselache*. In physical appearance it is similar to some of our present sharks.

Bony fish are the most successful vertebrate animals that evolved during the Devonian Period. Their successful development has been because of their ability to adapt to an aquatic environment.

After evolving from the placoderms, the bony fish divided into two groups: the ray-finned fish (Actinopterygii) and the air-breathing fish (Choanichthyes). The ray-finned fish did not give rise to any other vertebrate group. The air-breathing fish make up two important orders, the Dipnoi (lungfish) and the Crossopterygii (lobe-finned fish). The lobe-finned fish was never a dominant organism but is of particular significance because, like the modern lungfish, it was able to utilize oxygen from the air. Within this group we find the possible ancestors of the first land animals.

Fossils in late Devonian rocks from Greenland indicate that structural changes in the Devonian lobe-finned fish resulted in land animals. These modified animals are believed to have been the primitive amphibians. Further study of these Greenland fossils indicates that the primitive amphibians had weak limbs.

The Mississippian and Pennsylvanian Epochs were active periods in the development of the amphibian. Fossil remains indi-

FIGURE 15.26   Ichthyostega, the oldest known tetrapod (After Dunbar).

FIGURE 15.27   The *Seymouria* is an interesting intermediate animal that shows characteristics similar to both amphibians and reptiles (American Museum of Natural History).

cate a variety of amphibians, ranging in size from tiny salamanderlike animals to rather large crocodilelike organisms.

Since the Devonian Period, when amphibians began, they have changed structurally as well as modifying some of their natural processes. For example, the first amphibians laid their shell-less eggs in water. But as they became terrestrial animals, they changed structurally in order to survive on land. For one thing, they developed strong legs. For some, their eggs, no longer laid in water, developed a hard outer shell to prevent dehydration. The transition from aquatic animals to land animals resulted in the development of the reptiles. This change, made possible by the development of the shell-covered egg, occurred during the Car-

boniferous Period. The genus *Seymouria* (Figure 15.27) is one of the earliest reptiles. Although the oldest fossil egg found comes from the Lower Permian strata of Texas, scientists believe that reptiles must have existed for several million years before the development of the first land egg.

The most interesting reptile group of the Permian Period is the one that includes the ancestor of the mammals. This group includes the fin-backed and mammal-like reptiles. The fin-backed reptiles are represented by the genus *Dimetrodon* (Figure 15.28), a carnivorous reptile. The genus *Cynognathus* represents reptiles that possessed mammalian skeletal characteristics. It is believed that they might also have possessed hair and were warm-blooded. These mammal-like

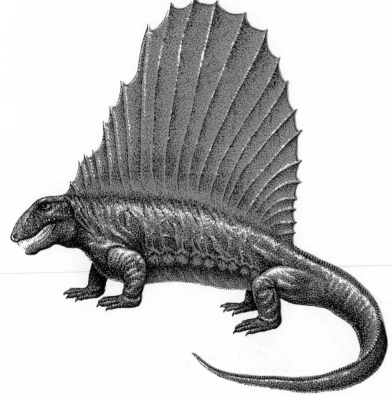

FIGURE 15.28 *Dimetrodon,* a fin-backed carnivorous reptile of the Permian Period (After Rudalph F. Zallinger).

reptiles developed in the Southern Hemisphere during the late Permian and Triassic Periods.

By the end of the Paleozoic Era several types of invertebrate life had disappeared, whereas others continued into the Mesozoic Era. Only those animals capable of surviving the different environmental conditions that developed endured to enter the Mesozoic Era.

Mesozoic Life

During the 140 million years of the Mesozoic Era many environmental changes occurred that affected the development of living things. As a result of these changes the land, sea, and air came to be dominated by the reptiles.

Dinosaurs, the most interesting reptiles, were the animal giants of this era. The only

FIGURE 15.29   A living cycad (The New York Botanical Garden).

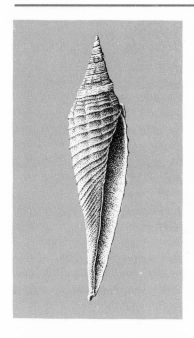

FIGURE 15.30 (*left*) *Volutoderma,* a representative gastropod of the upper Cretaceous Period.

FIGURE 15.31 (*below*) *Rhipidogyra,* a common coral of the Mesozoic Era.

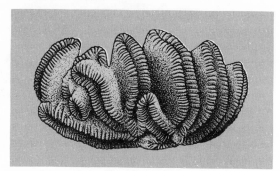

remains uncovered from this era are giant footprints and bones. With only footprints and bones to act as guides, the appearance of these giant reptiles has been left to man's imagination, except that some fossil eggs remain.

Mesozoic Plant Life

Mesozoic plant life begins with the survivors of the Paleozoic Era, among them ferns, bushes, conifers, and the earliest palm trees. The conifers and cycads (Figure 15.29) are the dominant plants that arose during this era. Evidence indicates that the famous petrified forest of Arizona was formed at this time.

One of the most important events in the evolution of plant life is the emergence of the angiosperm (seed-bearing plants). The origin of the angiosperm still remains a mystery. It has been suggested that they evolved from the cycads, but any final an-

swer must await a more complete record of the primitive angiosperms. The angiosperms gave rise to all of today's seed plants.

Mesozoic Invertebrates

The most successful invertebrates to continue developing during the Mesozoic Era were the mollusks. These invertebrates include the cephalopods, gastropods (Figure 15.30), and pelecypods. The ammonoids are the most characteristic group of this era, and underwent dramatic changes to become one of the most numerous groups of invertebrates of the Mesozoic Era. They continued to exist throughout the Mesozoic Era, as the abundance of ammonoid fossils shows.

During the Permian Period the ammonoids were very active. Toward the end of the Triassic Period changes in the marine environment resulted in the near extinction of the ammonoid group. The main survivor was the

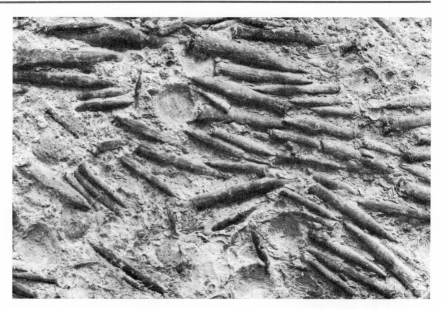

FIGURE 15.32 (*above*) The belemnoids were dominant during the Mesozoic Era. Fossil remains indicate that they were rod-shaped organisms. Their skeletal structure was composed of calcium carbonate. Because of their abundance, they are excellent index fossils (American Museum of Natural History).

FIGURE 15.33 (*right*) A Triassic pelecypod.

genus *Phylloceras*. This genus reproduced in sufficient numbers to survive the Jurassic and Cretaceous Periods. The ammonoids were most active during the Jurassic Period; their fossils are found in all parts of the world.

The corals (Figure 15.31) were present in certain parts of the sea during the Mesozoic Era, but some vanished by the end of the era. During the mid-Triassic Period a new coral appeared that possessed folded symmetrical septa. These septa still appear in today's coral.

The *Belemnites* (two-gilled cephalopods) developed successfully during the Mesozoic Era (Figure 15.32). Fossil remains indicate that they were rod-shaped and composed of calcium carbonate. Part of their skeleton is squidlike in structure. Geologists have concluded that the *Belemnites* were excellent swimmers, and therefore were dispersed throughout the world. Because of their abundance, the fossil remains of the *Belemnites* serve as good index fossils to Mesozoic rocks.

Pelecypods (axe-footed bivalve-shell mollusks such as clams, mussels, oysters, and scallops) were almost as numerous as the ammonoids (Figure 15.33). One species, the *Monotis subcircularis,* developed during the later Triassic Period, and fossils are found in all countries along the Pacific

Ocean. Another example, the *Exogyra* (oyster-like pelecypod) lived in North America along the Atlantic Coast to the Gulf of Mexico.

Mesozoic Vertebrates

The dinosaurs were the most abundant reptiles during the Mesozoic Era. Geologists originally believed that there was a single group of dinosaurs, but later discoveries proved that there were two major groups varying in height from 1 to 100 feet.

Dinosaurs are probably the most famous of prehistoric animals known to man. They originated in the Triassic Period, and are believed to have been comparatively small at

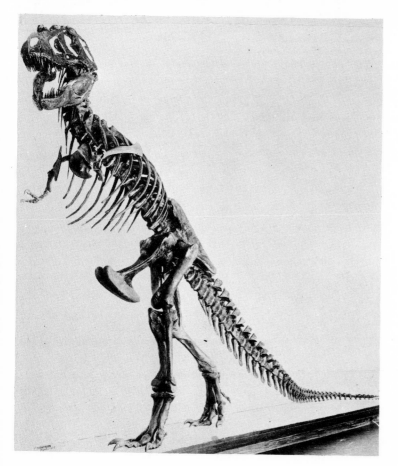

FIGURE 15.34 *Tyrannosaurus rex* was the largest, and is the most well known, of all carnivorous dinosaurs. At full growth it was between 30 and 40 feet long, 15 to 20 feet high, and weighed many tons (American Museum of Natural History).

FIGURE 15.35 The *Brontosaurus* was a herbivorous dinosaur. It was 50 to 80 feet long, had a long neck, and weighed about 50 tons. It lived primarily during the Jurassic Period and part of the Cretaceous Period (Peabody Museum of Natural History).

FIGURE 15.36 The *Triceratops* was a plant-eating dinosaur. It had two long horns projected forward from either side of its head and a third horn just above its nose. It was between 20 and 25 feet long (American Museum of Natural History).

FIGURE 15.37 The *Stegosaurus* was a large herbivorous dinosaur that had an armoured bony plate along its back (American Museum of Natural History).

first, reaching a maximum length of about 15 feet. In the Jurassic and Cretaceous Periods the most famous of the dinosaurs evolved: the larger *Tyrannosaurus, Brontosaurus,* and *Triceratops* (Figures 15.34 to 15.36).

During the Jurassic Period, reptilian dinosaurs having long tails and long necks were prominent. These lizardlike creatures walked on two legs, an adaptation that enabled them to move very quickly in the Jurassic plains and swamps.

Geologists classify the dinosaurs into two groups: the carnivores and herbivores. The *Tyrannosaurus rex* is probably the best-known carnivorous dinosaur. It is the largest of the carnivorous dinosaurs and, at full growth, was between 30 and 40 feet long, 15 to 20 feet high, and weighed many tons.

FIGURE 15.38 The *Trachodon* (duck-billed dinosaur) was a herbivorous, semiaquatic dinosaur. It walked on its hind feet and had as many as 2000 teeth in its mouth (American Museum of Natural History).

The carnivorous dinosaurs lived throughout the Mesozoic Era. But the well-known *Tyrannosaurus* did not make its appearance until the late Jurassic and early Cretaceous Periods. Carnivorous dinosaurs became dominant as the herbivorous dinosaurs declined near the end of the Mesozoic Era.

The herbivores (plant-eating dinosaurs) include the armored dinosaurs (*Stegosaurus,* Figure 15.37), horned dinosaurs (*Triceratops*). and duck-billed dinosaurs (*Trachodon,* Figure 15.38).

Fossil remains indicate that the *Stegosaurus* stood 30 to 40 feet long. Its body was covered by two long rows of great bony plates that stretched the length of its back.

The *Brontosaurus,* the heaviest of all the dinosaurs was between 50 and 80 feet long and weighed approximately 50 tons. The *Stegosaurus* and *Brontosaurus* lived during the Jurassic Period and the earlier part of the Cretaceous Period (Figure 15.35).

The *Triceratops,* a horned dinosaur, had two long horns projected forward from either side of the head and a third horn located above the nose. It was between 20 and 25 feet long (Figure 15.36).

The duckbills are known as the semiaquatic dinosaurs. They include the well-known *Trachodon,* the dinosaur that walked on its hind feet. Fossil remains indicate that the *Trachodon* had as many as 2000 teeth in their

FIGURE 15.39 (*right*) The *Plesiosaurus* was a marine reptile. It had a long neck and limbs that were modified flippers used for swimming (American Museum of Natural History).

FIGURE 15.40 (*below*) The *Ichthyosaurus* was a marine reptile. It had a streamlined body and a powerful tail to make it an efficient swimmer. The adult grew up to 10 feet long. The Ichthyosaurus resembles our modern-day porpoise (American Museum of Natural History).

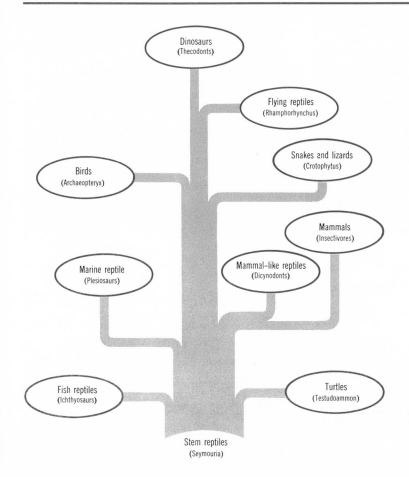

FIGURE 15.41 Reptiles, birds, and mammals arose from a common ancestor or stem reptile. From this ancestor the various types of each class evolved.

mouths. Both the *Triceratops* and *Trachodon* lived during the Cretaceous Period.

The Mesozoic Era was a very active stage for the dinosaurs. It was during this time that the pterosaurs (flying reptiles), plesiosaurs (those with long necks and limbs that were modified flippers), and ichthyosaurs (finned fishlike forms) evolved (Figures 15.39 and 15.40).

Toward the end of the Cretaceous Period the dinosaurs rapidly disappeared. Their decline has been attributed to a number of causes. Some dinosaurs were unable to cope with various diseases. Others, such as the herbivores, could not adapt to the rapid loss

of bush vegetation, while others declined because of climatic changes.

Not all of the dinosaurs were large. Some were only a few inches long. Most of man's attention, however, has been given to the larger dinosaurs.

Although birds seem to show little resemblance to reptiles, the two groups have many characteristics in common. Birds reveal reptilian ancestry in their features and in the manner in which they produce eggs (Figure 15.41).

One of the most valuable bird fossils is the *Archaeopteryx* (Figure 15.42). It possessed a skeletal structure similar to that of a reptile.

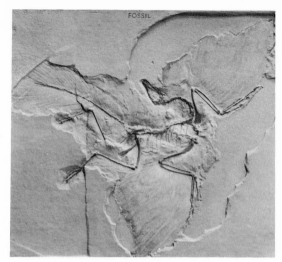

FIGURE 15.42   The Archaeopteryx is the earliest known bird. Its fossil is of particular importance to the geologist because it possesses skeletal structure very much like a reptile's (American Museum of Natural History).

FIGURE 15.43   Cretaceous landscape (American Museum of Natural History).

346

FIGURE 15.44 Many invertebrates that evolved during the early part of the Cenozoic Era are still present today. One of them, the sea urchin, is shown at left. Other examples appear in Figures 15.46 to 15.49 (Robert C. Hermes, National Audubon Society).

FIGURE 15.45 A "sand dollar" (C. G. Maxwell, National Audubon Society).

Geologists have not been able to find evidence that other birds existed during the Jurassic Period. But several fossils have been found which indicate that birds possessing modern structures existed during the Cretaceous Period. The information available to us regarding the evolution of birds during the Cenozoic Era is limited because of the lack of fossils.

Cenozoic Era

The Cenozoic Era makes up the last 70 million years of the Earth's history, and is the Era about which scientists know the most.

The Era has been divided into two periods and several epochs: the Tertiary Period contains the Paleocene, Eocene, Oligocene, Miocene, and Pliocene Epochs; and the Quaternary Period to date contains only one epoch, the Pleistocene. Man has only occupied about two million years of this era.

The great wealth of scientific knowledge about the existing life of this era comes from the fact that there has not been time enough for geologic upheavals to bury fossils very deeply in the Earth. And thus the evidence is readily available. Fossil remains provide more complete pictures of this era than any preceding time.

Cenozoic Plant Life

Many modern trees that developed during the Cretaceous Period were also present at the beginning of the Cenozoic Era. Then as now, the locations of trees depended on climatic variations, so that tropical vegetation grew in climates that were warm then but are no longer tropical today (Figure 15.43).

The diatom, a unicellular plant, was abundant in the Cenozoic seas. As they expired they sank to the bottom, and their shells accumulated in many places to form thick deposits of diatom ooze.

Cenozoic Invertebrates

Many invertebrates that are still represented today existed in the Cenozoic seas. Among them are many of the modern unicellular protozoans.

Sponges, corals, starfish, sea urchins, and sea dollars were common during this era, as they are today (Figures 15.44 to 15.48). Brachiopods and cephalopods were rare. The mollusks (clams and snails) were quite active during this era; they have increased both in number and type until the present day.

FIGURE 15.46 (*above*) Fire and antler coral (Carleton Ray, National Audubon Society).

FIGURE 15.47 (*top right*) Sponges (Robert C. Hermes, National Audubon Society).

FIGURE 15.48 (*right*) Starfish (G. Clifford Carl, National Audubon Society).

FIGURE 15.49   An adult lemur.
Notice the development of the paws
(Professor John Buettner-Janusch).

Insect remains are also abundant throughout the strata of the Cenozoic Era. The spiders, centipedes, and scorpions continued their development. Fields of vegetation provided food for moths, butterflies, wasps, bees, ants, and beetles.

Cenozoic Vertebrates

The fish of the Cenozoic Era resemble those of the Mesozoic. Sharks were abundant in the early and middle Cenozoic. Some were 70 to 80 feet long and had jaws up to 6 feet in width.

The amphibians during this era were numerous. They included frogs, toads, and salamanders—species which still exist today.

Reptiles were much as they are today, and included snakes, lizards, crocodiles, and turtles.

The birds became more numerous during the Cenozoic Era. They continued their structural development into the many modern forms.

Mammals

Mammals are warm-blooded animals that suckle their young, and their bodies are covered with hair. Primitive mammals first appeared in the Triassic Period. In the 100 million years that followed—the Mesozoic Era—evolutionary progress was slow. But, by the beginning of the Cenozoic Era, mammals had evolved from a position of obscurity to one of dominance. At that time, most mammals were small. Structurally, they are believed to have resembled our modern-day woodchucks. There were also small monkey-like mammals called lemuroids (Figure 15.49). We believe that they were the ancestors of our modern primates.

Many of the mammals that arose during the Eocene Epoch were rather small. These mammals include the rodents and monkeys. They, of course, were predominantly land animals but, before the Eocene Epoch ended, mammals evolved that lived in the sea. They were our first whales and porpoises. Other mammals such as the bats took to the air. The first horse, the *Hyracotherium,* evolved during this period. It was as large as a fox, had an arched back, front feet with four toes, and hind feet with three toes (Figure 15.50).

Many of the smaller animals of the Eocene Epoch that continued into the Oligocene Epoch began to grow during this epoch. The Camel, Horse, and Rhinoceros developed into large species. For example, the *Hyracotherium* preceded *Miohippus,* a larger form in which all of the feet had three toes. Monkeys gave rise to apes. New animals also developed during this epoch: pigs, dogs, cats, mice, beavers, rabbits, squirrels, and elephants.

Many animals that evolved during the Oligocene Epoch no longer exist. For example, the *Titanothere,* a distant relative of the *Rhinoceros,* and the *Brontotherium* (Figure 15.51), a slender-legged creature with a rhinoceroslike horned head and elephant-like body, lived for only a rather short period of time.

Toward the end of the Oligocene Epoch, animals about the size of sheep, resembling both deer and hogs, became extinct. These animals are known as creodonts and oreodonts.

During the Miocene Epoch grasslands developed and, as a result, grazing animals evolved. The Miohippus, Camel, and Rhinoceros grew to become very large animals that roamed the Earth. The *Miohippus* gave rise to the larger *Hypohippus.* Carnivorous dogs and cats became large. They continued to evolve to produce new species; the dogs produced wolves and foxes; and cats gave rise to leopards and saber-toothed tigers (Figure 15.52).

During the early part of the Pliocene Epoch, two types of horse evolved. First was the *Merychippus,* which was about the size of a pony. This animal had three toes of which the center toe was the only functioning one. Second was the one-toed horse, *Pliohippus.* The *Pliohippus* later gave rise to the *Equus* (modern horse) during the Pleistocene Period.

By the end of the Pliocene Epoch, the elephant and modern horse had made their way

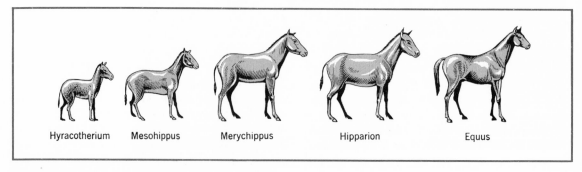

Hyracotherium   Mesohippus   Merychippus   Hipparion   Equus

FIGURE 15.50   The evolutionary development of horses. The horse arose from a much smaller animal about the size of a fox and has undergone several evolutionary changes to its present form.

FIGURE 15.51 The Brontotherium existed during the Oligocene Epoch. It had a very short life span. During its existence, it was one of the largest animals that roamed the continent of North America. It had a rhinoceroslike body and stood eight feet tall at the shoulder (American Museum of Natural History).

FIGURE 15.52 During the Miocene Epoch cats gave rise to the saber-tooth tiger. The saber-tooth tiger was the last of its species and became extinct during the Pleistocene Epoch (American Museum of Natural History).

FIGURE 15.53 The chimpanzee (a) and the gibbon (b) are classified as "manlike apes." Unlike the gibbon, the chimpanzee does not move swiftly in trees and spends most of its time on the ground. The chimpanzee walks in an upright position. The gibbon's forelimbs are longer than its hindlimbs, and it spends most if its time in trees. It is capable of moving rapidly in trees by swinging under limbs and by leaping from tree to tree (Bucky Reeves, A. W. Ambler, National Audubon Society).

into the pages of the evolutionary calendar. Each had finally succeeded the long line of ancestors that gave rise to the modern animals.

The Pleistocene Epoch marks the time of the Great Ice Age. During this period, many large animals became extinct: the great elephants, mastodons, imperial mammoths, wooly mammoths, giant ground sloths, giant armadillos, and saber-toothed cats.

Of all the mammals that have evolved to date, the primates are probably the most important (Figure 15.53). The anthropoid primates appeared some 30 to 40 million

years ago. They had well-developed limbs that enabled them to become good climbers. One branch of the anthropoid apes came to be ground dwellers. From this branch the apes evolved.

The early apes' forelimbs were equipped with thumbs for grasping and climbing. They learned to walk in an erect position, on two legs. As the apes encountered problems, their heads and brains increased in size. Before the beginning of the Pleistocene Period other primates had evolved that possessed brains two or three times the size of the ape's. Furthermore, these other primates were capable of using their hands and fingers to aid them in survival. Paleontologists believe that they, Ramapithecus, were the earliest example of man.

The Coming of Man

A complete understanding of anthropoid evolution is still to be developed. Human fossil remains 1¾ million years old have been uncovered in Tanganyika. Scientists hypothesize that the Ice Age may have been the prime factor in speeding the progress of man from the primitive cave dweller to present-day civilized man.

Unlike other primates, man is not adapted for swinging in trees like apes. He is not adapted for running or swimming; although he has these skills, they are not well developed. Nevertheless, with his intelligence to direct him, there is no other animal that can successfully compete with man.

It is man's brain that has made him more advanced than any other living thing. Man's intelligence led him to find shelter in caves. His power to reason lessened the chances of being trapped by larger animals. Moreover, his intelligence has allowed him to develop complex speech and to develop a written history.

Paleontologists can make some basic inferences about the evolutionary development of man. These inferences are based on the discovery of the remains of primitive animals and tools that have been located near a few skeletal structures of primitive man.

Eolith Man

The earliest artifacts of man are very crude flint utensils found in late Pliocene deposits in England. The crude flint utensils are called eoliths and the users of these utensils are known as Eolith Man. It has been concluded from the study of these utensils that the Eolith Man lived over one million years ago.

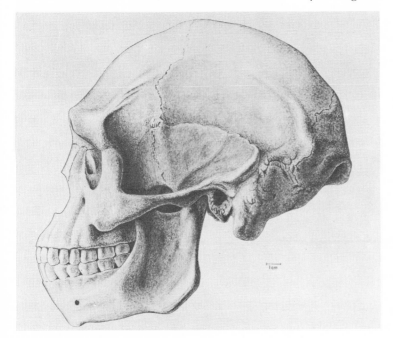

FIGURE 15.54 The Java man shows a low slanted forehead made of thick bone with protruding bones over the eye sockets. The facial features were definitely "apelike," but not as prominent as the ones found in apes (American Museum of Natural History).

Java Man (*Homo Erectus*)

In 1891 on the Island of Java, a Dutch Army medical officer discovered fragments of a skull, teeth, and bones which are believed to have been the remains of a man who lived more than 500,000 years ago. It has been proposed that Java may hold the answer to the development of man. Close examination of these bones show that this creature had a low slanted forehead made of thick bone with protruding bone over the eye sockets. These features are definitely apelike. But the protruding bones are not as conspicuous as those found in the apes. Further examination of the skull shows that the braincase was larger than that of an ape. Other bones indicate that Java Man walked in an upright position. Anthropologists conclude from this discovery that these are the remains of a primitive man. This specimen has been classified as *Homo erectus* and is more recent than Ramapithecus (Figure 15.54).

Peking Man (*Homo Pekingensis*)

In 1928 near Peking, China, anthropologists uncovered the remains of the *Homo Pekingensis* (Peking Man). The Peking Man's skull resembles the Ape Man's except that the braincase of Peking Man is larger. Rather sophisticated stone tools called paleoliths were found in this area, showing that Peking Man was more advanced than the Ape

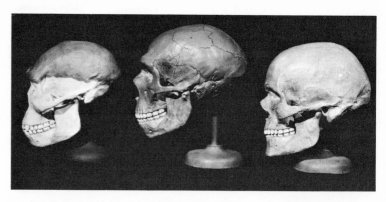

FIGURE 15.55 From the skulls of Pithecanthropus, Neanderthal, and Cro-Magnon man, scientists have been able to develop restorations of what these prehistoric men might have looked like. The apelike character of Pithecanthropus (left) is seen in the low forehead, heavy brow ridges, and chinless profile. The Neanderthal man (center) also had apelike characteristics, with a low forehead, heavy brow ridges, and a receding chin. A high forehead, prominent chin, and recessed brow ridges mark Cro-magnon man. The facial characteristics are very similar to those of modern man (American Museum of Natural History).

Man. It is from these paleoliths that the idea of an ancient Stone Age evolved.

Neanderthal Man (*Homo Neanderthalensis*)

The first primitive remains of man found in Europe were discovered in Germany, near the town of Neanderthal. Anthropologists believe the Neanderthal Man to be the most primitive of the true human race. The Neanderthal Man possessed more characteristics of modern man than the Ape Man, and was given the name *Homo neanderthalensis*. These were the first creatures classified as Man rather than Ape Man. Scientists believe that Neanderthal Man did not originate in Europe, but rather migrated to Europe from Asia.

It is believed that an advance of Pleistocene ice 100,000 years ago caused the extinction of Neanderthal Man.

Cro-Magnon Man (*Homo Sapiens*)

In the late 1800's, anthropologists discovered prehistoric skeletons in caves located in southern France. Since then, many similar skeletons have been discovered in France, Spain, and Italy. They have been classified as Cro-Magnon Man.

Like modern man, the Cro-Magnon Man was tall and erect. His skull was quite similar to that of modern man, showing the protruding forehead which indicated the presence of a large brain. This implies that Cro-Magnon Man had a brain similar to that of modern man. Moreover, Cro-Magnon Man has left other evidence of his intelligence: on the walls of his caves he made artistic drawings which still survive (Figure 15.59).

Cro-Magnon Man has been placed in the same species as modern man (*Homo sapiens*). And anthropologists assume that he is the direct ancestor of the people of southern Europe.

FIGURE 15.56 Pithecanthropus had a hairy "apelike" body (American Museum of Natural History).

Neolithic Stone Age

It has been proposed that Cro-Magnon Man made his appearance some time after Neanderthal Man, and somewhere between the last advance of the ice sheets and the end of the Pleistocene Period. It is believed that these men knew how to sharpen stones with abrasives. The manipulation of stone characterizes the Neolithic Period of man's history. Neolithic Man made many advancements over Cro-Magnon Man. During his period he began to train animals, built shelters, and learned to live within tribal communities.

Bronze Age

About 7000 to 5000 B.C., man learned to extract copper from its ore, and learned to manipulate and to work the copper into tools and weapons. Later man learned to make

FIGURE 15.57 **An artist's conception of a typical Neanderthal group (American Museum of Natural History).**

bronze, which is harder than copper. This marked the beginning of the Bronze Age.

When man learned to make iron, the Iron Age succeeded the Bronze Age. Iron is much harder than bronze and, therefore, is much more desirable for tools and weapons.

Undoubtedly, the discovery of these different processes did not occur at one time throughout the entire world. It is believed that Egypt was the first place they were discovered, and that from Egypt these processes were carried throughout the world.

Now modern man has passed through the Atomic Age and is standing on the threshold of the Space Age.

Modern Man

All human beings belong to the species *Homo sapiens.* Different types of humans are distinguished by different characteristics such as body build, bone structure, eye color, and skin color. Anthropologists recognize three basic racial groups of *Homo sapiens:* the Mongoloid, Negroid, and Caucasoid.

Mongoloid

The Mongoloid group is represented by people from Asia, the East Indies, and the Philippines. Mongoloids in the Western Hemisphere include the Eskimos and the Indians of North, Central, and South America. Anthropologists believe that the Mongoloid people came to North America from Asia by migrating through Alaska.

Negroid

Natives of North, South, and East Africa and the Congo are of the Negroid group. Negroid races are found also in New Guinea, the Philippines, and the islands surrounding Australia.

Caucasoid

People classified as members of the Caucasoid group are from Europe, southwestern Asia, and North Africa. Caucasoid people include types such as the Teutonic type of

FIGURE 15.58 Cro-Magnon man resembled modern man in that he was tall, erect, long-legged, had a prominent chin, double curvature of the spine, and a high forehead. His advancement over the Pithecanthropus and Neanderthal man can be determined from the Cro-Magnon paintings that still remain on the walls of caves (American Museum of Natural History).

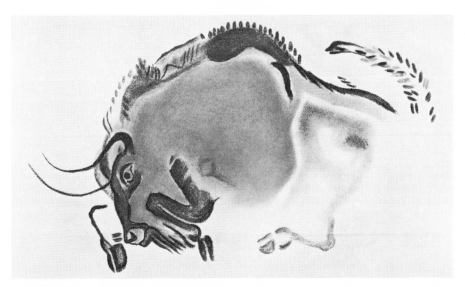

FIGURE 15.59 Modern man has been able to gain insight into prehistoric culture by studying the artwork left on cave walls, like this drawing of a bison found in a cave in Altamira, Spain (American Museum of Natural History).

Northern Europe, the Iceland type, the Southern Europeans, the Slavs, the Hindus, the Arabs, the Egyptians, and the Ethiopians.

Australoid

A fourth type of modern man, the Australoid, does not fit into any of the three major groups. This small group includes the original inhabitants or aborigines of Australia and Ceylon. These primitive people still live in Australia and maintain the life of the early Stone Age. A slanted forehead and prominent eyebrow ridges are characteristic of the Australoids, and suggest a relationship to ancient man.

CHAPTER 16

The Evolving Earth

THERE IS A GREAT DEAL OF EVIDENCE TO indicate that the Earth, just as the life forms which inhabit it, has changed. Some of this evidence is apparent from examination of fossils, but much has been obtained from other techniques available to paleontologists and geologists. As these techniques have been developed and refined, our understanding of the history of the Earth and its life has undergone many changes.

Modern science has developed methods of examining rocks and rock structures that delve into the most fundamental aspects of the Earth's history. These techniques go beyond the mere recognition and identification of fossils as remains of once dominant life forms. Today, the ancient climates, continental structures, magnetic fields, and their associated changes have been recreated—at least on a theoretical basis. Like fossils, the structural formations within rock, or the sediments within the oceans, and the continental masses themselves all yield clues about the Earth's past history.

The Earth not only underwent dramatic changes during its early formation but has changed slowly and steadily ever since. Our planet has evolved, just as its life forms have evolved, throughout the last four billion years.

The History of Fossils

Before geologists and paleontologists discovered fossils and studied them, the chronological history of the Earth was not well understood. As a result, people also did not understand how the Earth developed, how the Earth changed, how life developed, and how life changed. Only with the advent of systematic study of fossils came an understanding of the geological and evolutionary calendars.

From the discovery of fossils came methods and techniques for studying the processes of the Earth's history and the evolution of life.

Controversy over Fossils

Herodotus, a famous Greek historian, was probably the first to understand that fossils were the remains of marine life left from the ancient seas. He was able to correlate the relationship between rocks and the imprints found in these rocks. But Herodotus was very much alone. Even Aristotle, the great philosopher, could not see that these figures were the imprints of ancient animals that once lived in the ocean or roamed the land.

In the Middle Ages, it was commonly held

359

FIGURE 16.1 William Smith (Radio Times Hulton Picture Library).

that fossil remains were not imprints of living organisms that had existed on the Earth, but rather were the worlk of the devil, designed to confuse people and to destroy their belief in God.

During this time the church proposed another origin for fossils: that they were a message from the Creator. The church proposed that when the Earth was formed, fossil remains were also formed to give man an indication of what type of life was to be created in the future.

However, the church also suggested that fossils were an indication as to what life was like when animals and plants first evolved. This hypothesis is true to a point, and is the same hypothesis on which our scientists base their study of the Earth. There was one basic difference between the hypothesis proposed by the church and that proposed by today's scientists. The church believed that fossils were the first living things to be created by

God, but that he was discontented with what He had created and, thus, destroyed these animals and plants by replacing them with the animals and plants known at that time. It was not until the 1800's that the basic modern concepts about fossils were accepted. These modern concepts owe much to the work of an Englishman named William Smith.

The Law of Faunal and Floral Succession

William Smith, the son of a farmer, was born in 1769 in Oxfordshire, England (Figure 16.1). He grew up in an area where there was a great abundance of fossils, became interested in them, and began studying and recording information about them. His interest in fossilization continued throughout his life. He traveled throughout England, obtaining specimens and gathering information about them. He kept notes showing where he found the fossils, the type of fossil found, and what type of rock contained the fossil. He also made geological maps showing the rock strata in areas he had covered, and what types of fossils he found in them. From his work, he proposed the Law of Faunal and Floral Succession, which states that strata contain fossils in the same succession everywhere, and that fossils can be identified in different strata. Thus, he concluded that the strata found on the east coast of England were the same as those found on the west coast, a pattern that could be followed across England. He also made an important conclusion: that each layer of rock in the historical development of the Earth could be identified by the fossils associated with that layer. His conclusions opened up a completely new method of studying the Earth's history.

The Theory of Continental Drift

For many years the Theory of Continental Drift has been a matter of fierce debate. Some scientists believe that the continents have always been separated. Others believe that the continents were, at one time, joined together, and then eventually "drifted" apart. Within the past four years, the controversy has become more prominent, as new evidence has been found to support the idea of continental drift.

The idea that the continents in the Western Hemisphere at one time were joined to Europe and Africa is not a new one. It was proposed as far back as 1620 by Francis Bacon (Figure 16.2). A half century later, P. Placet proposed the idea that America was part of the other continents. In 1858, Antonio Snider proposed that the continents were all

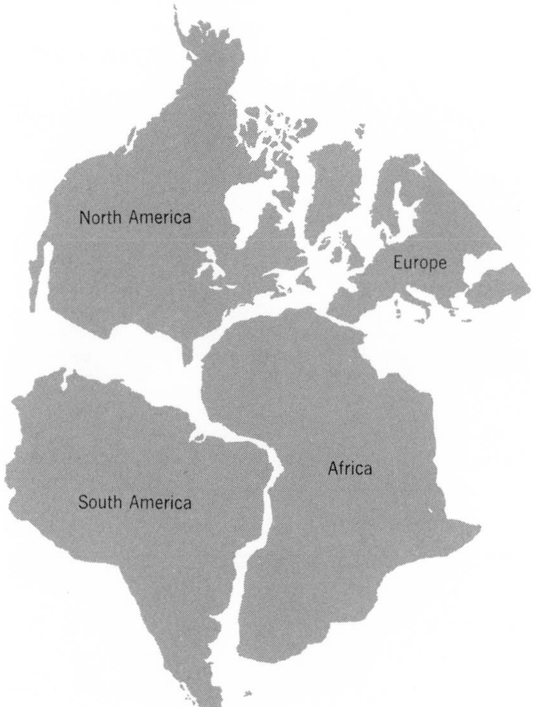

FIGURE 16.2 Francis Bacon developed the concept that the continents which are presently in the Western Hemisphere — South and North America — were once joined to Europe and Africa as one supercontinental mass.

joined at one time. His proposal was based on the similarities between North America and European fossils of the Carboniferous Period.

Toward the end of the nineteenth century, Edward Suess, an Austrian geologist, noticed that there was a close relationship between the geological formations in South America and Africa. He stated that the landforms of the Southern Hemisphere could fit together to form a single continent. He called this continent Pangaea.

In the early 1900's, F. B. Taylor, an American geologist, and Alfred Wegener, described ways in which lateral displacements of the Earth's crust could have occurred to move the continents apart.

Wegener's work has remained a point of lively discussion. He observed the similarities in geology and paleontology on both sides of the continents that face the Atlantic Ocean. Based on the observations and information he collected, he proposed that, at one time, the continents were all joined together to form one large land mass. He had

reached the conclusion that this large land mass existed 200 to 225 million years before the Mesozoic Era. Wegener referred to this large land mass as the *Pangaea,* after Suess.

Today, geologists disagree somewhat with Wegener's theory. They believe that instead of one large land mass, there were two large land masses, one in the Southern Hemisphere and the other in the Northern Hemisphere. The mass in the Southern Hemisphere is referred to as *Gondwanaland,* and that in the Northern Hemisphere as *Laurasia.*

However, evidence has been found in the Southern Hemisphere to support Wegener's theory. This evidence has been provided by the glaciers that occurred during the Permian and Carboniferous Periods. In the southern parts of Africa, Australia, South America. Madagascar, the Peninsula of India, and in Antarctica, the glaciers left behind tillite (old glacial material) and rare fossil forms of ancient life. Geologists have studied this tillite and found evidence to support the theory of one continental mass. (See Figure 16.3.)

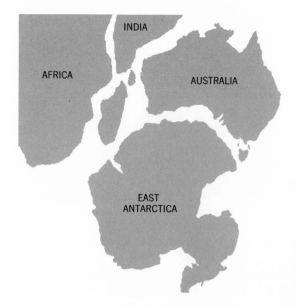

FIGURE 16.3 The Gondwana formation, a supercontinent envisioned by Wegener, contained the land masses that were to become India, Australia, Madagascar, Africa, and part of Antarctica.

FIGURE 16.4   A reconstruction of a *Glossopteris.*

Some Permian and Carboniferous tillite beds contain fossil remains of two plants that were in the peak of their development from the Devonian to the Triassic Periods. These two plants are the *Glossopteris* and the *Gangamopteris* (Figure 16.4). Scientists believe that it is almost impossible for similar Gondwana plant fossils to be located on different continents unless they were at one time all connected.

Thus the evidence provided so far not only indicates that continents of the Southern Hemisphere were joined together but also indicates that these land masses were connected to Antarctica.

Not all scientists believe in the Theory of Continental Drift. Many were put off by the arguments of Sir Harold Jeffreys, an English geophysicist. He based his opposition to the theory on the idea that the Earth's crust and mantle were too rigid to allow the movement of the continents. However, in the late 1930's, F. A. Vening-Meinesz, a Dutch geophysicist, contradicted Jeffreys' theory. He held that it would be possible for the continents to move apart under the influence of convection currents taking place in the Earth's mantle. His hypothesis was based on data that he collected while study-ing the Earth's gravity over the deep sea trenches and island arcs of the western Pacific Ocean.

Evidence of Continental Drift

The Earth's surface is divided into two levels: first, the continental surface and second, the oceanic surface. The continental surface is composed of six blocks of approximately equal area with an age of about 3000 million years. In Africa, there are older *cratons* (ancient nuclear areas) that are surrounded by belts of younger rocks. The young rock belts range in age from 600 million years to the present. Even though the younger rock belts have new material, they also contain older rock blocks of the same age as the cratons. The presence of the older cratons suggests that the Earth's surface has passed through periods of warping and folding around the ancient continental land masses, accompanied by the intruding and folding of young igneous rocks. Much of the older material has changed far beyond recognition, although some has remained untouched. These hypothetical young rock belts, referred to as *zones of rejuvenation,* are thought to have eroded, leav-

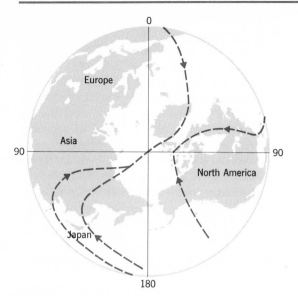

FIGURE 16.5 Rocks in which iron particles are lined up with the magnetic poles show evidence of polar wandering. This apparent wandering may be the result of continental drift (After Cox and Doell).

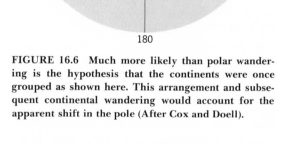

FIGURE 16.6 Much more likely than polar wandering is the hypothesis that the continents were once grouped as shown here. This arrangement and subsequent continental wandering would account for the apparent shift in the pole (After Cox and Doell).

ing bare the continental platforms. Close observation of the African cratons has revealed preexisting mountain belts that have been divided into segments by younger material intruding into older, structurally formed patterns.

The continents of the Earth are surrounded by earthquake zones, volcanoes, folded rock, and faults. Within recent years, a great number of displacements have been recorded on seismographs. In the future, the displacement of the Earth's solid surface can be recorded, traced, and the direction of the displacement detected. Thus it would be possible for the seismograph to determine large-scale motions of material in the Earth's interior, and evidence for or against the theory.

The ocean floor is covered by a thin layer of sediment whose deposition began in the Cretaceous Period, thus dating back about

100 or 200 million years as compared to the continental and oceanic history of 3,000 million years. The question is, why is there so little sediment? An answer to this question was put forward in 1960 by Harry H. Hess and Robert S. Dietz. They propose that the ocean ridge and rift systems were formed by increasing currents of material that spread out to form new ocean floors. They suggest that sediments were swept along the ocean floor, leaving the ocean floor with a relatively young appearance.

Paleomagnetism

In the last 20 years, research in the study of rock magnetism (paleomagnetism) has helped scientists investigate the Theory of Continental Drift. Rock magnetism is a property of rocks containing small amounts of iron. These rocks can be formed by crystallization from a melt, or by precipitation from an aqueous solution. When the rocks form, the iron deposits are magnetized in the direction of the Earth's magnetic field.

If the iron-bearing rock remains untouched from the time when it is first formed, its magnetism will indicate the direction of the Earth's magnetic field at the time it was formed. Thus, by determining the direction of the magnetism in the rock, a scientist can obtain the Earth's magnetic direction at the time the rock took shape.

Scientists have done much work in this area. They have found that, going backward in time, the rocks of each continent indicated a different position of the poles. (See Figures 16.5 and 16.6.) This finding has resulted in the idea that the continents have moved in respect to the present position of the Earth's magnetic poles. The fact that the magnetic poles of each hemisphere are different, suggests that the continents have moved apart. It is unlikely that the magnetic poles have changed greatly from the axis of the Earth's rotation, or that the axis of rotation has changed in relationship to the Earth's mass. The study of paleomagnetism strongly suggests that the continents have moved over the Earth's surface. This magnetic evidence supports the idea of continental drift. Since it suggests that the continents in the Southern Hemisphere have shifted southward,

this evidence specifically supports the Theory of Gondwana Formation.

Many scientists have thought that the information obtained thus far from the study of the Earth's paleomagnetic properties was not enough to establish the Theory of Continental Drift. In 1966, however, at the annual meeting of the Geological Society of America, dramatic new support was provided by a series of papers showing the relationship of ocean-floor spreading, continental drift, the cause of oceanic ridges, fault systems, and the direction and time of drift motions. (See Figure 16.7.)

New Evidence to Support Continental Drift

Newer evidence since 1966 has done much to substantiate continental drift. Discovery of mid-oceanic ridges and sea-floor spreading have been of dramatic importance, as have been the equally exciting successes of the drilling vessel Glomar Challenger. Significant pieces of biological evidence have appeared as well, chief among which was the discovery of the fossils of amphibians in Antarctica.

In 1964, J. E. Everett and A. G. Smith, from the University of Cambridge, presented some more support for the Theory of Continental Drift. They tried matching the continents on both sides of the Atlantic Ocean together, taking the central depth of the

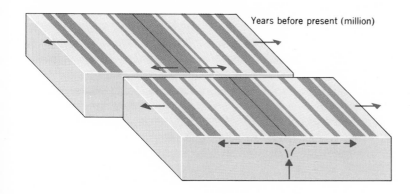

Years before present (million)

FIGURE 16.7 Evidence from the sea floor points to a welling up of material near the rift valley. As one nears the continental masses, the material on the floor becomes older, indicating a slow spread of the sea floor (After Hurley, 1968).

continental slope to represent the true edge of the continent. They gathered mathematical data that were programmed into a computer. After analyzing their results, they found that their margin of error was less than one degree.

A recent geological research project into the theory of drift took as its starting point the fact that a geological boundary line exists between the 2000-million-year-old geological province underlying parts of Ghana and the Ivory Coast, from these countries, and

FIGURE 16.8 Geological matching of South America and Africa as they looked 200 million years ago.

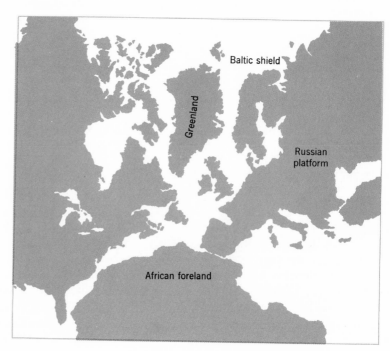

FIGURE 16.9 Although the North Atlantic continental masses are difficult to match, evidence appears that these regions were once part of the same continental mass.

the 600-million-year-old province under parts of Dahomey and Nigeria. The boundary turns south to the ocean near Accra in Ghana. The scientists on this project proposed that if South America had been connected to Africa, the boundary should enter Sao Luis in South America. After studying the rocks in Sao Luis, 2000-million-year-old rock was found on the west and 600-million-year-old rock on the east boundary. Thus, it was concluded that part of the 2000-million-year-old craton found on the west coast of Africa was present in South America. (See Figure 16.8.)

As Figure 16.9 shows, no such divided geological provinces have been found in the Northern Hemisphere. The breaking up of a continent appears to occur in areas of rejuvenation between cratons, as if these areas were zones of weakness. Only in the Southern Hemisphere does the break transect the continent through age provinces.

It appears that the Northern Hemisphere was transected obliquely by Paleozoic belts. These belts produced the areas of the Appalachian Mountains. The Maritime Provinces of North America, with an overlap along the west coast of Africa, then split into 2 belts. The first belt passes through the British Isles along the Atlantic coast of Scandinavia and Greenland, and the second belt turns east toward Europe. Close examination indicates the presence of four periods on both sides of the Atlantic Ocean, making the study of the Northern Hemisphere very difficult.

Gondwana Breakup

Drillings of the west coast of Africa have provided evidence as to when the Gondwanaland may have split up. These drillings indicate that the African sediment is quite young, possibly of middle Mesozoic origin of some 160 million years ago. If the south Atlantic

had been in existence since the beginning, Africa would have a large shelf of older sediment.

It appears that the rift in Gondwanaland started from the northern edge of western Africa in the mid-Triassic Period and slowly opened toward the south until the final separation occurred in the Cretaceous Period. Recent drillings indicate that the east coast of Africa opened in the Permian Period.

Narrow Land Bridge Concept

Some geologists still object to the Theory of Continental Drift. They cannot fully accept the theories that have been proposed to explain how it would be possible for great continents to divide. But even those who do not accept the total Continental Drift theory believe that some form of connection must have existed among the continents. There is no other way to explain the similarities and differences of life and geology on the different continents. These scientists have proposed that instead of a large land-mass connection, narrow land bridges were once present to connect the continents, probably during the early Paleozoic and late Mesozoic or early Cenozoic Eras (Figure 16.11). According to this hypothesis, Antarctica was connected to South America by a land bridge that extended northeast and included the Palmer Peninsula, South Orkney, South Sandwich, and South Georgian Islands. The bridge then proceeded in a western direction to the Falkland Islands and was connected to Patagonia. It has also been proposed that South America and Africa were also connected by a land bridge extending from Brazil to Africa, and that Africa and India were connected by a land bridge that followed a course northeast. Land bridges, they hypothesize, also connected Australia and Indonesia to the Asiatic continent.

The Migration of Life

If the continents were a solidified unit, or if they were joined together by land bridges, it is possible to understand how similar life forms could be found on different continents. If we assume that, at one time, the continents were a solidified structure, it is easy to understand that, when the continents broke apart, similar life forms would be living on different continents. It is also easy to understand that some organisms are found only on particular continents because they evolved after the continents divided. Similarly, if we accept the narrow land bridge as a possible explanation for similar life forms on different continents, it is also easy to consider that animals migrated and were scattered among the different continents when the bridges collapsed.

On the other hand, plants have no means of locomotion. In order to explain the appearance of the same plant species on different continents, those who accept the land-bridge theory suggest that animals or wind carried the pollen or spores from the plants.

One of the most interesting characteristics of plants is their worldwide distribution. Plant fossils similar to European types have been found in permocarboniferous rock of the eastern United States and Asia.

Fossil remains indicate that the land plants had their beginnings in the Silurian Period, and remained similar throughout the world from the Silurian Period through the early Carboniferous Period. It was during the late Carboniferous Period that plants began to diversify. The most recognizable plant of this period is the *Glossopteris*. It is the distribution of this plant in South America, Falkland Islands, Antarctica, South Africa, Madagascar, the Indian Peninsula, and Australia which suggests that all these land masses at one time may have been connected.

The Development of Fossils

The examination of fossils, allows scientists to trace the evolution of the plant and animal kingdoms back to the Cambrian Period, and almost to the beginning of life. But there are no remains from the time before the Paleozoic Era. All indications are that life began sometime during the Precambrian Period. However, only a few primitive Precambrian fossils have been located, and little is known about this period.

In spite of the lack of Precambrian fossils, geologists have gained insight into this period by analyzing the physical, biological, and chemical conditions under which

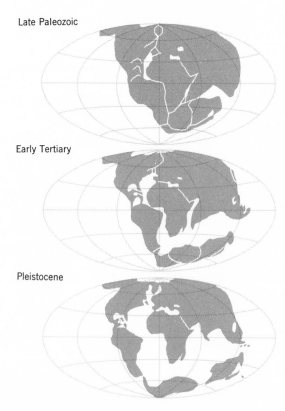

Late Paleozoic

Early Tertiary

Pleistocene

FIGURE 16.10   The drift of the continents as proposed by Wegener (After A. Wegener, 1912).

living things developed. They have produced hypotheses about the types of living things that may have evolved during the Precambrian Period.

Precambrian Fossil Records

Our knowledge about the Precambrian Period is derived by reasoning back from what we know about the Cambrian Period. During the Cambrian Period, all the major invertebrates existed, as did some of the more advanced forms. A large number of the living organisms of this period were so simple that they could have occurred by the conversion of inorganic chemical compounds to living organic compounds. This conversion would have taken place by the end of the

Precambrian Period. Fossils give evidence that multicellular life began no less than one billion years ago, although it is possible that simple life existed a billion or more years before that.

That life did exist in the Precambrian Period is shown by the fossil remains found in limestone structures called *stromatolites*. These fossils are the remains of one-celled plants called Cyanophyceae (blue-green algae) and can be found in all areas of the world. The Cyanophyceae evolved throughout the Paleozoic, Mesozoic, and Cenozoic Eras, and they continue to exist today. Some of these fossils have been discovered by Lake Michigan; some are encased in silica and have retained some of the complex organic compounds they formed millions of years ago.

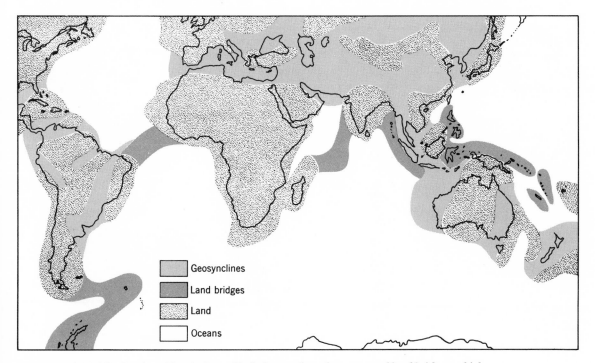

FIGURE 16.11   The Southern Hemisphere. Shaded areas show the concept of land bridges, which have been hypothesized as connecting the various land masses.

The Cambrian Period is characterized by the appearance of trilobites, which belong to one of the more complex invertebrate groups, the arthropods. The early Cambrian Period saw the appearance of other types of animals, including jellyfish, sponges, brachiopods, and arthropods.

The following is a brief description of the organisms that are valuable to geochronology as index fossils.

Protista Index Fossils

The protists are one-celled organisms, some of which possess characteristics that are both plant and animal. The foraminifers were the most abundant Protista during the post Precambrian Period. They first appeared in the late Ordovician Period, and did not become important until the late Mesozoic Era. One group of Protista, the foraminifers, provide scientists with information about the Precambrian, Pennsylvanian, and Permian Periods. Another group of the Protista, the *Globigerina*, makes up approximately 50 per cent of the modern ocean floor.

Invertebrate Index Fossils

The ancient record of the porifera is lengthy but not abundant. They have proved

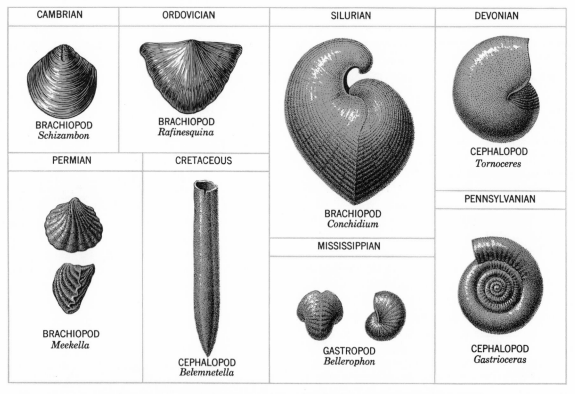

FIGURE 16.12 Brachiopods are the most common index fossils. Numerous examples of this class of animals remain because of the hardness of their shells. Some of the more common fossils from each period are shown.

useful in identifying Cambrian sponge reefs.

The formation of coral reefs was continuous throughout the Paleozoic and Mesozoic Eras; it began in the Ordovician Period and continues up until the present day. The fossil coral reefs have been studied extensively because they are frequently located near oil reservoirs.

There are three orders of corals: the Rugosa, Tabulata, and Scleractinia. The Rugosa and Tabulata were active reef builders during the Paleozoic Era. The Scleractinia appeared during the Triassic Period and reached the summit of their evolutionary development in the Triassic Period. The corals have been excellent index fossils in geochronology.

The brachiopods are especially useful for the correlation of the early Paleozoic Era. At that time, they were extremely abundant. Few brachiopods remain in our modern seas.

Mollusks serve as excellent stratigraphic tools. A series of mollusks that are often used in geologic dating are shown in Figure 16.12. The cephalopod fossils' zone is the marine strata of the Mesozoic Era throughout the world. This group of animals, which originated during the Cambrian Period, is still present today.

Other molluskan fossils are also useful. The pelecypods, which appeared during the Ordovician Period, have had wide distribution and are good paleoecological tools. At present, the gastropods appear to be at the peak of their evolutionary development, and are now facing the end of their evolutionary development. They first appeared in the lower Cambrian strata. Their fossils have served as good stratigraphic tools.

It was during the beginning of the Devonian and late Silurian Periods that the ammonoids evolved. Their evolutionary development was rather slow, and they experienced their decline toward the end of the Triassic Period. One group of ammonoids

that survived this near extinction of the group had a rapid evolution during the Jurassic and Cretaceous Periods, and then became extinct. These cephalopods serve as excellent stratigraphic tools of the Mesozoic Era.

The echinoderms include such organisms as the crinoids, blastoids, and cystoids. These organisms were prevalent during the Paleozoic Era. The cystoids, blastoids, and crinoids first appeared during the Ordovician Period. The cystoid and blastoid remains are also in strata of the Paleozoic Era. Several groups that evolved from the crinoids are still present today as are crinoids.

The starfish has evolved from the Ordovician Period up to the present. However, it has left few fossil remains and, therefore, cannot be used in geochronology. On the other hand, the sea cucumbers have left fossil forms, because of their calcareous plates, from the Ordovician Period up to the present day.

The sea urchin first appeared in the Ordovician Period, was especially numerous during the Jurassic Period, and is still present today. Their fossils have been used as stratigraphic tools in identifying the Late Mesozoic and Cenozoic Eras.

The oldest fossils of the arthropods are the trilobites. Remains have been found from the lower Cambrian Period to the Permian Period, when the trilobites became extinct. These fossils are important in identifying the Cambrian and Ordovician deposits.

Chordate Index Fossils

The most useful stratigraphic tool in identifying the Ordovician and Silurian Periods has been the graptolites, a group thought to be related to the hemichordates. They first appeared in the Middle Cambrian Period, and disappeared toward the beginning of the Carboniferous Period. These fossils have aided geologists in identifying the Ordovician and Silurian strata.

Vertebrate Index Fossils

The chordates, the first of the vertebrates, made their appearance during the Ordovician Period. It was not until the end of the Devonian Period that the amphibians first appeared.

The reptiles appeared in the early part of the Pennsylvanian Period. Late in the Paleozoic Era the Age of Reptiles was born; this was the time of great reptilian diversification when the reptiles became land, sea, and air travelers. Toward the end of the Mesozoic Era, many of the reptiles became extinct. The reptile fossils have aided in the identification of Mesozoic strata. Together with the fossils of mammals and birds, which first appeared in the Jurassic Period, reptile remains have played important parts in telling us something about continental formation during the Cenozoic Era.

Plant Index Fossils

The first simple aquatic plants evolved during the latter part of the Precambrian Period; more complex water plants and the first land plants appeared during the early Paleozoic Era.

The first land plants, the psilopsida, appeared during the Silurian Period. Structurally, they were not as complex as today's plants.

During the Devonian Period great forests evolved. Plants present at this time were the club mosses, horsetails, ferns, and gymnosperms. During the Mesozoic Era the gymnosperms were dominant. The angiosperms first appeared during the beginning of the Cretaceous Period and have remained until the present. The gymnosperms have since become prevalent as conifers.

Fossil variations have aided geochronology and acted as stratigraphic aids in telling geologists something about when rocks and fossils appeared. These fossils have tied together many loose ends to give man some insight into the evolution of the Earth and life.

Questions

CHAPTER THIRTEEN

1. How can the theory of abiogenesis be used to explain the origin of life from outer space?

2. How do the autotroph and heterotroph hypotheses differ?

3. What are the basic differences between the heterotroph hypothesis and the theory of abiogenesis?

4. Why are the nitrogen bases so important in explaining the origin of life?

5. What is the most important loophole in the autotroph hypothesis as a means of explaining the origin of life?

6. Explain how Dr. Miller's work supports the heterotroph hypothesis.

7. After reading this chapter, how do you think life began? What form of life do you think was the first to develop? Provide evidence to support your hypotheses.

CHAPTER FOURTEEN

1. In spite of the Law of Superposition, geologists encounter problems in learning about the Earth's history. Why?
2. How is it possible to tell the difference between marine and terrestrial land masses? What causes these differences?
3. Discuss some of the problems that geologists have encountered in geological correlation.
4. Discuss in order, the geological changes of North America during the Paleozoic, Mesozoic, and Cenozoic Eras.
5. What is a geosyncline? Explain its origin.

CHAPTER FIFTEEN

1. Why is there limited information about the Precambrian Era?
2. What are some reasons for the disappearance and extinction of various types of flora and fauna?
3. What types of fauna have been particularly helpful in unraveling the history of animal life?
4. Why have plants not been useful as stratigraphic tools in decoding the Earth's history?
5. What has been the main reason for man's ability to survive and become the most complex and highly developed organism in the animal kingdom?
6. Why have scientists proposed the idea that marine fauna was the first type of life to evolve?
7. Discuss, in chronological order, some of the changes that came about in the vertebrates.

CHAPTER SIXTEEN

1. Describe how fossils are formed. Give examples.
2. What is meant by the term "geochronology"?
3. How is the Law of Faunal Succession related to explaining the Earth's history?
4. Discuss the Theory of Continental Drift. Give evidence to support this theory.
5. How is paleomagnetism related to the Theory of Continental Drift?
6. Why has there been much difficulty in gathering evidence to support the Theory of Continental Drift in the Northern Hemisphere?

Bibliography

BOOKS

Buettner-Janusch, J., *Origins of Man: Physical Anthropology*, John Wiley & Sons, New York, 1966.

Carles, Jules, *The Origins of Life*, Walker, New York, 1963.

Clark, T. H., and Stearn, C. W., *Geological Evolution of North America*, 2nd Ed., The Ronald Press, New York, 1968.

Dunbar, C. O., and Waage, K. W., *Historical Geology*, 3rd Ed., John Wiley & Sons, New York, 1969.

Ericson, D. B., and Wollin, G., *The Deep and the Past*, Knopf, New York, 1964.

Grant, Verne, *The Origin of Adaptation*, Columbia University Press, New York, 1963.

Jukes, T. H., *Molecules and Evolution*, Columbia University Press, New York, 1966.

McAlester, A. L., *The History of Life*, Prentice-Hall, Englewood Cliffs, N.J., 1968.

Phinney, R. A., *The History of the Earth's Crust*, Princeton University Press, Princeton, N.J., 1968.

Stokes, W. L., *Essentials of Earth History*, 2nd Ed., Prentice-Hall, Englewood Cliffs, N.J., 1966.

Vening-Meinesz, F. A., *Earth's Crust and Mantle*, Elsevier Publishers, New York, 1964.

Woodford, A. O., *Historical Geology*, W. H. Freeman and Co., San Francisco, 1965.

PERIODICALS

Bullard, E., "The Origin of the Oceans," *Scientific American*, **221,** September, 1969.

Calvin, M., and Eglinton, G., "Chemical Fossils," *Scientific American*, **216,** December, 1967.

Emery, K. O., and Milliman, J. D., "Sea Levels During the Past 35,000 Years," *Science*, **162,** December, 1968.

Cloud, Preston, "Atmospheric and Hydrospheric Evolution on the Primitive Earth," *Science*, p. 729, May, 1968.

Ericson, D. B., and Wollin, G., "Pleistocene Climates in the Atlantic and Pacific Oceans," *Science*, **167,** March, 1970.

Herz, N., "Anorthosite Belts, Continental Drift, and the Anorthosite Event," *Science*, **164,** May, 1969.

Hurley, P. M., "The Confirmation of Continental Drift," *Scientific American*, **218,** April, 1968.

Kornberg, A., "The Syntheses of D.N.A.," *Scientific American,* **219,** October, 1968.

Leakey, L. S. B., "Exploring 1,750,000 Years into Man's Past," *National Geographic,* **120,** No. 4, 1961.

Lear, John, "The Bones on Coalsack Bluff, a Story of Drifting Continents," *Saturday Review,* February, 1970.

Miller, Stanley L., "Production of Amino Acids Under Possible Earth Conditions," *Science,* **117,** May 15, 1953.

Silver, E. A., "Late Cenozoic Underthrusting of the Continental Margin Off Northernmost California," *Science,* **166,** December, 1969.

Simons, E. L., "The Earliest Apes," *Scientific American,* **217,** June, 1967.

Vine, T. J., "Spreading of the Ocean Floor: New Evidence," *Science,* **154,** September, 1966.

Washburn, Sherwood L., "Tools and Human Evolution," *Scientific American,* September, 1960.

Scientific American, September, 1970.

    Articles by G. Evelyn Hutchinson, "The Biosphere."

               Abraham H. Oort, "The Energy Cycle of the Earth."

               George M. Woodwell, "The Energy Cycle of the Biosphere."

               H. L. Penman, "The Water Cycle."

               Preston Cloud and Aharon Gibor, "The Oxygen Cycle."

               Bert Bolin, "The Carbon Cycle."

               C. C. Delwiche, "The Nitrogen Cycle."

               Edward S. Deevey, Jr., "Mineral Cycles."

               Lester R. Brown, "Human Food Production as a Process in the Biosphere."

               S. Fred Singer, "Human Energy Production as a Process in the Biosphere."

               Harrison Brown, "Human Materials Production as a Process in the Biosphere."

Space

TO THE ANCIENTS, THE STARS REPRESENTED THE ABODE of the gods, and the constellations were representative of the gods' work in the sky. Eventually, the stars came to mean more than just representations of nature's wondrous works. Early sailors used the stars to navigate the uncharted seas. Religious leaders relied on the stars as accurate timepieces to forecast different seasons. Finally, the stars became the keys to unlocking the many secrets of the universe. Modern astronomers use stellar observations to study the various sources of energy and to determine the origin of the universe.

CHAPTER 17

The Science of Astronomy

ASTRONOMY IS ONE OF THE OLDEST SCI-
ences known to man. It began with
the ancients seeking ways to calculate the coming of seasons, to determine
when to plant crops, and to calculate the
dates of religious holidays. Often, early
astronomy was indistinguishable from astrol-
ogy and mysticism.

As men studied the heavens, their questions grew more general and entered into
more theoretical issues. These questions concerned the characteristics of the Earth and
the solar system. Many of the early myths
about the formations of the constellations
were attempts to explain the origin of the
universe.

As other sciences became more sophisticated, so did astronomy. Today, astronomy
is at a point in its development that enables
astronomers to seek answers to the most
fundamental questions about the origin of
the universe and life itself.

Ancient Science

We often make the mistake of placing
ancient science in the category of mere story-
telling. This is because many of the ideas
proposed by the ancients have proved to be
wrong. However, their science should not be
dismissed in such a sweeping fashion.

Although ancient science was filled with
legend and myth, much of it was informative,
and it laid the foundations for the future
courses of mathematics, physics, chemistry,
and astronomy. The idea that atoms make up
matter was first conceived by the ancient
Greeks. They were the first to develop the
theory of the Earth-centered universe, and
they were also the founders of the entire
science of mathematics.

Geocentric Versus Heliocentric Theories

Aristarchus, a Greek philosopher who
lived in the third century B.C., was the first
to propose that the universe is heliocentric
(sun-centered). However, his theory did not
gain favor. General acceptance instead was
given to the geocentric (Earth-centered)
theory, which was proposed by Hipparchus a
century later. The Greeks found it hard to
accept any idea that did not place man and
the Earth at the center of the universe.

In the second century B.C., Eratosthenes
successfully calculated the size of the Earth
and found it to be 24,000 miles in circumference. Ptolemy, in the second century A.D.,
also recognized the spherical shape of the
Earth, although he is better remembered
for his belief in geocentrism. Ptolemy established the geocentric theory which was accepted by the Church. As a result, this was
the official church doctrine for many centuries, and to disagree was heresy.

As more and more astronomical observations were made, the Ptolemaic system became quite unwieldy. In order to explain the observed motions of the planets, Ptolemy proposed that each planet whirled in loops about the Earth. These loops were called deferents. Smaller circles (epicycles) were added to the system: the planets were assumed to travel in a series of epicycles that became more and more complex (Figure 17.1).

The epicycles were necessary to explain the retrograde motion of the planets—the planets sometimes appear to slow up, re-

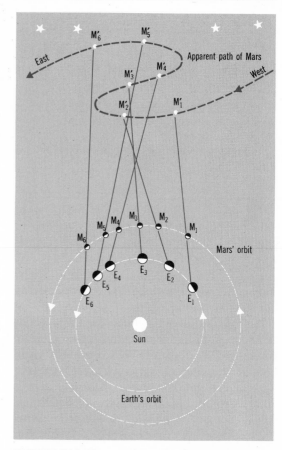

FIGURE 17.2 Retrograde motion is an *apparent* motion. The relative backward motion of some of the planets is because of the varying speeds of the planets in their respective orbits. As the Earth catches up to and passes Mars, the planet Mars will appear to slow up, move backward, and then once again move in a forward direction.

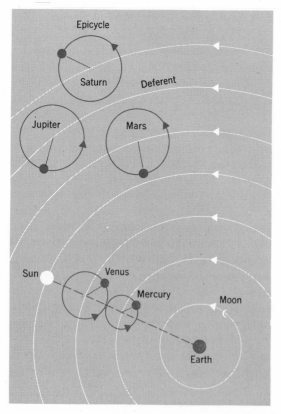

FIGURE 17.1 The Ptolemaic solar system was a complex arrangement of orbits and epicycles. The epicycles were necessary to explain the retrograde motion of the planets as they were viewed from the Earth.

verse their motion, and then move forward. The real reason for this appearance is that the planets move at different speeds around the sun. Thus, the Earth catches up to a slower-moving planet, and then passes it. (See Figure 17.2.)

In the sixteenth century, Nicolaus Copernicus (1473–1543) disputed the Ptolemaic concept of the universe. Copernicus, basing his work on Aristarchus, presented the Earth and other planets as revolving around the sun (Figure 17.3). According to Copernicus, the sun and not the Earth was the center of the solar system.

The Copernican system was given further substance by Galileo Galilei (1564–1642). Galileo made many astronomical observations with the newly invented telescope. He observed that Venus had phases like the moon, and that other planets were similar to the Earth in many respects. He also made many observations of planetary satellites— including our own moon.

In addition, Galileo investigated bodies in motion and the principle of the pendulum. His work dealing with bodies in motion led to further investigations that culminated in Newton's Laws of Motion.

In the sixteenth century, a Danish astronomer by the name of Tycho Brahe (1546–1601) built an astronomical observatory on the island of Hven. He devised astronomical instruments that were used to determine star locations. With these innovations he studied the heavens and developed exhaustive logs of his observations. In this manner, Brahe helped to set the stage for modern astronomy.

Astronomy as a Modern Science

Johannes Kepler (1571–1630), who is often called "The Father of Modern Astronomy," established the modern language of

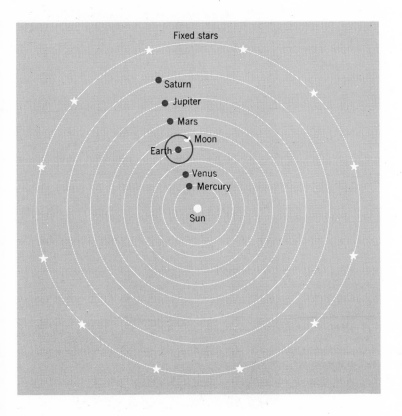

FIGURE 17.3 The Copernican view of the solar system placed the sun at the center of the solar system with the planets revolving about it. This concept of the solar system marked one of the most revolutionary aspects in the history of man's astronomical observations.

science and mathematics in astronomy. In 1609 he derived his first two laws of planetary motion (page 419) from the observations made by Brahe. From these, Kepler established astronomy as an exact science.

Late in the seventeenth century, Sir Isaac Newton advanced astronomy as a mathematical science. In fact, his laws of force and motion are the very basis for the science of physics and astronomy. Newton's Law of Gravity was derived by means of a careful mathematical analysis of the motions of the moon. Still further (as we will see later in this chapter) he invented the reflecting or Newtonian telescope, and also made great contributions to man's understanding of light.

The Characteristics of Light

The understanding of the characteristics of light is a basic need for astronomers. When we discuss the science of astronomy, we are really referring to the observation of light and how light travels.

The Wave Versus Particle Controversy

One can talk about light as traveling by means of either rays or waves. The word "rays" most often implies a stream of rapidly moving particles; whereas "waves" refer to energy and not individual and discrete particles. Which of these concepts, then, is correct?

FIGURE 17.4 The innovations that Brahe developed enabled him to compile an immense number of observations of the heavens. They were to lead to Kepler's laws, which in the 17th century placed astronomy in the forefront of the sciences (New York Public Library Picture Collection).

FIGURE 17.5 Sir Isaac Newton was one of the intellectual giants of science. There is hardly a single field in which his influence has not been felt during the last three centuries (National Portrait Gallery, London).

By the seventeenth century, two different theories had been proposed as to the structure and origin of light. Sir Isaac Newton proposed the particle theory. His theory was based on the hypothesis that light travels in straight lines and consists of individual particles, or "corpuscles." Since light cannot bend around corners as do other energies such as sound, his theory seemed acceptable.

At the same time, Christian Huygens (1629–1695), a Dutch scientist, proposed that light travels in waves as pure energy. By this time, it was recognized that light can bend or be refracted when it moves through denser substances such as water. Huygens was not successful in firmly establishing the fact that refraction is due to differences in the speed of light, and his wave theory was not accepted. He could not bring a great weight of evidence or prestige to support his idea as did Newton. The contro-

versy raged for several decades. Evidence slowly built up for both theories, but the wave theory began to accumulate more support. The fact that two waves can cancel out each other (interference) was discovered by Thomas Young (1773–1829) and this added further backing to Huygens' theory. Interference of varied waves is shown in Figures 17.6 and 17.7. By the beginning of the twentieth century, it was accepted that light does travel as a wave.

However, there was still some evidence upholding the particle theory which apparently disproved the wave theory.

In the late nineteenth century, it was discovered that light causes certain substances to emit electrons. This was called the photoelectric effect. After studying this effect, Max Planck, in 1900, proposed that light was transmitted in discrete packets of energy called quanta. These quanta were not a continuous band of energy; they were packets

FIGURE 17.6 The existence of waves that produce interference patterns such as these helped establish the wave theory of light. They demonstrate that wave energy is distributed uniformly in all directions and is not due to a stream of "corpuscles" or particles (Fundamental Photos).

of energy that were absorbed or emitted by electrons, enabling electrons to move from one energy level to another around the nucleus of an atom. Figure 17.8 shows the energy levels of a typical atom.

Planck's theory was still not as successful as it might have been. There was some disagreement about the many ideas regarding light transmission. In 1905, these ideas were finally reconciled by Albert Einstein, who suggested that light travels in photons. He defined photons as discrete packets of energy, like quanta, but he held that when photons travel closely together in large quantities, they act as if they are a continuous band of energy. This part of Einstein's ex-

planation satisfies the characteristics of reflection, refraction, and interference.

The photoelectric effect can be explained as photons traveling in space or in small quantities which spread out and become individualistic. In particle emission, individual photons give up their energy to electrons. If the energy is sufficiently strong, the electron is emitted from the atom and the photon disappears.

Modern scientists accept the wave theory of light with one important change, that light travels as discrete packets or photons of energy. These photons exhibit properties of energy waves or particles depending on different circumstances.

The Speed of Light

Several characteristics of light are of prime importance to the astronomer. One of them is the speed of light.

Originally, it was felt that the speed of light was instantaneous from one point to another: that light, when emitted from a source, would travel instantly to all observers in all directions. This belief was due, of course, to the fact that light traveled so rapidly that no precise measurement was possible.

Olaus Roemer (1644–1710), in 1676, demonstrated that light does have a specific speed. While observing a moon of Jupiter pass into an eclipse behind the planet, he showed that the eclipse time was different when observed from two points on the Earth's orbit. (See Figure 17.9.)

Roemer observed that it took longer for the light to travel across the Earth's orbital diameter than when the Earth was closer to Jupiter. Taking the Earth's orbit to be 192,000,000 miles in diameter, he calculated that the speed of light was about 192,000 miles per second. However, Roemer was using an inaccurate figure for the Earth's orbital diameter. Thus, his calculations were only fairly accurate.

Further measurements were made in 1849 by Armand Fizeau (1819–1896), who calculated the speed of light to be 194,675 miles per second. In 1862, Leon Foucault calculated the speed of light to be 185,177 miles per second. Both of these measurements were concerned with light passing through air at the Earth's surface.

In the 1920's, Albert Michelson (1852–1931) and his associates helped to further refine this measurement by determining the speed of light through a vacuum tube as

FIGURE 17.7 An interesting pattern resulting from light waves passing the teeth of a pocket comb (Fundamental Photos).

FIGURE 17.8 (*right*) Electron energy levels. The absorption of energy causes an electron to move to a higher energy level. The emission of energy allows the electron to move towards a lower energy level.

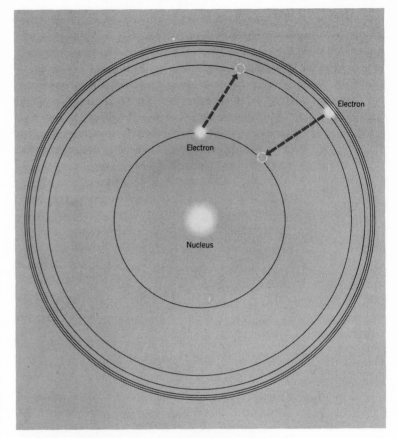

FIGURE 17.9 (*below*) Olaus Roemer developed one of the earliest systems for determining the speed of light. He measured the elapsed time for an eclipse of one of Jupiter's moons. His measurement showed that the light took a longer period of time to cross the Earth's orbital diameter (b) than to reach the Earth when it was closest to Jupiter (a). The determination of the speed of light showed it to be finite, and not instantaneous from one point to another (After Navarra and Strahler).

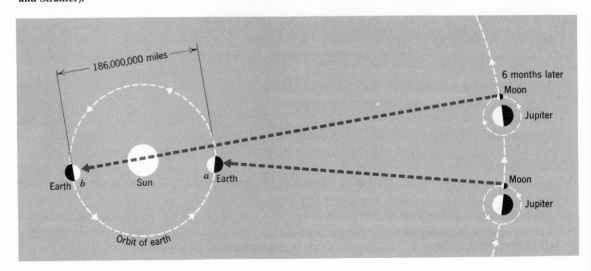

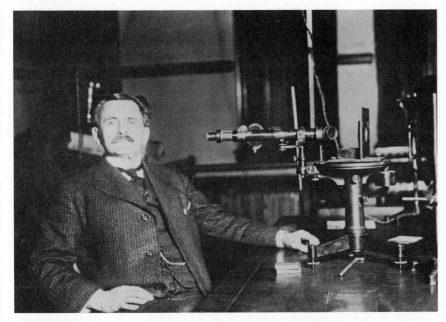

FIGURE 17.10 The long series of experiments carried on by Albert Michelson and his associates are prime examples of excellent scientific experimentation. They helped establish Michelson's reputation as an American research scientist of the first rank (Bettmann Archive).

186,284 miles per second. This figure agrees very closely with the now accepted value of 186,282 miles per second.

The speed of light is very important to the astronomer, since it gives him an important standard of measurement, the light-year. A light-year is the distance light travels in one year at a speed of 186,282 miles per second. This is actually a distance of about six trillion miles.

Spectral Lines

Because of the work originally begun by Newton, we know today that white light is actually a series of bands of energy. These bands of energy appear as various colors to the human eye. We also know that various gaseous materials, when heated to incandescence, produce characteristic colored lines called bright line spectra. Other materials in a liquid or solid state produce a rainbow effect, a continuous spectrum. When

a continuous spectrum passes through a transparent substance, a curious effect occurs. Dark lines are located where the transparent material would have yielded bright lines if it had been heated to the incandescent state. This is known as an absorption spectrum. Information from spectral analysis provides astronomers with much basic knowledge about celestial objects. The information is used to determine the brightnesses, chemical compositions, and temperatures of the stars, and also distances between stars.

The Doppler Effect

Spectral lines can provide important information about star motion. When a star is moving toward an observer, the spectral lines tend to pile up on one another. The wavelengths appear to shorten, and move toward the blue-violet end of the spectrum. Conversely, when the source of light moves away from the observer, the wavelengths ap-

FIGURE 17.11 A piece of sodium (wire) yields a bright-line spectrum in the flame of a Bunsen burner. The burner itself yields a continuous spectrum. On the left the spectrum from the sodium has been passed through a diffraction grating for analysis of the wavelength of light (Fundamental Photos).

pear to spread out and lengthen (Figure 17.12). Thus, the color of the star seems to move toward the red end of the spectrum. This shift in wavelength, due to the motion of celestial objects, is known as the Doppler Effect. The speed of a star and the direction it travels can be determined by the shift in the spectrum.

Basic Tools and Devices

Modern astronomers depend on many sophisticated tools and devices in order to observe the heavens. One of the oldest of these devices is the refracting telescope, which was first used by Galileo in his celestial investigations.

The Refracting Telescope

A simple refracting telescope consists of a front lens (objective lens) and an eyepiece. The objective lens gathers light that enters the telescope and brings it to a point of focus called the focal plane. (See Figure 17.13.) The eyepiece is another lens that magnifies the image for the observer. (A photographic plate may be substituted for the eyepiece.)

The images produced by a simple telescope like that used by Galileo were rather blurry and unclear. If a telescope is to give a clear picture, several lenses of various shapes must be used. These lens combinations eliminate bands of color (chromatic aberrations) and are said to be *achromatic*. The bands of color result when the various wavelengths of light are refracted unequally because of imperfections in the objective lens.

The light gathered by the objective lens increases proportionally with its size. As the diameter of the objective lens (called the aperture) becomes larger, the length of the telescope must increase if greater magnification is to be achieved.

The largest refracting telescopes are found in Yerkes Obervatory in Wisconsin (40 inches in diameter), Lick Observatory in California (36 inches in diameter), and Meudon Observatory in Paris (33 inches in diameter). The diameters of these lenses are not really very large. This is because the construction of lenses is not practical in large sizes. Beyond 40 inches in diameter, a lens bends under its own weight, since it can be supported only at its edge.

The Reflecting Telescope

The problem of lens distortion was solved in 1668 by Newton, who constructed the first reflecting telescope. Instead of relying on a lens to bend light, the reflecting telescope uses concave mirrors to catch the light and to reflect it to a point of focus. Mirrors, when carefully ground, need not be corrected for chromatic aberration and can be made larger than the lenses in a refracting telescope.

The reflecting telescope gathers light in a concave mirror. The concave mirror converges the light to a focal point called the prime focus. (See Figure 17.14.) At the prime focus, a small mirror is situated to throw the light through an eyepiece. Although the small mirror interferes with some light, it is a relatively small amount that does not greatly affect the efficiency of the reflecting telescope.

The mirrors of reflecting telescopes consist of aluminum-coated glass. They do not need to be as perfectly constructed as glass lenses, and only one surface, the reflecting surface, needs to be ground. The mirror can be supported at the back as well as the rim, and can be built much larger than 40 inches in diameter. Large reflecting telescopes can be found on Mount Palomar in California

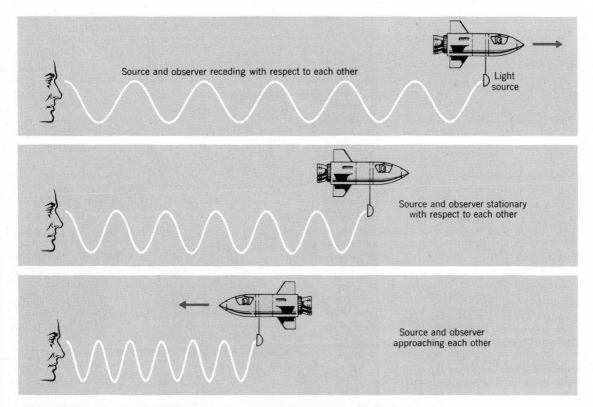

Source and observer receding with respect to each other

Light source

Source and observer stationary with respect to each other

Source and observer approaching each other

FIGURE 17.12   The Doppler Effect produces a change in the wavelength of light. As a light source approaches the observer the wavelength shifts toward the blue end of the spectrum; as the source recedes, the shift is toward the red end.

(200 inches in diameter) and the Lick Observatory (120 inches in diameter). Very large reflecting telescopes have small cages with observers riding at the prime focus.

For many years, the reflecting and refracting telescopes were the prime tools of the astronomers. Recently, however, other devices have come into use. The Schmidt

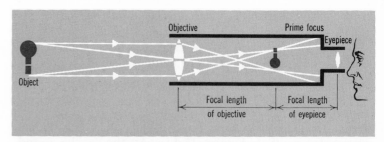

FIGURE 17.13 The refracting telescope produces an inverted image of the object in front of the eyepiece. The eyepiece then enlarges the image. Light is bent by the objective lens and focused at the prime focus. The 36-inch refracting telescope in the photograph is at Lick Observatory, University of California (Lick Observatory).

telescope, which is a reflecting telescope with a smaller lens to help bring a wide-angle picture into focus, is one of the most important variations of the optical telescope.

Radio Telescopes

In 1931, Karl Jansky of the Bell Telephone Laboratories discovered that radio waves can be detected from stars. From this discovery came a device known as the radio telescope. The radio telescope uses a reflecting metallic surface instead of a mirror, and has an antenna at the prime focus. This antenna picks up the reflected radio signals and sends them to an amplifier, from which they are sent to a recording device. (See Figure 17.15.)

Radio telescopes cannot detect signals from small areas of the sky, but are used to detect general or large areas in the sky that send out radio waves. One of the advantages of the radio telescope is that it reveals stars in parts of the sky in which optical telescopes reveal nothing. Radio-wave analysis has

given astronomers a new and very significant tool for examining the sky.

Other Important Devices

Photometry

Photometry has proved to be a very effective method in studying stars. Photometric methods are based on the principle that, when light strikes a photocell, an emission of electrons takes place. The emission or flow of an electric current is recorded and amplified by sensitive measuring devices. From this, an accurate computation of star brightness is recorded. (See Figure 17.16.)

The light sensitive portion of this measuring device, the photocell, is coated with a

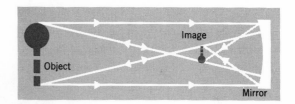

FIGURE 17.14 A reflecting telescope produces an inverted image which is then enlarged with an eyepiece. The light is focused by means of a concave mirror. The photograph shows the 200-inch Hale reflecting telescope at Mt. Wilson and Palomar Observatories (Mt. Wilson and Palomar Observatories).

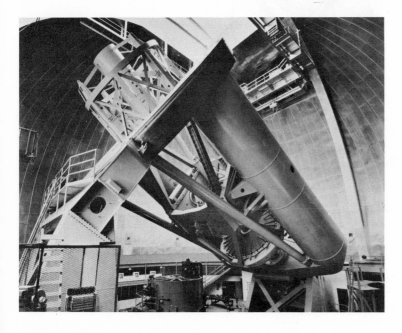

chemical that is very sensitive to minute changes in light. It gives researchers a picture of star brightness that is more accurate than photographic film or observation by the human eye.

The Spectroscope

The spectroscope provides information that gives clues to the intensity, temperature, and composition of stars. It is based on the

FIGURE 17.15 Radiotelescopes such as this one at Goldstone Tracking Station in Arizona are used to analyze wavelengths of energy other than visible light. Specially designed instruments of this type allow for more stellar observations than ordinary telescopes (Jet Propulsion Laboratory).

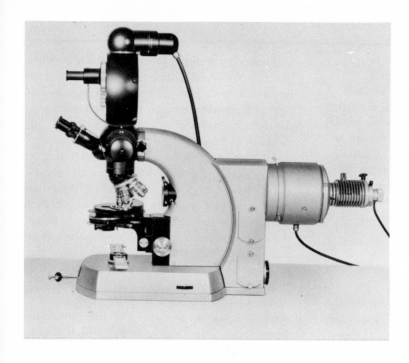

FIGURE 17.16 Photometers utilize photocells to produce a reading of the intensity of light (Carl Zeiss, Inc.).

familiar device known as a prism that breaks white light into its component colors.

In 1814, J. Fraunhofer, a German astronomer and optician, found that when the sun's rays pass through a slit prior to passing through a prism, dark lines are obtained. The spectroscope allows a ray of light to pass through a slit and, then, through a focusing lens called a collimator. The function of the collimator is to separate the original beam of light into separate parallel rays. (See Figure 17.17.) These rays then pass through a prism that separates the beam into a series of colors. This series then enters a second focusing lens where the colors are focused on an eyepiece or film. The spectroscope has an eyepiece whereas the spectrograph utilizes film.

Other Astronomical Tools

We have discussed the devices used to detect radio waves and visible light. Actually, these energies are only two forms of an entire range of energies known as the electromagnetic spectrum. (See Figure 17.18 and Chapter 19 – The Solar System.) In the future, a great deal of information about stars will be derived from observations of other energy bands, such as gamma rays, X rays, ultraviolet rays, and infrared spectra. Many elements give off such rays in addition to visible light. These bands of energy will enable astronomers to explore the processes of energy production by stars. Together with the subatomic particles found in space, these energies may shed some light on the origin of the universe.

Recent Theories of the Cosmos

To develop a basic understanding of the nature of the universe and the many interactions between energy systems that make up the universe, you must understand something about the basic unit of the matter that makes up the universe – the atom. It is the study of the atom, and the processes by which atoms form, that will eventually yield clues to the formation of the entire universe.

The Nature of Matter

The atom is composed of three kinds of particles—protons, neutrons, and electrons (Figure 17.19.) These are the fundamental particles from which the 92 naturally occurring elements are composed. Actually, some 30 additional kinds of particles have been discovered, but most of them arise in special cases. Our study of matter will be limited to only those particles that are basic to the understanding of astronomy.

The Atomic Nucleus

The atom is composed of a central mass called the nucleus that includes various combinations of protons and neutrons. The proton is a positively charged particle with a mass that is nearly the same as the neutron's. The neutron has no charge, and thus is said to be neutral. The only exception to this general scheme is the hydrogen nucleus, which is composed of a single proton. For this reason, protons are at times referred to as hydrogen nuclei. Surrounding the nucleus at varying orbital distances are the negatively charged particles called electrons. The mass of an electron is $1/1836$ that of the proton.

All the structural material of the atom is held together by electrical forces of attraction. These forces are produced by the positively charged nucleus and the negatively charged electron. The electron determines the chemical activity of the atom, since it is the electron that is involved when bonds are formed or broken between atoms during chemical reactions.

A mass of 1 is assigned to the proton and a similar mass is assigned to the neutron. Therefore, a hydrogen atom, which has a single proton in its nucleus, has an atomic mass of 1. The number of protons present inside the nucleus determines the atomic number. Thus, the hydrogen atom has an atomic number of 1.

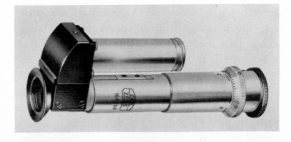

FIGURE 17.17 The spectroscope separates light into its colors in a manner similar to a simple prism (Carl Zeiss, Inc.).

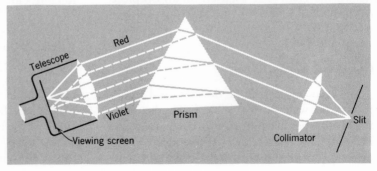

The Electron

Electrons are found revolving around the nucleus of an atom in a fashion that has, at times, been compared to the planets orbiting around the sun. This comparison must not be taken too literally. It is true that the electrons spin or rotate on their axes. But the electron is not held in place by simple gravitational and centrifugal forces like those act-

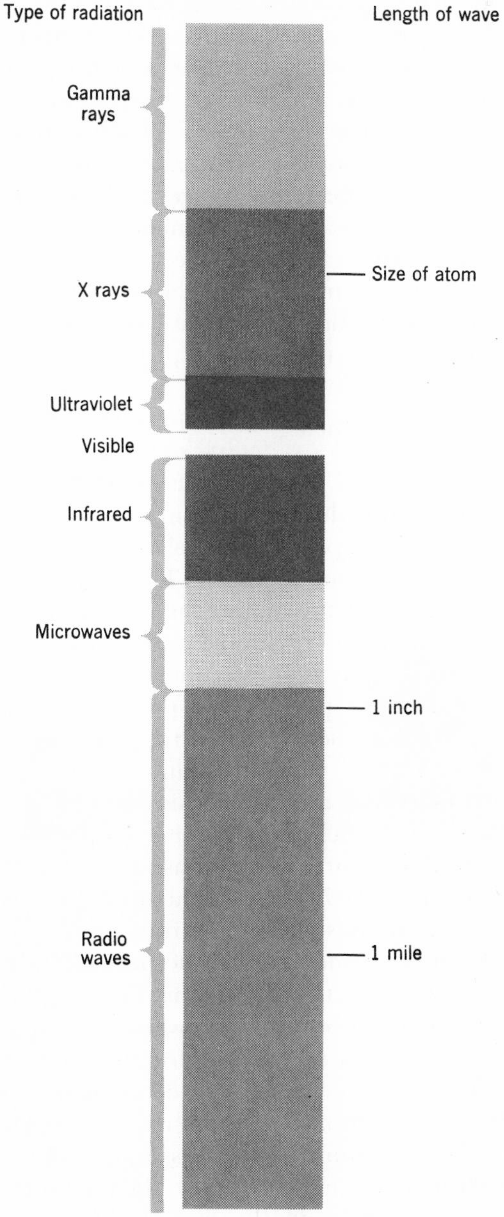

Type of radiation
Length of wave
Gamma rays
X rays
Size of atom
Ultraviolet
Visible
Infrared
Microwaves
1 inch
Radio waves
1 mile

FIGURE 17.18 (*left*) The electromagnetic spectrum and the relative lengths of the various wavelengths.

FIGURE 17.19 (*below*) A generalized view of the atom showing the positions of protons, neutrons, and electrons. The view is of a typical helium atom.

Electron
Neutron
Proton
Electron

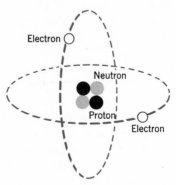

ing on the planets. Its orbit is determined by several other factors: the size of the nucleus of the atom, the distance of the electron's orbit from its nucleus, and the shape of the orbit, which is determined by the quantity of energy contained in the electron. Electron orbits are really more properly described as energy levels.

It requires more energy for an electron to move away from the nucleus of an atom than it does to move into an orbit closer to the nucleus. The closest orbits are thus the lowest energy levels. They are known as ground states.

Electrons are described in terms of the states or levels in which they are found. An electron that absorbs extra energy from some source moves to a higher energy level and is said to be "excited." When it absorbs a sufficient amount of energy, the electron is lost by the atom. Such an unbalanced atom is called an ion.

A specific amount of energy is required for the move to each higher orbit; thus, energy levels are not haphazard arrangements. Similarly, an electron moving to a lower energy level does so by releasing a specific amount of energy. These are the quantities of energy that Einstein called photons. They are absorbed or emitted in the form of one of the frequencies of the electromagnetic energies—electrical, infrared, ultraviolet, etc.

The spectra with which astronomers are so deeply concerned are the results of electron transitions from one level to another. Since the energies given off by excited atoms in various elements are specific for each level and element, the examination of spectra reveals two things to the astronomer. First, the energy levels of the electrons in a star indicate the temperatures and activity. Second, the kind of spectrum emitted indicates which element is emitting the energy as photons.

The outer electrons of atoms, those involved in reactions, require different amounts of energy to become excited. In cooler stars, most hydrogen lines found in the spectra are neutral and not ionic because the energy required to excite the outermost hydrogen electron is less than that required by helium. As the stars become hotter, more hydrogen ion lines can be found, but it is only in the hottest stars that lines of helium are present. Since scientists know the approximate temperature required to excite helium electrons and, thus, to produce strong helium lines, such spectra give astronomers clues about the temperatures of stars. It is also known that metallic elements are excited rather easily. Thus a spectrum showing rather strong metallic lines, while hydrogen and helium lines are almost completely absent, indicates a very cool star.

The Periodic Table

The periodic table (Figure 17.20) lists the elements in order of increasing atomic number. For example, helium, element number 2, has a mass number of 4. This means that the nucleus of the atom contains 2 protons; thus helium has an atomic number of 2. (Its mass is 4, thus it must contain 4 particles. Since 2 of the particles are protons, the remaining 2 particles are neutrons.) Helium is followed in the table by lithium, which has a total mass of 7 and an atomic number of 3. (We know from lithium's atomic number that the nucleus contains 3 protons; since the total atomic mass is 7, the remainder of the mass must be composed of 4 neutrons.)

Further examination of the periodic table would show that the elements in any given horizontal row, or any given vertical column, have similar properties. Since the electron is that part of the atom involved in the formation of different chemical compounds, it stands to reason that there is some order to electron arrangement. In fact, the total number of electrons in the outer orbits of an

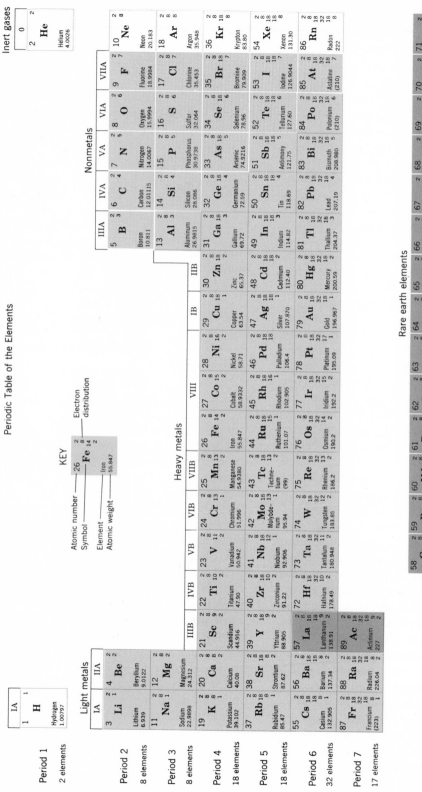

FIGURE 17.20 The periodic table of the elements.

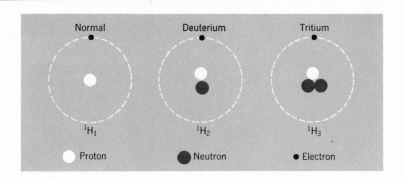

FIGURE 17.21 The isotopes of hydrogen. "Normal" hydrogen contains one proton in its nucleus and has an atomic weight of one. Deuterium contains one proton and one neutron and has an atomic weight of two. Tritium contains one proton and two neutrons and has an atomic weight of three. All have the same atomic number—one—as determined by the lone proton.

atom is equal to the number of protons in the nucleus. The atom is said to be neutral, since each positive charge is balanced by a negative charge.

It must be remembered that the number of electrons of an atom varies, since they are lost or picked up very easily. It is for this reason that the atomic mass numbers given in the periodic table are not whole numbers, but fractions. The fractions do not mean that the atom contains pieces of protons or neutrons. That would be impossible. The fractions are *average* figures, because there are different forms of the same atom. These forms, called isotopes, all have the same atomic number but have different atomic masses. Hydrogen, for example, has three forms (Figure 17.21). "Normal" hydrogen has an atomic number of 1 and a mass of 1. Thus, it has a single proton in the nucleus, and a single electron. Another less common form of hydrogen, deuterium (from the Greek word *deutero,* meaning "second"), has an atomic number of 1 but a mass of 2. It has 1 proton, and 1 neutron in its nucleus, with a single electron in its orbit. The third form of hydrogen is tritium. It has an atomic number of 1 but, mass of 3, with a single proton, 2 neutrons, and 1 electron. This form of hydrogen is not very abundant and is radioactive. That is, it will decay and send particles out of its central nuclear mass.

These 3 forms of the hydrogen atom are called hydrogen *isotopes.* Isotopes have the same atomic number because they have the same number of protons, but they differ in mass number because of a different number of neutrons. Thus the mass number for hydrogen shown on the periodic table is an average figure calculated from the presence of each isotope of hydrogen in a sample of hydrogen gas.

Nuclear Transformations

It is the atomic nucleus, not the electron, that is involved with changes that produce the temperatures in the stars. The extraordinary temperatures found in stars can be produced only by nuclear reactions and not by chemical reactions. When scientists seek answers as to the origin of the universe, how atoms and compounds came into being, and how the processes of the universe itself are maintained, they are not at all concerned with the electron but with the buildup and breakdown of the atomic nucleus.

Scientists have proposed that, in the beginning of time, the universe consisted mostly of neutrons with a small percentage of electrons and protons. (It is known today that, under the proper conditions, neutrons decay or break down into electrons and protons.) Primitive material then began to build up the elements by a successive series of complex processes.

Neutrons and protons could initially have

joined together to form hydrogen isotopes. Protons, we recall, are already hydrogen nuclei. Thus, if a proton and a neutron join together, they form deuterium, the second isotope of hydrogen.

Atomic mass
  number$\longrightarrow 1$ ^{1} ^{2}
Hydrogen$\longrightarrow$ H + n → H + energy
Atomic number$\longrightarrow 1$ $_0\uparrow$ $_1$
                    Neutron

If this particle now captures another neutron, tritium is produced.

$$_1\text{H}^2 + _0 n^1 \rightarrow _1\text{H}^3 + \text{energy}$$

Tritium is radioactive. When one neutron decays and gives off an electron, 2 protons and a neutron are left in the nucleus. If 2 protons are present, the structure is no longer considered a hydrogen nucleus but is now helium.

$$_1\text{H}^3 \rightarrow _2\text{He}^3 \leftarrow (\text{Isotope of helium})$$

$$+ _{1-1}e^0 \leftrightarrow \text{positron (one positive charge and a negligible mass)}$$

This is an isotope of helium with an atomic number of 2 and a mass of 3.

Some scientists propose that the universe began with a series of reactions similar to the above. This series would produce a number of helium and tritium atoms that could then begin joining to form the heavier elements: lithium, carbon, etc. About 99% of the universe's mass is hydrogen even at the present time.

As this process continued, these scientists maintain, most of the elements in the earlier part of the periodic table formed quite easily and rapidly. As temperatures increased in the primitive stars that were being formed, another series of reactions began that formed the very heavy elements nearer the end of the periodic table. These reactions were a series of steps, including the buildup of nuclei and periodic decay such as the series described, except that they involved heavier nuclei as the temperatures increased.

The Development of the Solar System

It must be remembered that the atomic processes outlined in the last section were

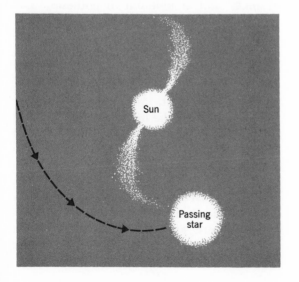

FIGURE 17.22 The tidal hypothesis proposes that the solar system originated in a near collision between our sun and a passing star. Material lost by the sun coalesced into the present planets.

not discovered until recently. Therefore, most earlier theories of the origin of the solar system do not attempt to include any nuclear buildup or development. In any discussion dealing with solar system origins and development, it is valuable to examine how man's thinking in this area has changed.

The Sun Forming the Solar System

Theories dealing with the origin of the solar system were proposed as early as the eighteenth century. Scientists hypothesized that the sun formed first and then gave rise to the present planetary system. This theory has several variations. One theory holds that a star passed by the sun closely enough to pull out a long streamer of material by gravitational attraction. This streamer of material broke into a series of globules, each of which became planets or satellites. The other theory, which is essentially the same, holds that another star actually collided with the sun, causing an explosion of material. This material then formed the planets.

One theory of this sort has been put forth in the twentieth century by Sir Harold Jeffreys and Sir James Jeans. Their theory, called the *tidal hypothesis,* proposes a near-collision between the sun and a star (Figure 17.22). The star presumably pulled out a streamer that coalesced into planetary globules with a rotational motion about the sun. The main problem with this theory is the difficulty in trying to explain how the planets began rotating on their axes. The theory also fails to explain the difference in densities between the inner planets (Mercury, Venus, Earth, and Mars) and the outer planets (Jupiter, Uranus, Neptune, Saturn, and Pluto). Jeffreys and Jeans were able to explain rather successfully how the cores of planets accumulated heavy elements. However, they could not explain why the outer planets accumulated the lighter elements.

Another variation of the tidal hypothesis is the *planetesimal theory,* which was actually proposed by Thomas Chamberlin and Forest Moulton just prior to Jeans' and Jeffreys'. This theory states that a series of streamers left the sun to form the planets.

It has also been suggested that the sun was originally a double star, and that the second star exploded, sending its mass outward as the material that formed the planets. So many double stars have been discovered that this hypothesis is not at all farfetched. However, it too presents problems in explaining how the solar system came to be as it is today. All the planets revolve around the sun in nearly the same plane, and an explosion would not have taken place only in one plane.

Various Nebular Hypotheses

A more realistic series of hypotheses suggests several refinements in the formation of the solar system. These are the *nebular hypotheses.* The first nebular hypothesis was presented in the eighteenth century; its descendant has replaced the tidal, planetesimal, and double-star hypotheses. The modern nebular hypothesis assumes that the entire solar system began at the same time. This hypothesis is more in line with present thinking dealing with the origin of the universe.

Immanuel Kant, an eighteenth-century philosopher, was the first to suggest the idea of a nebula, or dust cloud, as the origin of the solar system. Kant assumed that there was once a huge flattened disk of hot gases that were rotating about a common center. As the gases continued to rotate, the outermost portions of the disk broke off and became planets. The inner mass of the nebula became the sun. In the latter half of the

FIGURE 17.23 Immanuel Kant drawn from life in 1789. This German philosopher developed an early nebular hypothesis of the origin of the solar system (Bettmann Archive).

eighteenth century and early nineteenth century, this theory became a true nebular hypothesis with the ideas of Pierre Laplace.

Laplace felt that an explosion of the sun would not have produced the nearly circular orbits of the planets. Instead, he assumed that there was a whirling sun that exploded, leaving a flat, whirling disk of gases. As the disk cooled, it began to contract. This contraction left a series of rings forming around a central mass. Laplace held that these rings eventually consolidated and became the planets and their moons. This theory successfully explained the flat plane in which the planets orbit and the fact that all the planets revolve in the same direction (Figure 17.25).

Laplace's idea was upheld until the twentieth century, when it fell under attack. His hypothesis could not explain all

the motions found in the solar system— including those of the planets and their various moons. Thus, the Kant-Laplace nebular hypothesis was replaced by the tidal hypothesis and later by the planetesimal hypothesis. However, recent theories have once more moved back toward nebular ideas.

Modern nebular ideas developed from the work of an American astrophysicist, G. P. Kuiper, and a German mathematician, Carl Von Weizsäcker. Essentially, their ideas also begin with a nebula of particles, probably atomic in nature. As a cloud of cosmic matter began to condense, smaller eddies or currents formed. Clouds of material developed, each with its own circulation, revolving about the central mass. The smaller clouds gathered more particles by accretion or clumping together, and formed the beginnings of primitive or protoplanets. Although Kuiper

**FIGURE 17.24 Laplace, the French
scientist-philosopher (New York
Public Library, Picture Collection).**

and Von Weizsäcker differed somewhat in the theories they independently proposed, each suggested that the smaller eddies or protoplanets formed within the larger nebula. Each man also showed mathematically that the outer eddies increased in size, forming the large planets. Lastly, Kuiper showed that the lighter materials could be pushed away by radiation pressure from the sun. (This is the same process that pushes the gaseous tail of a comet away from the sun.) Most of the lighter material was pushed away from the inner planets so that the outer planets gained most of it. This hypothesis would satisfactorily explain why the outer planets are less dense than the inner planets.

Fred Hoyle, the most provocative astronomer today, has agreed with the proponents of the nebular hypothesis. Hoyle points out that the materials in the nebula would re-quire specific temperatures in order to coalesce. The inner planets are thus quite dense because they are composed of materials that could coalesce at the fairly high temperature near the sun. The lighter elements, being pushed outward by radiation pressure, could not coalesce until they reached the temperatures found in the region of the larger planets. Thus, the low density of the outer planets is easily explained.

The theories dealing with the origin of the solar system began with a collision hypothesis and moved toward the nebular hypothesis, which was in turn replaced by a collision or near-collision hypothesis. Now we have come one full circle to a nebular hypothesis. Undoubtedly, further refinements will be made in this theory in the future, but it seems to be a more satisfactory answer than the various collision hypotheses.

Origin and Structure of the Universe

Within recent years, man has begun to increase the range of his theoretical syntheses to include the entire universe. With the advent of the radio telescope and other highly sophisticated measuring devices, the movements of galaxies and other structures in the universe have been determined rather accurately. These new instruments have provided additional information so that scientists are now able to formulate general theories about the nature and development of the universe.

Estimates as to the age of the Earth and solar system vary widely. The oldest known rocks in the Earth's crust are about 3 billion years old. It is estimated that the solar system, assuming that it all formed at nearly the same time, is about 5 billion years old. The age of the universe is highly conjectural. It has been estimated at anything from 10 billion to 25 billion years. The difficulty with estimating the age of the universe lies in the fact that no one theory of origin has managed to solve all the questions dealing with all the situations found in the universe.

At present, there are two main theories vying for acceptance. One, called the "Big-Bang" theory, holds that the universe is going through only one of a series of evolutionary stages. The second, known as the Steady-State theory, proposes that new matter is constantly being created to replace old matter, and that evolutionary changes do not occur to any great extent.

The Big-Bang Theory Versus the Steady-State Theory

Both the Big-Bang and Steady-State theories propose that we are living in an expanding universe, in which the essential parts—the galaxies—are rapidly moving away from one another. When the galaxies are examined for spectral lines, a noticeable red shift appears. This red shift, based

FIGURE 17.25 Kant and Laplace developed the hypothesis that the solar system was produced by a whirling cloud of dust. Material near the center contracted and heated, forming the sun. Smaller eddies of gas formed the planets now revolving about the sun.

on the Doppler Effect (p. 387), implies strongly that the galaxies are moving away from each other.

The question that arises is, will this expansion continue indefinitely or not? Some answers must be derived in order to explain how this motion was initially imparted to the structures of the universe.

The Big-Bang or Evolutionary Universe theory has been expounded by George Gamow, a leading theoretical physicist from the University of Colorado. Gamow proposed that the universe originally consisted of a hot "primeval soup" consisting of neutrons, protons, and electrons. The "primeval soup" began to contract very rapidly. There was an explosion, the "Big-Bang," that created the initial material of the universe. This took place eight to ten billion years ago. Through a series of neutron captures in the hot gaseous mass, the lighter elements formed. Gamow hypothesizes that the initial temperature of the mass was about one billion degrees. Within the first hour, it dropped to several million degrees Fahrenheit. Then most of the lighter elements formed. Thus, the formation of elements took place during an infinitesimally small period of time.

During the next quarter of a billion years, as the expansion continued, the elements cooled to the temperatures at which they condensed to form the planets, stars, and galaxies. At this point, the general temperature fell to about 6000°K, which is the present temperature of the sun's surface.

The major opposition to the Big-Bang theory comes about because it has difficulty in accounting for the formation of the heavier elements. The theory proposes that the heavier elements formed within the cores of hot stars. This theory, of course, implies that there was a beginning of time, and that the entire development of the universe can be traced.

There is another school of thought that has refined the Big-Bang theory to propose a universe that is oscillating between repeated periods of expansion and contraction. This theory holds to no actual beginning or end of time, but maintains that we are, at present, in one cycle of expansion. In time, the momentum imparted to the expansion will decrease, and the universe will begin to slow down, contract, and eventually be annihilated. The annihilation will cause another "Big-Bang" when the mass of the universe comes together and, thus, the process will repeat itself. One cycle of expansion and contraction may take 100 billion years to be completed.

In complete opposition to these two theories is the Steady-State theory. Fred

FIGURE 17.26 Fred Hoyle, the English astronomer, is the major architect of the steady-state hypothesis of the universe (UPI).

FIGURE 17.27 The late George Gamow, Professor of Physics at the University of Colorado. This prolific and popular science researcher and writer was the chief early proponent of the "Big-Bang" hypothesis (Wide World).

Hoyle, the chief proponent of this theory, proposes that the concentration of matter remains constant, and that the mass of the universe is uniform throughout its entire volume. Hoyle is then faced with the problem of reconciling this condition with the observed phenomenon of an expanding universe: How can matter be moving apart in the universe and remain uniform in density at the same time? Hoyle and his supporters believe that new matter is constantly being created. They hold that the processes described earlier are still going on in various parts of the universe, with lighter elements like hydrogen being created from basic particles, and heavier elements being synthesized in the stars. If this theory is correct, then the distances between galaxies actually are not increasing, since new structures are forming to take the place of those moving apart. Hoyle's theory requires large amounts of hydrogen gas to be present in the universe. In fact, scientists have found large quantities of hydrogen moving into the outer fringes of the Milky Way Galaxy, our galaxy, and out of its core.

At present, scientists disagree over the nature of the universe. Recent investigations have provided information that seems to favor the Big-Bang theory, particularly when spectral shifts are closely examined. But it is too early to make any predictions as to which theory will finally be accepted. Answers to many questions dealing with the origin of the universe will begin to appear when tools and techniques allow man to examine structures presently beyond his limits.

Recent examinations of the heavens with radio telescopes have revealed rather exciting structures called quasars. Quasars—

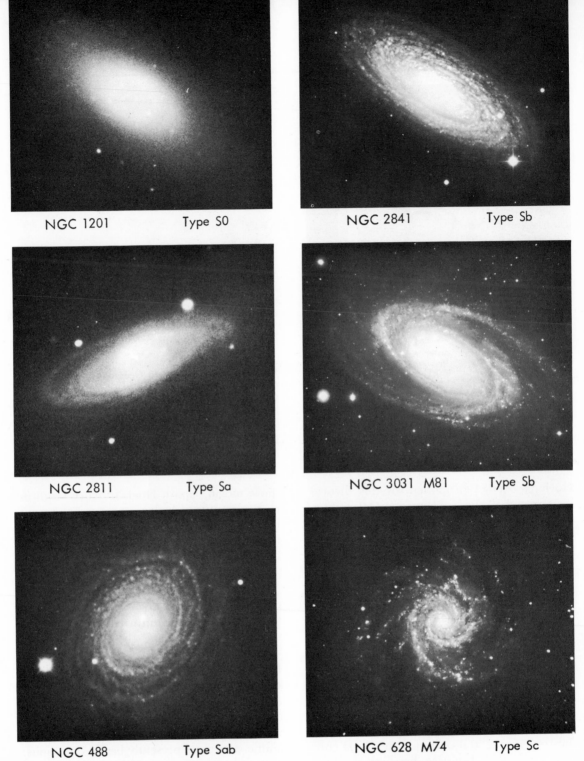

FIGURE 17.28 Normal galaxies as photographed by Mount Wilson and Palomar Observatories. These galaxies are all flattened and disk-shaped; some have arms radiating from the central mass (Mount Wilson and Palomar Observatories).

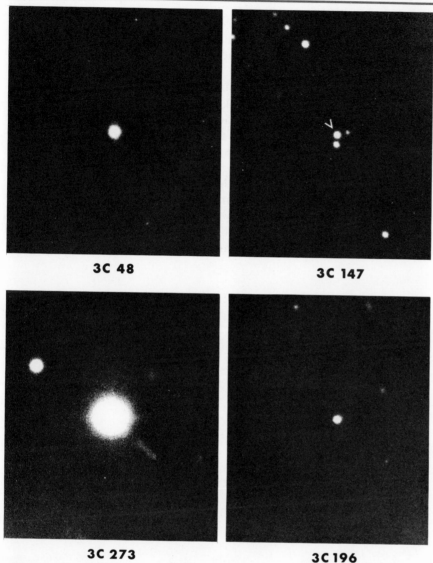

3C 48

3C 147

3C 273

3C 196

FIGURE 17.29 Quasars, or quasistellar radio sources, are extremely energetic structures. They may represent galactic births or simply superenergetic star types. The discovery of these structures has caused more excitement and controversy in astronomy than almost any other event during this century (Mount Wilson and Palomar Observatories).

more properly termed quasi-stellar radio sources—are highly energetic sources of radio emission believed to be billions of light years away, although this fact has not been definitely established. These quasars are larger than stars, but are not large enough to be galaxies. They produce energy at rates greater than any star or galaxy, but it is only recently that we have been able to examine them.

The existence of quasars has excited astronomers because of their apparent great distances from our solar system. The light that we receive from these structures may have begun its journey six or more billion years ago. It is possible that we are looking at some of the highly condensed primeval material that coalesced into the galaxies and stars we know today. Thus, it is highly probable that we are observing light that left a galaxy when it was in the process of birth.

At present, many theoreticians are leaning toward the idea that the age of the universe may not be very different from that of the Earth and the solar system. In effect, they are looking toward a theory of the formation of the universe that will also answer the questions about the origin of the solar system. In the near future, great changes will undoubtedly be made in man's theories of the solar system and the universe. Apparently they will become part and parcel of a single theory. As evidence begins to accumulate, the idea of an oscillating universe, although it is disconcerting to the self-centered human race, seems to become more and more plausible. Furthermore, this idea would make the Earth, solar system, and all the structures of our universe of the same age and place of origin.

CHAPTER 18

The Planet Earth

THE PLANET EARTH IS ONLY ONE SMALL body out of the countless billions of objects in our universe. But it is our home. For this reason it is an object of great concern and interest to us. Only through an understanding of our planet and its characteristics can we develop standards of measure against which to compare other planets in the solar system.

The study of the Earth as one small part of a great, balanced universe will give us insight into the mechanics of the universe, for the same physical laws govern both. As we study our Earth to learn how and why it moves, we will begin to understand the workings of our solar system and our universe.

Terrestrial Characteristics

As Chapter 1 pointed out, the Earth is not a perfect sphere; it bulges at the equator and is flattened at the poles. Furthermore, the Arctic is farther from the equator than expected by about 25 miles, and the Antarctic is closer by 25 miles. The Earth is actually slightly pear-shaped; it is technically an oblate spheroid (Figure 18.1).

The actual diameter of the Earth when measured from pole to pole is about 7900 miles. The equatorial diameter measures 7927 miles. Thus, the difference is very small—only 27 miles when the equatorial diameter is compared to the pole-to-pole diameter.

The Earth's Shape

Very sensitive devices known as gravimeters have been used to determine irregularities in the Earth's surface. These devices show a stronger gravitational attraction for objects located on flattened areas and a lesser attraction for objects located where the Earth bulges. The difference is accounted for by the fact that as the object moves away from the center of the Earth less gravitational attraction is exerted on it. For example, a man who weighs 200 pounds will notice a weight difference of approximately one pound when weighed at the poles and the equator. The pole weight is larger because the poles are flattened and the man is closer to the Earth's center.

Newton's Law of Universal Gravitation has provided us with valuable information about the Earth's characteristics. We can accurately calculate the attraction between the Earth and a known mass. Because we can determine this relationship, we are able to calculate the mass of the Earth. For example, if the gravitational attraction for a mass of metal that is accurately known is measured, simple substitution in Newton's formula gives us the mass of the Earth. It has been found that the mass of the Earth is 6.0×10^{21} tons (or 6.0×10^{27} grams). Once we know the mass of the Earth we can then derive other pieces of information, most notably the Earth's density. Density is the mass per unit

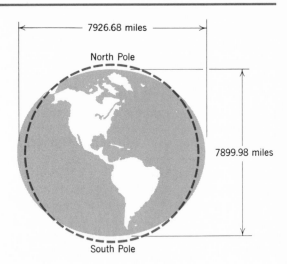

FIGURE 18.1 The rotation of the Earth causes an equatorial bulge, with an equatorial diameter approximately 27 miles more than the pole diameter.

volume—usually expressed in grams per cubic centimeter. The volume is easily calculated once the diameter or circumference of the Earth has been determined. The density formula is

$$D = \frac{M \text{ (mass)}}{V \text{ (volume)}}$$

The density of the Earth is about 5.5 grams per cubic centimeter, or 5.5 times that of an equal volume of water. This is an average density for the entire Earth, and does not hold true at all locations.

Earth Motions

As the Earth travels through space, it undergoes several motions. The Earth's motions are important because they affect our seasons as well as other characteristics of our planet.

Two commonly known motions of the Earth are its rotation on its axis, and its revolution around the sun. Although these motions seem obvious today, they were not recognized until recently, when proofs of these motions and their consequences were established.

Rotation and Its Consequences

To the casual observer, the heavens look as if they are moving across the sky while the Earth remains stationary. However, this phenomenon can be shown to be a consequence of the Earth's rotation. So can the oblateness of the Earth. If the Earth did not rotate, it would be a spherical structure instead of an oblate spheroid. (See Figure 18.3.) According to Newton's First Law of Motion, a moving object tends to move in a straight line. Thus, as the Earth rotates on its axis the material of the Earth tends to be thrown off its surface. Therefore, the equator is thrown out in a line away from the surface of the Earth. Gravity holds the Earth together, so that the material cannot escape from the Earth's gravitational pull. As a result, a bulge appears around the waist of the Earth.

If you were to take time exposure photographs of the heavens from different points on the Earth, you would obtain very different results (Figure 18.3). A picture taken at the equator would show the stars moving in a series of straight lines across the film. A picture taken at either pole would show the stars forming circular paths directly above the camera. It is more than obvious that the stars cannot move in straight lines and in circular paths at the same time. Therefore, it can be concluded that the rotation of Earth is the cause of such a series of pictures.

In the nineteenth century, Foucault, a French physicist, provided the most convincing experimental evidence of the Earth's rotation. His experiment used a pendulum made of a heavy bob suspended by a wire. This pendulum was able to swing freely, independent of the Earth. Around the plane of the pendulum, a series of pegs was set up in a circle. Foucault was aware that a pendulum will always oscillate back and forth in

one plane unless acted on by some outside force. Thus, if the Earth were stationary, the pendulum would only swing in a particular direction and knock down a peg at either end of its swing. However, if the Earth rotated underneath the pendulum, the pendulum would swing back and forth in a wide circle and knock down succeeding pairs of pegs.

Foucault's pendulum did knock down the pegs in succeeding pairs. The plane of oscillation of the pendulum was checked against the stars, and the bob always swung toward the same place in the heavens. Since the change of plane was apparent only against the Earth, it was apparent that the Earth must have changed position relative to the pendulum and the stars. A reconstruction of Foucault's experiment is shown in Figure 18.4.

The rotation of the Earth on its axis produces our night and day. As the Earth rotates, a shadow slowly moves across its face,

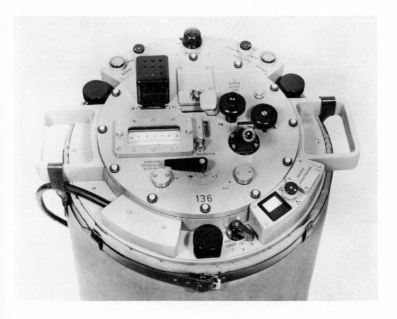

FIGURE 18.2 Gravimeters accurately measure the minute differences in gravitational attraction at various points on the Earth (Askania-Werke).

plunging us into the darkness which we call night, and departing to leave us in the light of day.

Revolution and Its Consequences

Revolution around the sun is the Earth's second major movement relative to the sun and stars.

When we observe stars from different positions in the Earth's orbit, those stars closest to the Earth shift position in relation to those farther away. This can be seen in Figure 18.5. When the Earth is in position A, star number 1 lines up in front of star number 2. However, when the Earth is in position B, star number 1 lines up with star number 3. Each succeeding year this apparent star motion is repeated.

FIGURE 18.3 A time-exposure photograph of stars taken at the North Pole (A) (*above*) shows a series of concentric circles. A similar photo taken at the equator (B) (*right*) shows the stars moving across the sky in relatively straight lines. The only reasonable explanation for both occurrences is the rotation of the Earth beneath the stars (Lick Observatory).

FIGURE 18.4 As the Earth rotates beneath a Foucault pendulum, successive pairs of pegs are knocked over (California Academy of Science).

Those who believed in the Ptolemaic concept of an Earth-centered system became increasingly entangled in trying to explain how a star could jump back and forth through the heavens in this manner. It is much easier to show that the Earth itself produces this motion by changing its position relative to the stars in question.

Further proof of revolution can be obtained from consideration of the Doppler Effect (Chapter 17). When the Earth is in position A in Figure 18.5, it is moving toward the star; when it is in position B, it is moving away from it. The shift in the spectrum of the star toward the blue, in the first case, and toward the red, in the second case, cannot be explained by the star alternately changing color, but is easily explained by the Earth's revolving in an orbit about the sun.

Revolution about the sun is the major cause of our change in seasons (Figures 18.6, 18.7). Because the Earth's axis is tilted some 23½°

from its perpendicular toward the sun, a portion of the Northern Hemisphere receives direct rays of the sun in the summer. These rays tend to be more perpendicular to the surface of the Earth; therefore, they are less scattered, and a given amount of radiation covers a smaller surface area. In addition, the tilting of the Earth's axis causes the Northern Hemisphere to remain in daylight for a longer period of time and, thus, to receive these direct rays for a longer time. In the Southern Hemisphere, the situation is reversed.

In the winter, the Northern Hemisphere receives rays of solar energy which enter the atmosphere at a greater angle. Therefore, they are more diffused by the atmosphere and are less intense. Moreover, the winter days are shorter, so that the Northern Hemisphere receives this less intense energy for a shorter period of time. In the Southern Hemisphere again, the situation is reversed.

Precession

Precession is a motion of the Earth that is not as obvious as the two just discussed. The tilt of the Earth's axis enables the moon's gravitational pull to cause the Earth to spin in a wobbly fashion, like a top. The "wobble," or precession, is in a slow clockwise direction and causes the North Pole to draw out a great circle in the heavens (Figure 18.9).

Precession is an extremely slow process; 26,000 years are required to complete a cycle. The present pole star, Polaris, will not always be the pole star. It happens to be the star to which the North Pole most nearly points at the present. In about 13,000 years,

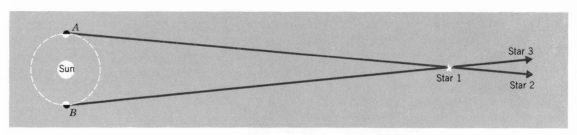

FIGURE 18.5 The Earth's orbital motion will cause stars to appear to move across the heavens in a repetitive pattern. This apparent motion can only be explained by the motion of the Earth around the sun.

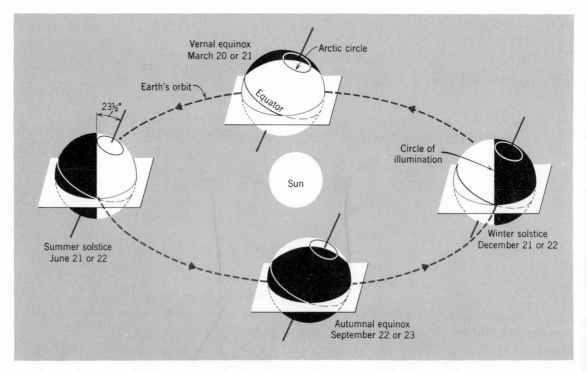

FIGURE 18.6 The relationship of the Earth and sun in each seasonal position (After Longwell, Flint and Sanders).

precession will have caused the North Pole to point in the opposite direction in the sky. In fact, it will point near the star known as Vega. In another 13,000 years, the North Pole will have come around once again to its present position near Polaris.

It has been suggested that precession causes complete changes of the seasonal patterns of the Earth. Although this is a logical suggestion, many other considerations make a simple answer to seasonal changes highly conjectural.

Axis precession is slightly irregular and, thus, the axis describes an irregular circle. This variation of precession is known as nutation.

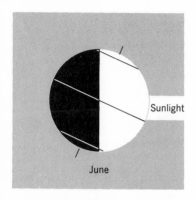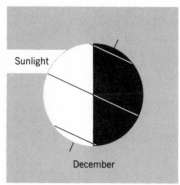

FIGURE 18.7 Relationship of the Earth's tilt and sun's rays in winter and summer. The rays of the summer sun strike the Earth directly at nearly a 90° angle and are scattered less. Thus, they are more intense than the oblique winter rays that cover a large surface area and are less intense.

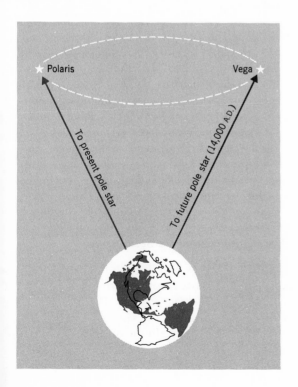

FIGURE 18.8 The Earth's axis describes a wide circle in the sky over a period of 26,000 years. This motion constantly changes the point in the sky to which the North Pole is pointed. This motion determines the pole star, which is Polaris at present.

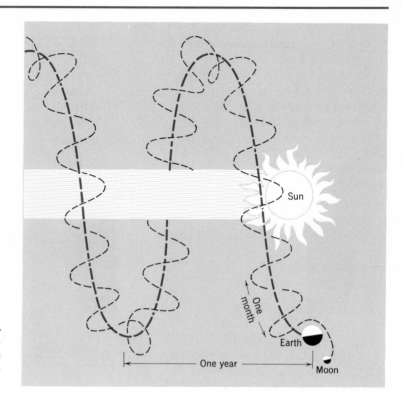

FIGURE 18.9   The various motions of the Earth. The Earth revolves around the sun, the sun moves through space, and the Earth-moon system accompanies it.

Other Motions of the Earth

There are several other Earth motions that require discussion. It has been found that the sun is moving, along with the rest of the solar system, toward the star Vega in the constellation Hercules. The rate of this motion is calculated to be 12 miles per second.

Like the rest of the galaxy, the sun also revolves around the center of the galaxy. At its present rate of 200 miles per second, this rotation takes 200 million years to complete one revolution. Finally, our entire galaxy is moving with a small local group of galaxies at a rate of about 50 miles per second.

The Celestial Sphere

To locate celestial objects in respect to the Earth, an imaginary system of celestial co-ordinates has been developed to coincide with surface features of the Earth. Figure 18.10 shows the development of this sphere.

When the Earth's poles, equator, and prime meridian are projected into space, the result is a celestial sphere that allows convenient reference to the positions of stars. On this sphere may be drawn the apparent path of the sun around the Earth during the year.

The Equinoxes and the Solstices

The sun does not follow the equatorial regions but moves north and south of the equator at different seasons (Figure 18.11). During the summer, the sun is north of the equatorial line. In the winter, the sun is south of the equatorial line. In the spring, the sun's apparent path—called the *ecliptic*—

crosses the equator and moves north. This event is the *vernal equinox*. When the sun moves south and again crosses the equator, the *autumnal equinox* has occurred. (All of these references are for the Northern Hemisphere. For the Southern Hemisphere the situation is the reverse.) During the spring and autumn equinoxes, the Earth is equally illuminated in both hemispheres, and the days and nights are of equal length.

When the sun is at its most northern position, the Northern Hemisphere experiences its *summer solstice*. When the sun is in its most southern position, the north experiences *winter solstice*. Again, these positions are reversed for the onset of the various seasons in the Southern Hemisphere. These motions of the sun are only apparent motions. They are produced by the changing position of the Earth in its orbit relative to the sun.

The vernal equinox occurs on March 21 to 22, and the autumnal equinox about September 21 to 22. Summer solstice is June 21 to 22, and winter solstice is December 21 to 22.

Declination

The declination of a star is its number of degrees north (+) or south (−) of the celestial equator. Thus, declination is analogous to the north or south latitude on the Earth. A declination of +41, for example, places a star directly over 41° latitude north. However, since the Earth rotates beneath the heavens, that star may be anywhere along the 41° latitude line. Thus, we also need a system to describe the east and west movement of the star.

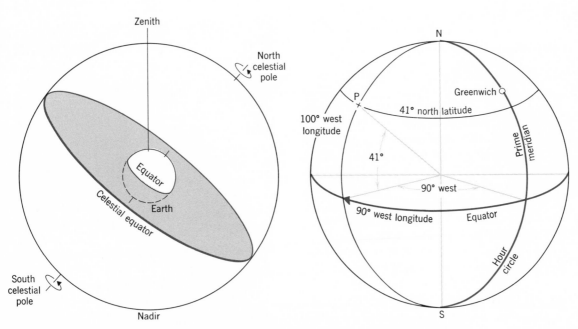

FIGURE 18.10 The celestial sphere is a system of star location similar to the longitude and latitude lines on Earth. The poles, equator, and horizon are extensions of these points on Earth.

Hour Circles

As a first step in developing such a system, a series of great circles is drawn through the celestial poles similar to meridians on the Earth. These are called *hour circles*. An arbitrary starting point has been chosen that coincides with the constellation Aries and is noted by the symbol γ. This is the 0^h circle. It is the point of the vernal equinox.

Every 15° east marks another hour of time across the globe. Thus, 15° east of 0^h is called 1^h and 30° east is 2^h, etc. This is continued all the way around the Earth back to the origin, 0^h. The star position is then described in hours and is referred to as the *right ascension* of the star.

Reference to a star having a right ascension of 21 hours means that the star will cross your specific location 21 hours after the vernal equinox crosses your meridian. One would then know when to look for the star.

The declination indicates the location of a star. For example, New York is about 41° north. Its zenith point is the point on the celestial sphere directly overhead and is +41°. If a star has a declination of +41°, it will cross directly over New York City. If its

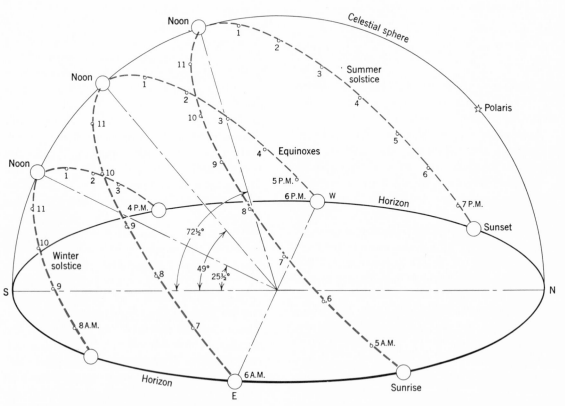

FIGURE 18.11 The sun's position in the sky changes throughout the year. The position of the sun above the horizon is determined by the relation of the Earth's axis to the sun. (Refer to Figure 18.7.)

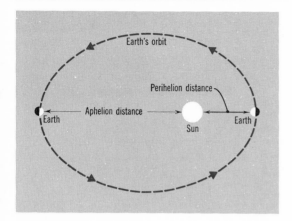

FIGURE 18.12 An elliptical orbit showing perihelion
and aphelion. Perihelion is that point on an elliptical
orbit where the satellite is closest to the sun. The sun is
located at one focus of the ellipse. Aphelion is the position of the satellite when farthest from the sun.

declination is +45°, it will be +45° minus
41° (or 4°) north of zenith; a star of +38°
would be 3° south of zenith.

In order to locate a star, three pieces of information are necessary; the latitude, the
star's declination, and its right ascension.

These reference points tell us where and
when to locate a particular star. Of course,
some stars pass overhead during daylight
hours and cannot be seen. Seasonal changes
in the sky also affect observation times.

The system of celestial coordinates has
been extended to include the sun and moon.
An almanaclike text called an ephemeris contains the declinations and right ascensions
for the stars, the moon, the planets, and the
sun.

The Physical Laws of the Universe

No study of astronomy or the Earth itself
can be complete without an understanding
of the physical laws that govern the motions

of celestial objects. A multitude of celestial
and planetary motions must be considered if
one is to be versed in astronomy.

Kepler's Three Laws of Planetary Motion

Johannes Kepler established astronomy on
a mathematical basis. He set down a huge
quantity of calculations based on the logs
kept by Tycho Brahe for nearly a generation
and, from these enormous undertakings, he
established his three laws of planetary motion. These laws explain how planets move
through the heavens.

Kepler's First Law states that each planetary orbit is an ellipse with the sun at one
focus. An ellipse is a geometrical figure that
is drawn around two foci (a circle has only
one focus). If you place two pins in a board at
some distance from each other, loop a string
around the pins, and draw a path with a
pencil by using the string as a guide, you
will obtain a flattened circle. This is an ellipse. An ellipse is the figure that the planets
and other objects in the solar system produce
during their revolutions around their parent
bodies. Actually, the planetary ellipses are
very slightly flattened circles barely distinguishable as ellipses if drawn to scale.
Figure 18.12 exaggerates the distortion
greatly.

The point in the Earth's elliptical path
closest to the sun (one of the foci) is called
the *perihelion* of that orbit. The *aphelion* is
that point which is farthest from the sun.

Kepler's First Law of Planetary Motion
further refined the Copernican Heliocentric
Theory, which assumed that the planets had
circular orbits. Kepler also helped to eliminate the idea of planets whirling about their
orbits in epicycles, which Copernicus was
forced to carry over from Ptolemy to explain
the retrograde motion of the planets.

Kepler's Second Law then carried his concept a step further. The Second Law relates

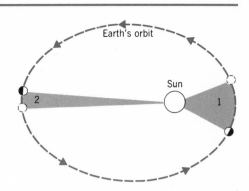

FIGURE 18.13 Each shaded area has the same area, and each arc is traversed in the same amount of time. The Earth must travel fastest to cover Area 1 and slowest to cover Area 2.

the idea of orbits to the velocities of the planets. Essentially the law reads, "the radius vector of a planet sweeps out equal areas in equal times." The areas shown in Figure 18.13 appear to be different sizes. However, if the base lines (the part of the orbit forming the base of the area) and the sizes of the legs are taken into consideration, it becomes clear that each area is equivalent to the other. In order to cover area number 1, near perihelion, in the same time as area number 2, near aphelion, the Earth must move faster near perihelion than it does near aphelion. Thus, as a planet moves around the sun, it moves at different velocities. Therefore, in the summer of the Northern Hemisphere, when the Earth is at aphelion, it travels at a slower speed than in the winter when it is at perihelion. All planets vary their orbital velocities in accordance with this law. Although the Earth's variation in velocity is slight, the velocity has been found to vary greatly with objects such as comets, which have gigantic elliptical orbits.

Kepler's Third Law, which took several years longer to develop than the first two, relates the length of time for a single orbit to the planetary distance from the sun. Kepler's Third Law states, "the squares of the periods of revolution (P_1 and P_2) of any two planets are proportional to the cubes of their mean

distances (D_1 and D_2) from the sun." Stated algebraically it reads:

$$\frac{P_1{}^2}{P_2{}^2} = \frac{D_1{}^3}{D_2{}^3}$$

In other words, if we take the period of a planet (year), square it, and divide by the cubed mean distance from the sun, a figure appears that is the same for all planets. This constant demonstrates that there is a relationship between the length of time it takes a planet to orbit the sun once and the distance between the sun and the planet.

Although these laws are of importance for the astronomer, they still leave a basic question: Why do the planets move as they do? Kepler showed the world how they move, no small accomplishment in itself, but it took another genius to show why all this takes place as it does. The question as to why planets move as they do was finally answered by Sir Isaac Newton.

Newton's Laws of Motion

Newton's contributions to modern experimental science are almost without peer in any field of endeavor. He developed an entirely new field of mathematics, the calculus, on which all modern science depends. We have already discussed some of his investigations into the nature of light, and his invention of the reflecting telescope. The final achievement that we shall discuss is his derivation of the laws of motion. Newton carried to their culmination Galileo's initial investigations into the motions of bodies, and Kepler's into planetary motion.

Newton's First Law of Motion explained the phenomenon known as inertia, the tendency of a body to resist a change in state. Newton's First Law states that "a body in motion or at rest tends to remain in motion or at rest *in a straight line*, with a constant velocity, unless acted on by an unbalanced outside force." Thus, if a satellite or planet is moving in a straight line, it *remains* moving in a straight line, unless acted on by an outside force. (See Figure 18.14.) The planets, very obviously, are not moving in straight lines but in ellipses, changing their directions as well as their speeds. Since they do not move at a constant velocity nor in a straight line, there must be some outside force affecting them to cause this unusual motion.

Newton's Second Law of Motion explains how motions are related to their causes. This law is often referred to as the force equation. It states that "the acceleration imparted to a body is in the same direction, and is directly proportional to the force applied and inversely proportional to the mass of the body." In other words, the motion given to a body is greater if the push or pull that is applied is greater. At the same time, the more massive the object, the greater the push or pull must be to impart a certain velocity to the object. Mathematically, the law is stated

$$F = m \times a$$

where m equals the mass, measured in kilograms, and a is the acceleration measured in meters per second squared. A force that imparts an acceleration of 1 kilogram-meter/second^2 is called 1 newton, in honor of the man who devised this scheme.

Newton's Third Law is one that everyone assumes he knows but very few people really understand. The law is called the "Action-

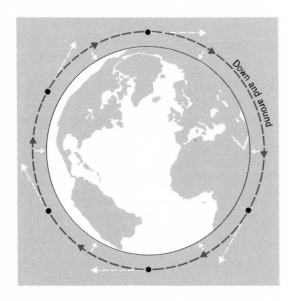

FIGURE 18.14 A satellite tends to move outward in a straight line (Newton's First Law of Motion). The gravitational pull of the parent body tends to pull the satellite toward the center of the parent body. The orbit traveled by the satellite is the figure that balances these two forces.

Reaction Law." Simply stated it reads: "for every action there is an equal and opposite reaction." All forces, therefore, really act in pairs. Notice the phrase "equal and opposite." This is the part that is rather difficult to visualize, because the opposite force is often not easy to see. For example, many people are under the impression that a rocket moves off the Earth because the hot, escaping exhaust gases push against the Earth. This idea has come about because photos of accelerating rockets show gases billowing up as they strike the surface of the Earth. However, if this is the driving force of a rocket, how can one explain the rocket's flight in space where there is nothing to push against? Actually, the rocket has a chamber in which the gases explode and expand. Some of the gas escapes

out of the chamber. But the rest of the gas within the chamber is pushing against the rocket itself (Figure 18.15), and the rocket opposes the gases pushing against its inner chamber. The gases pushing agianst the walls of the inner chamber are in the opposite direction from the escaping gases and drive the rocket forward.

In examining planetary orbits, it is important to remember inertia. Planets tend to move away *in a straight line*. Yet, they still revolve around the sun. The tendency to move away from the sun in a straight line is one force acting on the planets; the sun pulling the planets toward it (gravity) is the other one of this pair of forces (Figure 18.16). Thus the elliptical orbits are created. The orbit followed by the planets is the motion which ex-

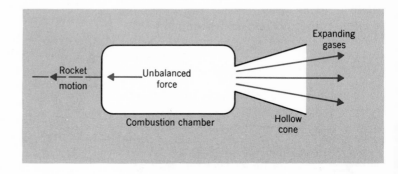

FIGURE 18.15 The explosive force of rocket fuel is expended in all directions. The force escaping from the rear allows an unbalanced force to push toward the front of the rocket. The rocket moves in the direction of the unbalanced force.

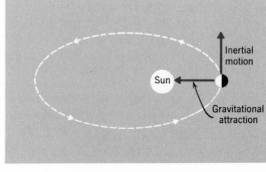

FIGURE 18.16 Inertial force tends to cause a satellite to move outward from the parent body.

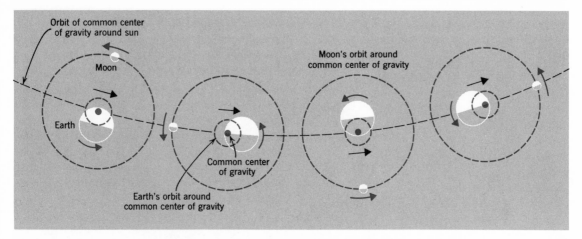

FIGURE 18.17   Earth-moon system. The Earth and moon form a common system. Both bodies actually revolve about a common center of gravity which is located below the surface of the Earth.

actly balances the inertial velocity of the planet and the gravitational pull of the sun.

Kepler's laws explained how planets move, and then Newton's laws related motion to all moving bodies. Finally, Newton explained why the planets behave as Kepler described them.

Newton's Law of Universal Gravitation

Everyone has heard that Newton, while sitting under a tree, was struck on the head by a falling apple. This supposedly gave him the idea for his Law of Gravity. Whether this story has any basis, it has very little to do with Newton's calculation of the force of gravitational attraction.

The planets, or any moving objects, tend to remain in motion at constant velocities in straight lines. Yet the planets do not follow straight lines, but are pulled from their theoretical straight lines in elliptical orbits around the sun. The moon also revolves around the Earth, so something must also be forcing the

moon to deviate from a straight-line path. Consideration of the Earth-moon system actually led Newton to formulate the Law of Universal Gravitation. The moon deviates from a straight-line path more than the Earth. Newton reasoned that the force responsible must be due to the massive size of the Earth compared to the much smaller mass of the moon. He later refined this idea when he realized that deviation from straight-line paths was also because of the distances between objects. Stated simply, the Law of Universal Gravitation says: "every object in the universe attracts every other object with a force directly proportional to the square of the distance between them." This means that the larger the object, the greater the force. However, as objects separate, the force of gravity decreases rather sharply. Force decreases by the square of the distance and, therefore, if the moon were twice as far from the Earth as it is now, the force would decrease to ¼. If the moon were four times as far away from the Earth as it is now,

FIGURE 18.18 Albert Einstein, the great 20th-century physicist-philosopher. Although Einstein is usually thought of as the man who developed the theory that led to the creation of the atom bomb, his contributions to our understanding of the universal laws of gravity and the characteristics of light are more numerous and equally important (Wide World).

the force would decrease to $\frac{1}{16}$, and so on. Newton's Law of Universal Gravitation can be stated algebraically as

$$F = G\,\frac{m_1 m_2}{d^2}$$

Gravity (G) is a constant used in all problems. Its value is 6.67×10^{-11}. m_1 and m_2 are the masses in kilograms of the two bodies under consideration, d^2 is the distance between the two objects measured in meters,

and F is the gravitational force between the two masses.

With Newton's work, Kepler's laws were finally explained. Scientists were at the point where they understood how and why planets move as they do. Scientists have now come to know what forces keep the solar system, in fact the entire universe, in its state of motion. Kepler's and Newton's laws have helped man to discover new planets, to measure celestial masses, to develop understanding about the

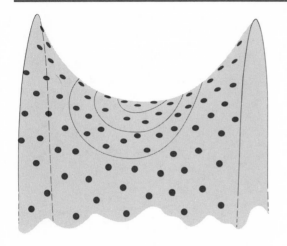

FIGURE 18.19 Space may be a saddle-shaped, infinitely curved structure with the galaxies randomly distributed throughout (After Inglis).

shape of the Earth, and even to understand the motion of atomic particles. But astronomy is by no means at an end, since there is much more to be understood about the motions of celestial objects.

Newton's Law of Universal Gravitation led to the further refinement of Kepler's Third Law. He showed that, in addition to the periods of the planets, the masses of the sun and the planet in question should be considered. However, the error is so slight that the law is usually still written as first stated by Kepler.

Einstein and Concepts of the Universe

Most people are under the misapprehension that once a scientific law is formulated, like Newton's Law of Universal Gravitation, it is unquestionable. Furthermore, they believe that scientists completely understand gravity. Unfortunately, this is not the case.

In the early part of the twentieth century, Albert Einstein further refined the concept of gravitational forces. There was still a question in the minds of many scientists as to what gravity actually is. To be sure, it is

a force, but what makes this force emanate from objects? Einstein held that planets move because of an actual alteration in space that results from a distortion created by the sun. Therefore, planets move in a space-time relationship, and follow a path dictated by spatial distortion.

Einstein predicted several events on the basis of his theory. He had already established the absolute motion of light; that is, that light moves at a speed independent of the motion of the observer—approximately 186,000 miles per second in a vacuum. Einstein believed that light from distant stars bends as it passes the sun because of the spatial distortion of the path the light follows. Furthermore, he stated that this alteration of space would cause a slight shift in the orbit of Mercury—a shift that had been observed and had puzzled scientists for years until Einstein's explanation.

Careful measurement of starlight passing close to the sun during eclipses shows Einstein to be correct. The light bends slightly toward the sun. Thus, the apparent straight line produced by starlight is not a straight line, and is misleading.

Einstein also predicted a shift in spectral

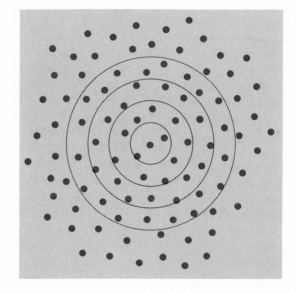

FIGURE 18.20 Euclidean Space. A two-dimensional view of the universe places the galaxies in an ever-spreading arrangement on a flattened circle.

lines similar to the Doppler Effect. He believed that light produced by a massive star would shift toward the red end of the spectrum from its place of origin. He hypothesized that light traveling from a strong gravitation field (or spatial distortion) set up by a very massive star slows down as it travels and shifts toward the red end of the spectrum.

The Einsteinian theory led to further refinements of our concept of the shape of the universe. Man has conceived of many types of universes since his thoughts first turned in this direction. From the ancients' ideas of a canopy of light covering the Earth to the idea of a boundless infinite universe, ideas and concepts of our universe have ranged far and wide.

If light does bend as it travels through space, it is conceivable that space is a curved, finite structure with specific boundaries. Our universe now presents us with a riddle. It may be, as some think, a curved, saddle-shaped structure with an infinite size (Figure 18.19), or a closed system with finite boundaries (Figure 18.20). The galaxies and intergalactic material may be located on and around the saddle, or they may be like points on, and within, a balloon.

In Einsteinian space, time is considered to be a coordinate of the system—the so-called fourth dimension. Thus, the relationship referred to as the space-time continuum takes into consideration the usual planes considered in geometry plus time.

To date, a satisfactory picture of the universe has not been developed. Scientists are still trying to arrive at a complete picture of the universe that correlates classical Newtonian mechanics and the Einsteinian modifications of classical physics. The finite versus the infinite universe controversy still goes on. However, as we gather more information, it is hoped that an improved concept of the universe will develop.

CHAPTER 19

The Solar System

SCIENTISTS HAVE TURNED THEIR EXPLORA-tions and thoughts to those objects in our immediate vicinity that make up the solar system. The sun, with its family of planets, forms a far more complex system than people usually imagine. This chapter describes, as accurately as present knowledge permits, the physical characteristics and interrelationships among the objects found in our solar system.

The Planets

The planets are roughly divided into two general groups; the Inferior planets, those that make up the inner sequence of smaller planets, and the Superior or large, outer planets.

The Inferior Planets

Mercury. Mercury is the planet closest to the sun. It is the smallest planet, with a diameter of 3008 miles, and it lies 37,200,000 miles from the sun. Mercury's orbit varies from 28.6 million miles at perihelion to 43.4 million miles at aphelion; it is the *mean* distance from the sun that is used to specify the position of the planet. Because astronomical distances are so large, it is cumbersome to talk in terms of miles. For this reason the *astronomical unit* (A.U.) has been adopted. One A.U. is equal to 93,000,000 miles, which

is the average distance of the Earth's orbit from the sun.

It is known that Mercury requires 88 days to revolve around the sun. However, there is some question about its rotation. Mercury is an extremely difficult planet to observe because of its proximity to the sun. For a long time it was thought that Mercury did rotate on its axis, and completed a cycle in 88 days. Both of these beliefs have since been questioned. At present it is believed that Mercury does rotate, although this has not as yet been firmly established. The period of rotation was recently estimated to be 60 days.

Temperature readings of the planet's surface show a range from +640°F on the lighted side to −415°F on the dark side. This tremendous difference results from the fact that Mercury lacks a protective atmosphere because its gravitational attraction is only about 1/4 that of the Earth's. With this minimal amount of gravitational attraction, molecules of gas can escape from the planet's surface. When the planet is closest to the sun it reaches a temperature of 780°F. This extreme heat is also believed to contribute to the atmospheric deficiency.

Venus. To the Greeks, Venus was known as the evening or morning star, depending on when it was sighted. Its brightness is due to the reflection of light by its atmosphere. The amount of reflected light, the albedo, is dependent on the covering of

427

gases on a planet. For example, Mercury has an albedo of approximately 6 to 7 per cent, very similar to our moon, whereas the planet Venus reflects nearly ¾ of its incident light for an albedo of 75 per cent. With a diameter of about 7600 miles, Venus is nearly as large as the Earth.

Venus revolves around the sun once every 225 days. It is not known how long its rotation takes, but it has been estimated that, perhaps like Mercury, Venus may rotate once for every revolution. In addition, although all the other planets rotate counterclockwise, Venus may rotate clockwise. Further examination is required before these questions can be answered.

Venus is so close to the sun (about 0.7 A.U.) that observation is difficult. Close observation is further complicated by the planet's dense cloud covering. Spectrographic analyses and space probes have detected large quantities of carbon dioxide in the atmosphere, with very small amounts of oxygen and nitrogen. The surface temperature, as measured by Explorer II, is in the neighborhood of 600°F. The planet appears to have almost no magnetic field.

As Venus is viewed from the Earth, it presents phases similar to those of the moon, although at its farthest point, on the other side of the sun, it appears as a small disk. It exhibits crescents of various sizes as it moves closer to the Earth.

Mars. Immediately beyond the Earth is Mars, a brilliant red planet. Mars is 1.52 A.U. from the sun and moves nearly as close to us as Venus does at times. Because it is brilliantly lighted by the sun, we know more about the surface of Mars than that of any other planet.

The diameter of Mars is only 4200 miles, nearly half that of the Earth. It is the only

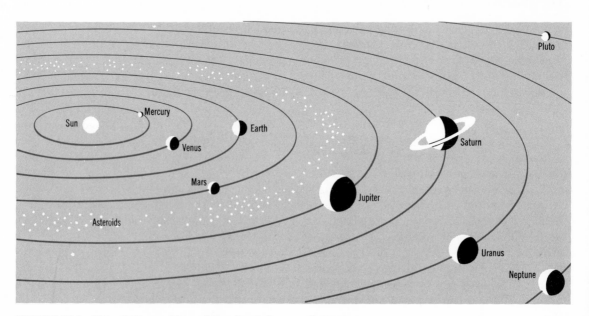

FIGURE 19.1   The relative positions of the planets in our solar system.

FIGURE 19.2 (*left*) Mercury (dark spot) making its transit across the sun. The proximity of Mercury to the sun makes observations of this planet extremely difficult (Russ Kinne, Photo Researchers).

FIGURE 19.3 (*below*) When viewed from the Earth, Venus exhibits phases much like the moon; a fact discovered as early as the time of Galileo (Lowell Observatory).

FIVE PHASES OF VENUS

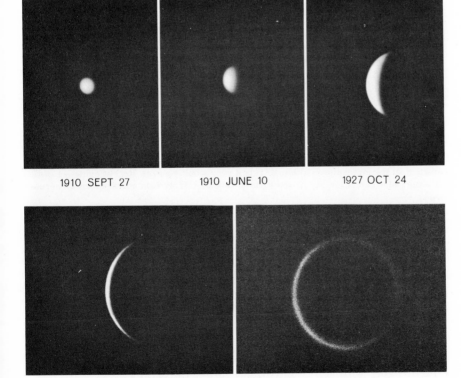

1910 SEPT 27   1910 JUNE 10   1927 OCT 24

1919 SEPT 25   1964 JUNE 19

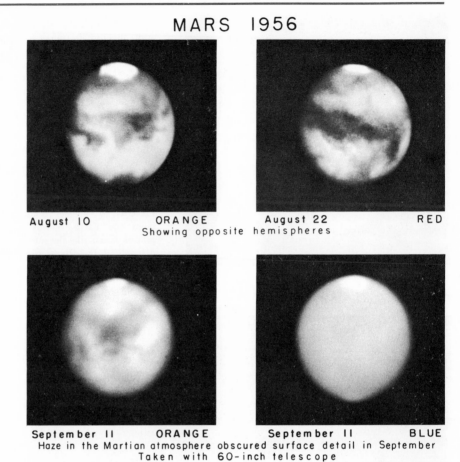

MARS 1956

August 10   ORANGE   August 22   RED
Showing opposite hemispheres

September 11   ORANGE   September 11   BLUE
Haze in the Martian atmosphere obscured surface detail in September
Taken with 60-inch telescope

FIGURE 19.4   The two hemispheres of Mars. The presence of an ice cap denotes some moisture in the atmosphere, although it is a very small amount. Recent evidence indicates the polar cap is largely frozen carbon dioxide (Mt. Wilson and Palomar Observatories).

other Inferior planet that has satellites like the Earth. The two moons of the planet are named Phobos and Demos.

The temperatures on Mars, although severe at times, would not prevent the support of life. Its daily surface temperature ranges from about −25°F to about 85°F. Its atmosphere is mostly nitrogen with a small percentage of carbon dioxide and possibly some oxygen. Because of its small

mass, Mars has a lower gravitational pull and thinner atmosphere than the Earth.

Although Mars has polar caps of frozen water which change in size from winter to summer, there is a scarcity of water vapor in the air. These caps, therefore, are not vast regions like the polar caps found on the Earth, but rather thin and sparse.

In 1877, an astronomer by the name of Schiaparelli noticed a series of lines on the

Martian surface. This finding caused great excitement. For many years these lines were thought to be a series of man-made canals. These canals, and larger areas called maria (seas), change from greenish hues in summer to brownish colors in the winter. As a result, it is now believed that these areas may harbor plant life similar to the lichens found on Earth.

The reddish color of the planet Mars is attributed to the presence of iron oxides. Further evidence of this mineral seems to be given by the fact that great dust storms periodically move over the planet.

Laboratory experiments demonstrate that primitive life forms such as bacteria can grow in Martian conditions. It is possible then, that life of some sort may be found on Mars, a possibility that is slight, but not to be discounted.

The Superior Planets

Beyond Mars lie the five outer planets. These planets, with the exception of Pluto, are massive, cold, lifeless, and hostile according to Earth's standards.

Jupiter. Jupiter is the largest of all the planets. It has a volume greater than all the eight planets put together, and it is over 300 times more massive than the Earth (Table 19.1 and Table 19.2).

FIGURE 19.5 Jupiter as seen through the 200-inch Mount Wilson telescope. The Red Spot is very obvious (Mt. Wilson and Palomar Observatories).

Although this planet has an immense volume, it is not very dense. In fact, it is only slightly heavier than water. This curious feature of the planet indicates that it is composed of the lighter elements and hydrocarbon compounds. Observations of the planet lead to the conclusion that its "atmosphere" consists of ammonia, helium, and large quantities of hydrogen. The atmosphere may be several hundred miles deep. In addition, methane, similar to the gas used in cooking stoves, has been found. All these gases are probably resting on a thick sea of ammonia and liquid hydrogen that was formed by the great pressures developed on the planet. Whatever solid rock exists probably makes up only 1/300 of the planet's mass.

Jupiter has several other unique characteristics. For instance, it has a huge Red Spot, 30,000 miles long and 7000 miles wide, which moves rapidly over the surface. The nature of this spot is unknown. It is thought not to be on the surface, but in the atmosphere.

Observation of Jupiter shows that the planet rotates faster at the equator than at the poles. In addition, the rotational period is faster than any of the other planets'. The rapid rotation, plus the unusual composition of the planet, causes an appreciable flattening at the poles, so that Jupiter is much more oblate than the other planets.

Jupiter receives less heat than the other planets and its albedo is high. That is, it reflects a great deal of the light back to space. As a result, its surface temperature is quite low, about −185°F.

The planet has 12 satellites, more than any other planet. Three of these moons revolve in counterclockwise fashion, while the others move clockwise about the planet.

The chances for life on this planet are, indeed, remote. Any water is surely in the form of ice, and the idea of a living creature drinking ammonia, breathing hydrogen, and made of compounds other than carbon-based organic compounds is rather unthinkable for us at this time.

Saturn. Saturn is unique among the planets in having a system of rings. It is one planet with which most people are familiar.

Table 19.1

NAME	DISTANCE FROM THE SUN (ASTRO-NOMICAL UNITS)	REVO-LUTION	ROTA-TION	RADIUS	VOL-UME	MASS	DEN-SITY
Mercury	0.39	88d	60d	0.38	0.054	0.054	5.4
Venus	0.72	225d	247d?	0.96	0.88	0.81	5.1
Earth	1.0	365.25d	23h56m	1.0	1.0	1.0	5.5
Mars	1.5	687d	24h37m	0.53	0.15	0.108	4.05
Jupiter	5.2	11.86Y	9h55m	10.8	1260	318	1.4
Saturn	9.5	29.5Y	10h15m	9.5	734	95.2	0.71
Uranus	19.2	84Y	10h50m	3.8	536	14.6	1.56
Neptune	30	164.8Y	16h	3.4	391	17.2	2.5
Pluto	39.5	248.4Y	6.4d	0.45?	0.09?	?	?

FIGURE 19.6 Saturn and its rings. Cassini's division can easily be seen separating the bands of debris (Mt. Wilson and Palomar Observatories).

Whirling about Saturn's equator are countless particles of stone and dust similar to meteorites. These bands are tilted at times so that we can see them on edge; at other times we see them in almost full view. They present a striking picture.

Saturn, whose surface temperature is on the order of −240°F, is less dense than Jupiter and, in fact, is less dense than water. In other words, if we had a large enough pool in which to place the planet, it would float! Saturn (like Jupiter) is composed of an im-

mense ammonia and methane atmosphere with a small solid core. Saturn has more methane than ammonia, just the reverse of Jupiter.

In addition to the ring system, Saturn also has nine moons. One of the moons moves in a counterclockwise direction while the other eight revolve about the planet in a clockwise fashion.

Saturn's period of rotation is 10 hours and 30 minutes, which is fairly rapid compared to the other planets. The rapid rotation produces an oblate figure. Because it is 9.5 A.U. from the sun, Saturn has a year equal to 29½ Earth years.

A great deal of research has been done on the rings of Saturn. It has been found that the inner ring, which is separated from the outer ring by a gap called Cassini's division, rotates more rapidly than the outer ring. In addition, when we see Saturn pass in

Table 19.2 **Inclinations**

Mercury	7°	Saturn	27°
Venus	6°	Uranus	98°
Earth	23.5°	Neptune	29°
Mars	25°	Pluto	?
Jupiter	3°		

FIGURE 19.7　Uranus and its three moons (indicated by arrows) (Lick Observatory).

FIGURE 19.8　Neptune and one moon (arrow) as seen by the Lick Observatory telescope (Lick Observatory).

front of a star, the star remains visible. Thus, the rings are not solid but are made of ice-coated rocks and pebbles. It is thought that these particles may be the debris from an exploded moon.

Saturn's atmosphere occasionally exhibits belts of cloudlike material that move across the surface of the planet. Scientists are unclear as to the origin and composition of these belts or white patches.

Uranus. All of the planets we have described thus far were known to the ancients.

Uranus was the first to be discovered with the aid of the telescope. It was discovered in 1781 by William Herschel, one of the greatest of the early modern astronomers.

Uranus is large, and is 19 A.U. from the sun. It is not brilliantly lighted and, thus, is difficult to see. Its surface temperature is about −300°F, with an atmosphere similar to the ones of the other large planets, possessing ammonia and large quantities of methane.

Although no distinct surface features have

FIGURE 19.9 A pair of photographs showing the motion of Pluto. The sequence covers a period of 24 hours (Mt. Wilson and Palomar Observatories).

been observed, Uranus does have one unique feature. Its axis is tilted at an angle of 98° to the perpendicular of its orbit. It rotates backwards to the plane of its orbit, the reverse of all other planets. The reason for this is unclear, although a satisfactory explanation is still being sought. This feature affects Uranus' five moons as well; they also revolve in a backward or retrograde motion.

Neptune. Neptune was discovered in 1846 after discrepancies in the orbit of Uranus were observed. These discrepancies could be explained only by the presence of another planet near the edge of the solar system.

Although little is known about the planet, its surface temperature is about −330°F, and its atmosphere is predominantly composed of methane.

The planet has two satellites which move in a retrograde manner. Although its period of rotation is short (16 hours), Neptune requires 165 Earth years to make one revolution about the sun.

Pluto. After the discovery of Neptune, it was noticed that there still were discrepancies in the orbit of Uranus that were not explained by the presence of Neptune. It was assumed, as in the case of Neptune, that still another planet existed. Pluto, a small planet about 40 A.U. from the sun, with a year comparable to 248 Earth years, was found. Its surface temperature is on the order of −350°F. It was discovered in 1930.

This planet has a rather eccentric orbit. At times it travels inside the orbit of Neptune. This fact leads some observers to conclude that Pluto may once have been a satellite of Neptune which broke away and went into an orbit of its own.

From time to time one hears stories of a new planet that will shortly be discovered beyond Pluto. There is no evidence to uphold this idea at present.

Other Bodies in the Solar System

There are several other bodies within our solar system that are under the gravitational influence of the sun. Among these bodies are comets, meteors, and the asteroids that lie between Mars and Jupiter.

Comets

Comets are one of the most spectacular phenomena in the solar system. They are actually loose masses of solid earthlike material mixed with frozen ammonia, water, carbon dioxide, and methane. As a comet passes near the sun (Figure 19.11), the solar heat causes the material to become gaseous. The main body—the head or coma—increases in size and develops a surrounding corona of gases. This gaseous material extends to form a long tail, which always faces away from the sun, and is millions of miles long. After swinging around the sun, the coma and tail begin to shrink and disappear. When the comet is traveling at a great distance from the sun, it is reduced to a small nucleus of material.

Although comets are considered part of the solar system, their orbits are highly elliptical and very long. Thus, comets usually take a long time to reappear since they pass beyond the solar system when approaching aphelion.

Of all the comets that have been observed and described, the most famous is Halley's Comet (Figure 19.12), which returns to the vicinity of the sun every 76 years. It is due to appear once again in 1986. As it returns to our vicinity in the solar system, it makes a brilliant appearance because of the reflection of solar light. Although it has the look of a fiery ball in space, Halley's Comet, like all comets, is nonluminous.

FIGURE 19.10 Comet Ikeya-Seki with its enormously long, wispy tail of debris (Lick Observatory).

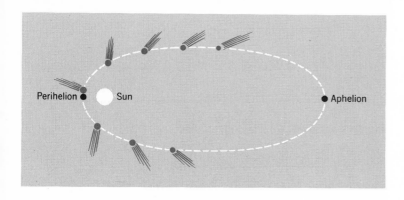

FIGURE 19.11 As a comet nears the sun, the tail increases in size as the particles are pushed from the comet's central region. The tail always points away from the sun, as the material is forced from the comet by the sun.

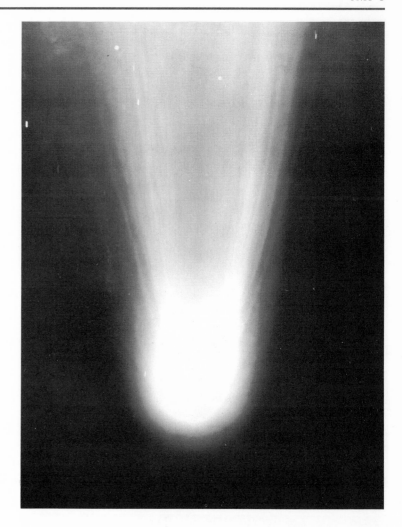

FIGURE 19.12 Halley's comet was photographed on its last visit (1910) by Mount Wilson and Palomar Observatories. The relationship of the head and tail of the comet can be seen (Mt. Wilson and Palomar Observatories).

Meteors

In a manner of speaking, meteors are the debris of space. They consist of rocky particles that travel through space. On occasion, one of these structures moves close enough to a planet like the Earth to be captured by gravity.

Meteoroids, the name given to meteors when still in space, become fiery masses because of friction with the air. These "shooting stars," as they are erroneously called, are visible as they burn in the atmosphere. Most meteors burn brightest between 70 miles and 40 miles above the Earth's surface.

Millions of meteors enter the Earth's atmosphere every day of the year. However, the friction between meteors and the atmosphere rather effectively burns up most of them before they strike the Earth.

The meteors that do reach the Earth's surface, called *meteorites,* yield a great deal of information about the problems of spaceship reentry into the atmosphere. They also produce a great deal of speculation about their origins. Some researchers think they come from outside the solar system; others believe that they originate in the belt of asteroids or as debris from comets. More than likely, meteorites begin within the solar system.

Three types of meteorites have been found on the Earth's surface (Figures 19.13 and 19.14). *Iron-nickel* meteorites consist of iron with a small percentage of nickel and a smaller percentage of earthy materials. *Stony-iron* meteorites are composed of iron with a high percentage of other earth materials. *Stony* meteorites are very much like terrestrial rock with a mixture of various minerals.

Although meteors are generally isolated phenomena, there are times when meteor showers occur, and huge swarms of meteors pass through the heavens. In early August every year (approximately August 9 to 13)

FIGURE 19.13 An iron-nickel meteorite that fell in Mexico. This specimen weighs 258 pounds (American Meteorite Laboratory).

FIGURE 19.14 This stony meteorite is one of a group of 26 that fell at Leedey, Oklahoma, in 1943 (American Meteorite Laboratory).

meteor showers can be seen in the Northern Hemisphere, from the general vicinity of the constellation Perseus. The August shower lasts for several evenings.

Meteors are credited with producing huge craters on the Earth (Figure 19.16). Barringer Meteor Crater in Winslow, Arizona, is 4150 feet in diameter. Chubb Crater in Quebec, Canada, is two miles in diameter. In June of 1908, a meteor explosion occurred in Siberia leaving a ring of ten craters. Trees were blown down 20 miles away, and the effects on homes were felt 50 miles away. Fortunately, no large-scale meteorites have landed on populated areas, and damage from the very few large meteorites has been negligible.

The Asteroids

The asteroids, located between Mars and Jupiter, are sometimes called the minor planets. They may be the remains of an exploded planet. More than 3000 of these so-called planetoids have been observed thus far, with diameters ranging in size from 480 miles (Ceres), 300 miles (Pallas), 240 miles (Vesta), 120 miles (Juno), down to those only a few miles in diameter.

Because of their small sizes, the asteroids have no gravity and, thus, lack an atmosphere. The accompanying dust boulders and rock that are found in this belt may give rise to meteors and other solar system debris.

Interplanetary Dust

Scientists used to think that space was an empty vacuum. Recently, however, evidence has been found to prove this belief false. Actually, space contains interplanetary dust or micrometeoroids. These particles of dust are no more than a few thousandths of an inch in diameter. In addition, there are no more than a few of these particles in a volume that would be equivalent to the Earth itself. Nevertheless, these particles are sufficient to

FIGURE 19.15 A shower of meteors photographed from Kitt Peak Observatory. Although seemingly far out in space, these showers are the result of meteors burning up in the atmosphere of the Earth (Dennis Milon, Sky and Telescope).

FIGURE 19.16 The Barringer Meteor Crater near Winslow, Arizona. The crater is 570 feet deep and 4150 feet wide. The rim rises above the ground to a height of 150 feet (American Meteorite Laboratory).

cause the scattering of sunlight. This faint *zodiacal light* can be observed under ideal conditions before sunrise and after sunset. The micrometeoroids are probably created by the slow disintegration of the comets during their orbits around the sun.

Natural Satellites

The largest planet, Jupiter, has the most known satellites—twelve in number. Saturn, next in size to Jupiter, has nine. There are five satellites of Uranus, and both Neptune and Mars have two. As far as we have been able to tell, Mercury, Venus, and Pluto have no moons. The Earth is the only planet with a single moon. Furthermore, the Earth's moon is the largest moon in proportion to the size of its planet.

The Moon

The moon revolves around the Earth in an orbit that varies from 220,000 to 253,000 miles. Actually, it is incorrect to say that the moon revolves around the Earth because both bodies revolve around a common center of mass, which is about 1000 miles from the Earth's center. This places the center of mass of the Earth-moon system within the Earth.

The moon's diameter is 2160 miles or about 1/4 that of the Earth. Its mass is only 1/81 that of the Earth's. As a result, the gravitational attraction of the moon is only 1/6 that of the Earth.

Because of the low surface gravity of the moon, there is little atmosphere and very little protection from the sun's heat. Since the Earth's atmosphere acts like a greenhouse, the Earth has a rather moderate change in temperature from day to night. On the moon, however, daylight surface temperatures approach the boiling point of water, and at night the temperature plunges to below −200°F.

As the moon moves about the Earth, it passes through a series of phases that we see as apparent changes in the shape of the moon (Figure 19.18). When the moon is between the Earth and the sun, the side facing

FIGURE 19.17 The Sea of Tranquility on the moon as photographed during the Apollo 8 mission. The shadows in the craters reveal that they are relatively deep. Scarps can be seen across the moon's face (NASA).

us is in darkness; this is called the new moon. As the new moon moves around the Earth, increasing small sections of its face move into the light, and we see it as a crescent. In about a week, we see ½ of the lighted face. This is called the first quarter of the moon. At the end of approximately two weeks, the full face is opposite the sun and is fully lighted. This is the full moon, or second quarter. In the third week (or third quarter), ½ of the lighted face is seen. The section of the moon that is lighted during the first quarter is not the same portion of the face that appears during the third quarter.

Day and night on the moon are different from Earth's. Each day and each night is about 14 Earth days long, for a total moon day of 29½ Earth days. By using the stars as

a reference, the *sidereal month* on the moon requires only 27⅓ days (27.3217 days) to complete a cycle. However, during this time, the Earth will have moved in its orbit, and two extra days are necessary for the complete cycle of phases to be seen from the Earth. Therefore, the complete revolution in respect to the Earth is 29½ days (29.53 days). This is called the *synodic month,* and is the time necessary to return the moon to the same position in respect to the Earth. (See Figure 19.19.)

The Lunar Surface

Because the moon has no atmosphere, it reflects less light than the Earth. Nevertheless, the amount of reflected light is sufficient

FIGURE 19.18   The various phases of the moon as viewed from the Earth (Lick Observatory).

to allow fairly easy examination of the lunar surface. The moon is pockmarked by high-rimmed craters, which vary from a few hundred feet to nearly 150 miles in diameter. Until recently, it was thought that lunar craters were volcanic in origin. However, current investigations indicate that most are meteor craters like the ones found on the Earth. Radiating outward from the craters are long streaks of material called rills. These rills may be either cracks in the surface or material thrown out by impact when meteors formed the craters.

The lunar surface is also marked by large smooth plains called maria (seas). They are not seas, in any sense of the word, but are flat unmarked plains of hardened lava covered by dust.

The moon revolves around the Earth once for every rotation that the satellite makes on

its own axis. As a result, the moon always keeps the same side toward the Earth and the other turned away. However, the moon is tipped at an angle of 6°. This enables us to see past the sides, top, and bottom at different times of the month. We have actually seen about 55 per cent of the lunar surface because of this unusual tilt.

The Tides

As we pointed out in Unit 3, the oceanic tides of the Earth are the result of the gravitational attraction between the masses of the Earth and moon. On the side of the Earth that faces the moon, the waters bulge toward the moon. On the opposite side of the Earth, the water is pulled slightly toward the moon, thus, moving away from the Earth on that side. On the two sides of the Earth where the

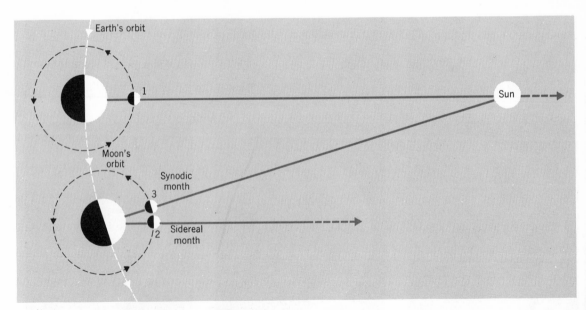

FIGURE 19.19 Synodic versus sidereal month. A sidereal month measured in relation to a star requires 27.3 days. When the moon's position is measured relative to the sun the length of one month is 29½ days. The additional time is required by the shift in the Earth's orbit causing the moon to travel the two extra days to fully complete one cycle.

bulges appear, we experience high tides. The low tides appear on the sides of the Earth at right angles to the high tides.

As the Earth rotates and the moon moves around the Earth, these tidal bulges move around the Earth at a rate of about 1000 miles per hour. Tides may vary anywhere from a few feet along shorelines to as much as 50 feet in deep channels like the Bay of Fundy.

There are usually two high tides and two low tides every day. It actually takes nearly 25 hours to complete a full cycle, since friction with the Earth retards the tides by nearly 50 minutes. This friction also affects the rotation of the Earth, slowing it down by approximately 0.001 of a second per century.

Although the moon produces the greatest tidal effects, the sun also plays an important role in producing tides. When the moon is at its full or new moon position, the sun's effect on the tides is added to the moon's, and thus a higher than normal tide occurs. This is called a spring tide. Low tides are also reduced at this time. Neap tides occur when the sun, moon, and Earth are at right angles to each other. Then the sun partially cancels out the effects of the moon, causing lower than normal high tides.

Although we have discussed tides only in relation to the waters of the Earth, tides have also been observed in the atmosphere, and even in the solid Earth. The air, being less dense than the water, reacts quite well to gravitational forces. The lithosphere, of course, does not react as obviously, but nevertheless does react to the pull of the moon.

The Sun

Solar Characteristics

The sun is the nearest star to the planet Earth, and it is the only one about which we have gathered much information. The sun

contains nearly 99.8 per cent of the mass of the entire solar system. The rest of the mass (0.2 per cent) makes up the planets, the asteroids, and the other interplanetary material that we have previously discussed. The sun is a massive structure. It has 330,000 times the mass of the Earth. Nearly 1,300,000 planets the size of the Earth would fit into the volume occupied by the sun.

The sun is approximately 93,000,000 miles away from the Earth. You will remember that this distance has been established as the astronomical unit. Actually the A.U. has been measured much more accurately as 92,950,-000 miles, but the other is sufficient for our purposes.

Because the Earth rotates around the sun in an elliptical orbit, at perihelion it is only 91,500,000 miles from the sun, and 94,500,-000 miles at aphelion. Thus, the 93,000,-000-mile figure for the A.U. is an average.

Knowing the solar distance allows certain other measurements to be made. We know today that the diameter of the sun is 865,000 miles, or about 35 times the diameter of the Earth. This figure has been calculated by geometrical methods, by employing the orbit of the Earth around the sun.

Although the sun is 330,000 times as massive as the Earth, its volume is more than a million times that of the Earth. This huge space occupied by the hot gases of the sun has a density similar to the larger planets'. It is only about 1.4, as compared to the 5.5 density of the Earth. Volume for volume the sun is just slightly heavier than water.

The sun is made of an overwhelming abundance of hydrogen, and a smaller percentage of the heavier elements. It is the huge volume of light elements that causes the sun to have a smaller density than the Earth.

The mass of the sun produces a gravitational attraction on the order of 28 times that of the Earth. In other words, a man weighing 200 pounds on Earth would weigh 5600

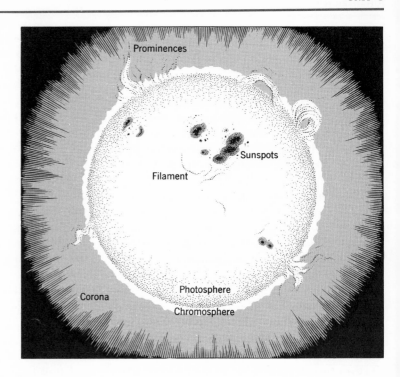

FIGURE 19.20 A cross-sectional view of the sun.

pounds on the surface of the sun. In order for a rocket to escape from the sun's gravitational pull, it would have to travel about 400 miles per second, as compared to the escape velocity of 7 miles per second on the Earth.

Wavelength analysis of solar light reveals that the surface temperature of the sun is in the vicinity of 6000°K or about 10,000°F. Further analysis also reveals that the core of the sun may have temperatures in the range of 20 to 25 million °F. The spectrum of the sun is in the slightly yellow range, placing it near the midpoint of star temperatures.

Analysis of the solar spectrum is complicated by the fact that the Earth's atmosphere interferes with and scatters solar energy. As a result, scientists must be sure to allow for refraction and scattering in producing figures regarding the spectra of celestial objects. This sort of spectral analysis also allows scientists to gather additional information.

For instance, astronomers can determine the different elements that make up the sun. Since the different elements produce different colors when heated, the spectrograph yields distinctive colors for each element. To date, nearly ⅔ of the 104 elements known on the Earth have been detected and identified in the solar spectrum. It is believed that further analysis will reveal the presence of still other elements in small quantities.

Energy Production

If we assume that the sun has been emitting energy for at least five billion years, then enormous amounts of energy have been released. It has been calculated, for example, that the sun emits 89,570 calories of heat per square centimeter every minute. To produce these tremendous quantities of energy, a very unusual energy-producing reaction must be involved. In fact, two kinds of re-actions take place on the sun to satisfy these energy requirements. One reaction is the carbon cycle, and the other the proton-proton reaction. The carbon cycle utilizes heavier elements than the proton-proton reaction and, although the carbon cycle may play some role in energy production within the sun, it is less likely to occur on the sun than on hotter stars.

The carbon cycle follows the pattern of these equations:

$$_6C^{12} + {_1}H^1 \rightarrow {_7}N^{13} + \text{gamma ray (energy)}$$

$$_7N^{13} \rightarrow {_6}C^{13} + {_1}e^0 \text{ (positron)}$$

$$_6C^{13} + {_1}H^1 \rightarrow {_7}N^{14} + \text{gamma ray}$$

$$_7N^{14} + {_1}H^1 \rightarrow {_8}O^{15} + \text{gamma ray}$$

$$_8O^{15} \rightarrow {_7}N^{15} + {_1}e^0$$

$$_7N^{15} + {_1}H^1 \rightarrow {_6}C^{12} + {_2}He^4$$

The net result of this cycle is the combination of four hydrogen atoms, which are single protons, into an atom of helium. Notice that

the original carbon ($_6C^{12}$) is returned at the end. It therefore acts as a catalyst in the cycle.

The proton-proton reaction, which is more likely to occur on the sun, is simpler. It causes this general sequence of events:

$$_1H^1 + {_1}H^1 \rightarrow {_1}H^2 \text{ (deuterium)}$$

$$_1H^2 + {_1}H^1 \rightarrow {_2}He^3 + \text{gamma ray}$$

$$_2He^3 + {_2}He^3 \rightarrow {_2}He^4 + 2{_1}H^1$$

Observe that the net result here is also the combining of four protons or hydrogen nuclei into a nucleus of helium. Each hydrogen nucleus has an atomic weight of 1.008; thus, four hydrogen nuclei would weigh 4.032. But helium weighs only 4.004, so that, in these reactions, 0.028 atomic mass units have disappeared. But where have they gone? The missing mass has been converted into energy. The amount of energy released is calculated according to the Einsteinian formula

$$E = mc^2$$

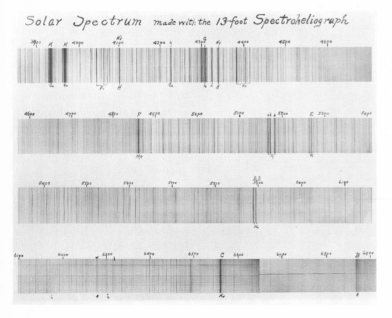

FIGURE 19.21 A spectrograph of the sun (Mt. Wilson and Palomar Observatories).

where *m* equals the mass that has disappeared during the reaction, and *c* equals the speed of light. Approximately four million tons of the sun are converted into energy every second. This is a tremendous quantity of material. Remember, the sun has been carrying on this reaction for at least five billion years and is likely to continue for another five to ten billion years. According to available knowledge, the proton-proton reaction easily accounts for the sun's production of energy throughout the ages.

Solar Structure

The sun is composed of several distinct and separate areas, each with its own set of characteristics and composition. First and foremost, the sun's interior is believed to be composed of hot gases under a tremendous amount of pressure. The density of the material in the interior is well over 100 times that of water. A box of this material with a volume of one cubic foot would weigh over 7000 pounds.

It is from within the interior that the sun's energy is produced. From here it escapes to the outermost layers and, finally, into space.

The Photosphere

The visible surface of the sun is called the photosphere (sphere of light). It is this part of the sun that actually radiates energy. Unlike the Earth's surface, the sun's is gaseous. This portion of the sun is visible because it is a high-pressure and temperature region that produces a continuous spectrum of light.

When the edge of the sun (called the *limb*) is examined, it is observed that the edges of the photosphere are cooler than the central region. Apparently, the temperature of the sun decreases in the region that can properly be termed the solar atmosphere.

The photosphere, when examined by techniques using the Doppler Effect, shows regions of granular structures slowly moving upward to the surface from the interior. Some scientists suggest that these structures represent convection currents moving from the interior to the photosphere. This may be one method by which energy emerges to the surface of the sun. But no one really knows what these structures are.

Sunspots

The curious phenomena known as sunspots have been under investigation ever since they were first discovered by Galileo. Sunspots often appear in clusters and are apparently concentrated in the region from the equator to latitudes 35° north and south on the face of the sun. At times they appear at other latitudes in each hemisphere, but the

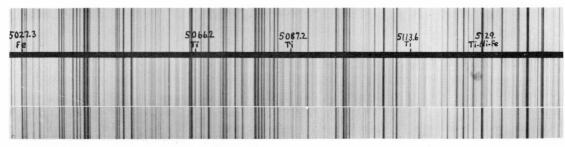

FIGURE 19.22 A continuous spectrum obtained from the photosphere of the sun (Yerkes Observatory).

former are the areas of greatest concentration.

The sunspots vary in size from a few hundred miles across to diameters so large that the Earth could easily be lost in one of them. They consist of a dark central region, the umbra, surrounded by a less dark penumbra. A sunspot usually lasts for about two weeks, although some individual spots have been observed for several months. The fact that a sunspot looks darker than the solar background implies that the sunspot is cooler than

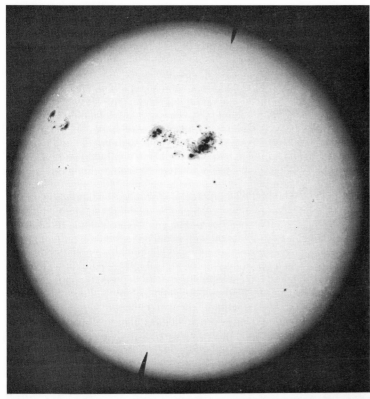

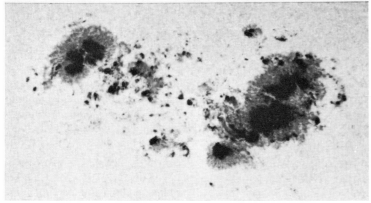

FIGURE 19.23 Sunspots photographed on the sun on April 7, 1947. A detailed view of this storm appears below (Mt. Wilson and Palomar Laboratory).

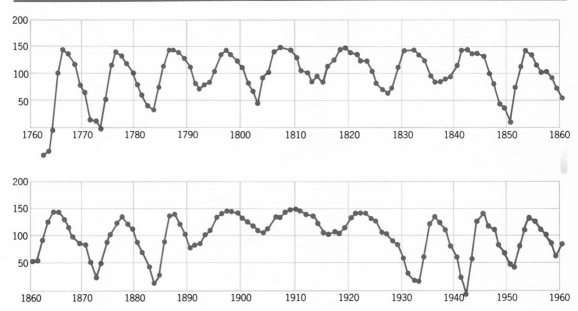

FIGURE 19.24 The sunspot cycle for the last 300 years. The high points correspond roughly to an 11-year cycle.

the surrounding area. During its brief lifetime it cools rapidly before disappearing.

It is thought that sunspots are electrical storms of hot gases emanating from the sun's interior. Examinations of sunspots reveal that they are bipolar; that is, they have a magnetic polarization with north and south ends. They align themselves in specific positions in each hemisphere, much like a series of freely moving magnets or compasses.

Sunspots appear to occur in cycles; on the average, eleven years pass between maximum numbers of sunspots, as Figure 19.24 shows. Between these events the number drops significantly. In fact, the International Geophysical Year of 1957–1958 was chosen to coincide with a sunspot maximum. The reason for the cycle is not clear, but it may be related somehow to the magnetic field of the sun. The relationship of the magnetic nature of the sun and the magnetic nature of the

sunspots has been under intense investigation since the IGY of 1957–1958.

Sunspots give rise to streamers of hot gases extending outward from the sun. These streamers are called faculae, and are larger than the normal spicules or streamers found on the sun's surface.

Telescopic examinations of sunspots have given rise to a greater understanding of the nature of the sun. The sun rotates faster at the equator than at higher latitudes. The movements of sunspots across the face of the sun show the rotation of the sun to be about 25 days at 60° north or south latitudes and 31 to 33 days as one approaches 75° north or south. Successive spectroheliographs (solar photographs) reveal these variations in movement. It is conceivable that these differences in rotation produce the disturbances in the hot gases that initiate the production of sunspots.

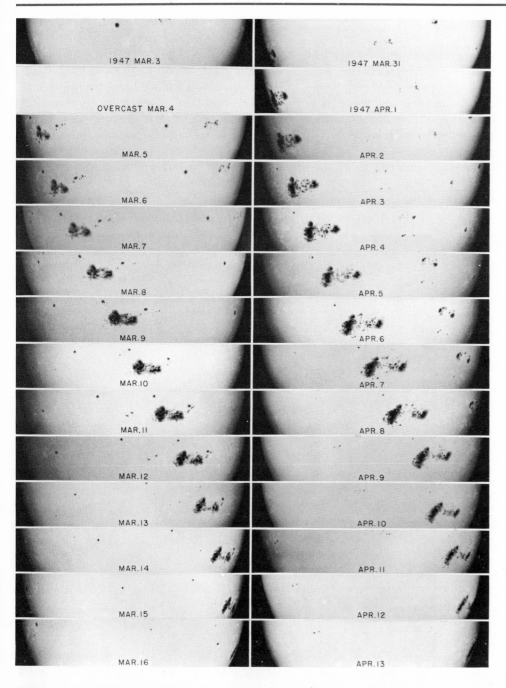

FIGURE 19.25 This sequence shows the motion of a series of sunspots across the face of the sun (Mt. Wilson and Palomar Observatories).

The Chromosphere

The photosphere blends into a layer of gases marking the boundary between the photosphere and the chromosphere. This layer is also called the reversing layer because the continuous spectrum of the photosphere suddenly becomes a dark-line spectrum. This spectrum is produced by the cooler gases that are encountered above the photosphere.

The chromosphere is the actual atmosphere of the sun. When the sun is examined through a coronagraph, a device used to block out the central disk of the sun, leaving only the edge or corona in view, large flares (solar prominences) are seen extending into the chromosphere. These solar prominences are actually cooler gases falling back to the surface from the upper reaches of the atmosphere. As the gases approach the higher temperatures of the sun's surface, they begin to glow and become visible.

Although there are several types of flares,

FIGURE 19.26 **A total eclipse of the sun. Only the photosphere can be seen (Lick Observatory).**

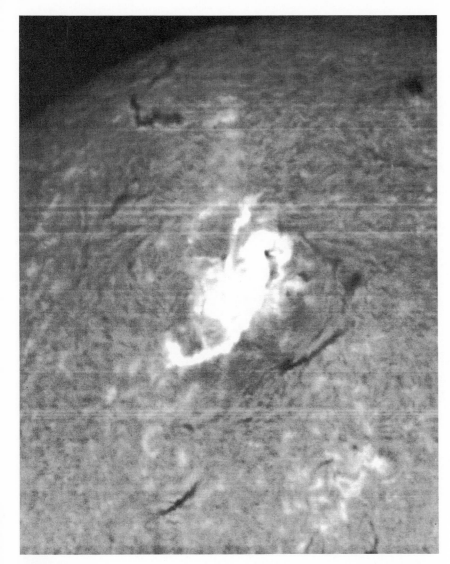

FIGURE 19.27   A solar flare on the surface of the sun (Mt. Wilson and Palomar Observatories).

they are all spectacular, brilliant filaments of gases. They rank as one of the most dramatic solar displays.

The Corona

The outermost portion of the sun is a thick layer of highly ionized gases called the corona. The continuous spectral lines and

bright-line absorption spectra obtained from this layer suggest temperatures of 1,000,-000°F.

Solar Wind

The sun emits a stream of particles (mostly protons, neutrons, and electrons), which is referred to as the solar wind. This stream is

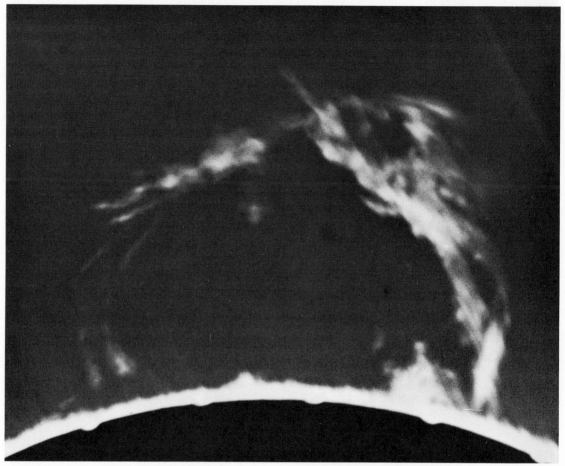

FIGURE 19.28   A solar prominence 205,000 miles high (Mt. Wilson and Palomar Observatories).

what pushes comets' tails away from the sun, for these particles produce enough energy to push the minute particles lost by comets along behind the coma. It has been observed that the solar wind increases during periods of unusual solar activity.

Electromagnetic Energy

The energy emitted by the sun covers a wide range of electrical and magnetic fre-quencies. It has been known for more than a century that electric and magnetic energies are similar in origin and are closely related. However, today, we know that there are sev-eral other types of electromagnetic energy with similar origins and characteristics. These electromagnetic energies include radio waves, infrared waves, visible light, and ultra-violet energy. One type gradually blends into another to form a continuous spectrum.

Radio waves include both long- and shortwaves. Radar and television communications bands are the longest radio wavelengths, varying from 10^6 m for true electric waves down to 10^{-4} for radar and microwaves.

Infrared waves, or heat waves, make up the next shortest range of energies. They have wavelengths from 10^{-4} m to 10^{-6} m.

The visible light bands are shortwaves about 10^{-6} m long. This is the only energy that the eye can see; human vision perceives red at the long end of these bands, and violet at the short end.

Ultraviolet energy is made up of bands of energy just below the range of vision, being on the order of 10^{-8} m long. Although we cannot see these frequencies, human skin pigments are sensitive to them.

The shortest end of the electromagnetic spectrum varies from 10^{-8} to 10^{-14} m. They are the most penetrating types of energies known. They include X rays, gamma rays, and cosmic rays. These shorter wavelengths can penetrate solid materials, severely damage living tissue and, particularly in the atmosphere, cause the ionization of atoms.

The energies found in the electromagnetic spectrum have several characteristics in common. They all have the characteristics that are usually attributed to light. First, they travel at a speed of 186,000 miles per second in a vacuum. Second, they are all capable of being refracted and reflected. Finally, they are similar in all respects to visible light with the exception that none are able to excite the retina of the eye.

Eclipses

Eclipses occur when the shadow of one object is thrown on another. Although any celestial object may be in eclipse in respect to another, we are really concerned with only two examples of this phenomenon — solar and lunar eclipses.

As the moon orbits the Earth, it periodically crosses the plane of the ecliptic of the Earth — Sun — path and passes in front of the sun. When this occurs, the solar disk is effectively blocked from view, and a solar eclipse occurs.

The possibility for an eclipse of this type occurs twice a year, when the moon crosses the ecliptic. The moon is most often located

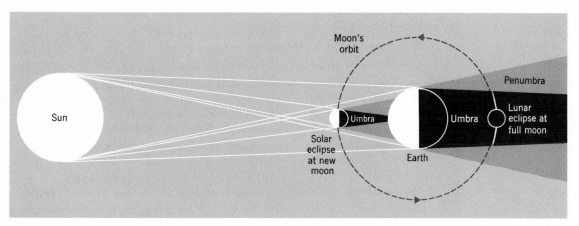

FIGURE 19.29 A solar eclipse occurs as the moon passes between the Earth and the sun, throwing its shadow on the Earth's surface.

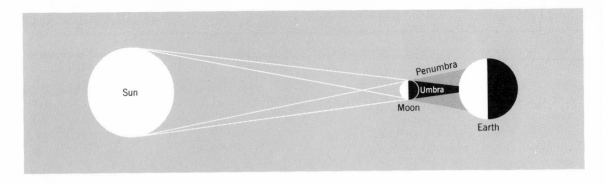

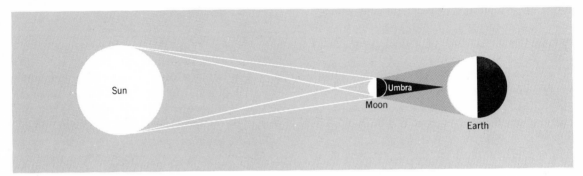

FIGURE 19.30 A lunar eclipse occurs as the moon passes behind the Earth and is darkened by the Earth's shadow.

above or below the plane joining the Earth and the sun, and an eclipse can only occur when the Earth, sun, and moon are in direct line, on the plane of the ecliptic.

When the moon's shadow falls on the Earth, the darker part of the shadow or umbra places a portion of the Earth in total eclipse. The lighter penumbra of the shadow produces a partial eclipse over the other portions of the Earth.

Since a large portion of the sun is blocked from view during a solar eclipse, a great deal of valuable information has been obtained while observing this phenomenon. Investigations of the corona and the sun's limb have given us the information we have regarding spectra, temperature, and the solar prominences, as well as other solar displays.

As the moon revolves around the rotating Earth, its shadow moves across the Earth's surface. The eclipse moves out of view in one portion of the Earth and into view in another.

Solar eclipses cover a relatively small area and last only a few minutes—seven at the most. Once the moon moves out of the ecliptic, the eclipse ends.

Lunar eclipses (Figure 19.30) occur when the sun, Earth, and moon are lined up with the moon opposite the sun at what is normally the full moon position. As the moon passes into the shadow of the Earth, the moon is darkened. Again, the umbra of the Earth's shadow produces a total eclipse, and the penumbra, a partial eclipse.

Lunar eclipses are better known because they are seen by a large number of people. Actually, lunar eclipses are rare and more infrequent than solar eclipses.

CHAPTER 20

The Universe and Its Structures

OUR SOLAR SYSTEM IS ONLY A SMALL part of the entire universe. The sun is only one star out of a probable 100 million in the galaxy, and the galaxy, in turn, is only one of millions of galaxies in the universe. And the various positions and periodic motions of these stars have enabled man to develop his concept of time and to measure its passing.

Star Classification Systems

Under ideal conditions, with clear skies and absolute darkness, we can see about 2500 stars. With telescopes, we can see many times this figure. Eventually, we come to the point where some method of classifying stars is needed. The method that is in use today classifies the stars on the basis of differences in temperature and color.

Stellar Temperature and Color

Star temperatures fluctuate from 1,500°K for the coolest star to nearly 100,000°K for the hottest. Our sun has a surface temperature of about 6,000°K. The temperature of a star determines the type of spectrum that it emits.

One way that star temperatures and colors are defined is according to their spectral lines (Figure 20.1). Several groups have been established, each denoted by a letter— O, B, A, F, G, K, M, R, N, and S.* The hot-

est stars are in the O class; the coolest are the S type. In addition, each letter group ranges from zero to nine, for a further subdivision of ten groups within each spectral class. This allows for the fine grades of spectral lines within each major group.

Spectral types vary according to those element lines that are found in the spectrum of the star. The cooler stars in the K, M, R, N, and S types show strong lines of metals present with weak bands of hydrogen. In the K, G, and F classes, temperatures increase, and strong metallic lines are present with increasing lines of calcium. Hydrogen becomes stronger and is highly excited; that is, the electrons are found at higher energy levels. In the A type stars, ionized hydrogen bands are the strongest, and the metallic elements are weaker. Ionization increases until it is quite strong in this class. Finally, in the B class, hydrogen and helium are dominant and ionized oxygen and silicon atoms are strong. The hottest stars, the O type, have weak hydrogen lines with strongly ionized lines of helium, nitrogen, oxygen, and silicon.

This scheme of classification is known as the Draper classification in honor of Henry Draper (1837–1882), one of the first Americans to study the spectra of stars. In the

---

* Students of astronomy find it easier to memorize this series if they use the following mnemonic—"Oh Be A Fine Girl, Kiss Me Right Now, Smack!"

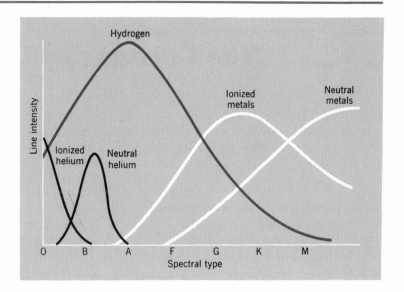

FIGURE 20.1 The spectral line produced by a star is determined by its temperature; different intensities will yield varying lines of ionized atoms.

spectral lines of any star, temperature and color are closely associated.

The O type stars contain mostly helium, nitrogen, oxygen, and silicon in the ionized state, although some ionized hydrogen is also present. These stars have temperatures of 50,000°K to 100,000°K, and exhibit a blue-white color. Class B stars have strong helium lines with hydrogen, oxygen, and silicon in great abundance. The temperature range of Class B stars is approximately 20,000+°K. They exhibit a white color. Class A stars exhibit strong hyrogen lines, and are in the 10,000+°K temperature range. They exhibit a blue-white color. Calcium lines begin to increase in F (8,000°K), G (6,000°K), and K (4,500°K) stars, and the colors range through white, yellow, and orange, respectively. The last spectral classes, M through S, begin to show strong metallic lines and some carbon compounds. They exhibit temperatures downward from 3,500°K, and are in the red end of the spectrum.

By observing the types of elements and compounds that are present in a stellar spectrum, scientists can determine the tem-

perature necessary to produce these lines. Further, spectral lines, temperatures, and approximate colors are correlated. Actually, the way we commonly describe the color of a star is not completely satisfactory. Star color is more properly expressed algebraically as the function of its effect on photographic devices. This mathematical expression is known as the color index, and is the description used by astronomers.

Stellar Distances

To determine true brightness, luminosity, and mass, the distance of the star must be known. It is the distance that determines the quantity and intensity of light received on Earth.

Stellar distance is calculated by a system of triangulation known as parallax (Figure 20.2). Parallax is based on the fact that a star appears to shift position in respect to background stars. This is because of the changing of the Earth's position in its orbit. Because of this apparent shift, astronomers can measure the angle of the star's position in relation to some known line, called the base line. The base line used for these measurements is the diameter of the Earth's orbit—93,000,000 miles.

Parallax is determined by first taking a series of pictures over a period of at least two years. The angle of shift of the star in respect to background stars is then measured. A triangular diagram develops with the Earth's orbital diameter as the measured base and with two imaginary legs drawn out to the star.

One leg runs from the sun to the star, and the other leg is drawn from the Earth in different orbital positions to the star. The angle of the star's apparent shift in position is measured with respect to the Earth and the sun. This determines the star's distance.

Star shifts are rather small because the stars are so far away. For this reason, the shift is stated in terms of a unit of measure called the parsec, which is equal to one second of arc of a circle. The farther away the star, the smaller the angle of the shift.

One parsec is much greater than one A.U.; it is equivalent to 20 trillion miles. The sun is the only star at a distance of less than one parsec.

The most widely known unit of astronomical measurement is the *light-year*—the distance light travels in one year at a speed of 186,000 miles per second. This distance is approximately six trillion miles. One parsec is equal to 3.26 light-years.

A knowledge of stellar distances is important to astronomers. It is a useful tool for

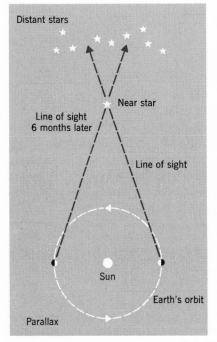

Distant stars

Near star

Line of sight
6 months later

Line of sight

Sun

Earth's orbit

Parallax

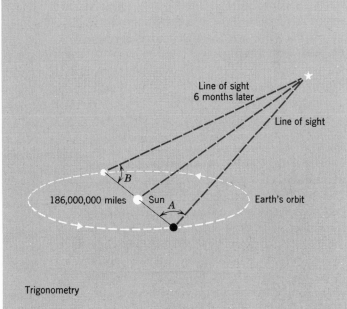

Line of sight
6 months later

Line of sight

B

186,000,000 miles   Sun *A*   Earth's orbit

Trigonometry

FIGURE 20.2 Star distances are determined by measuring the angle of parallax by using trigonometric techniques. The base line used for these measurements is the diameter of the Earth's orbit. The angles of the star position measured relative to the Earth's orbital diameter allow for determination of the star's distance from the sun (After Navarra and Strahler).

further consideration of stellar characteristics. In particular, it allows a true measurement of stellar brightness.

Star Brightness

The true brightness of a star depends on its temperature, its size, and the amount of energy it radiates. However, a star's apparent brightness is affected by its distance from the Earth. Consider a star at a particular distance from the Earth that has a certain brightness. If this star were twice as far away, its apparent brightness would be only $1/4$ of what it is in its original position. This decrease in brightness follows what is known as the *inverse square law*. The inverse of two (the increase in distance) is $1/2$. The square of $1/2$ is $1/4$. Therefore, the same size star at twice the distance from the Earth would be $1/4$ as bright. At three times the distance, the star would be $(1/3)^2$ or $1/9$ as bright. As these calculations show, it is essential to know the stellar distance in order to determine the star's true brightness.

Apparent Magnitude

The first scale established to classify stars according to brightness was the scale of *apparent magnitudes* of stars. This term denotes the brightness of a star as it appears to an observer on Earth.

When the scale was originally set up over 2000 years ago, it included six magnitudes of stars. However, with the advent of sophisticated measuring devices and the telescope, the scale has been extended all the way to very faint stars of the twenty-third magnitude. In addition, objects have been found that are brighter than the original first magnitude stars. A negative scale was devised to describe these very bright objects. Thus, magnitudes of 0, −1, etc., have been added to the scale. Each negative number repre-

sents an object about 2.5 times as bright as the preceding number. The brightest object on this scale is the sun, with a brightness of nearly −27.

The apparent magnitude system is not satisfactory, since the brightness is only apparent and is a consequence of distance. According to this scale, the sun, which is a rather small star with much less energy than most, is brighter and emits more energy than any other object in the universe. Even the moon with an apparent magnitude of nearly −13 is brighter than all the stars; yet, it is only a reflecting surface and emits no energy of its own.

Absolute Magnitude

To rate stars according to their true brightnesses, an arbitrary scale has been set up. This scale, the *absolute magnitude,* sets up an arbitrary distance from which to measure star brightness. Absolute magnitude is defined as the magnitude a star would have if it were 10 parsecs or 32.6 light-years from the sun. The true distance of the star must be known to calculate its theoretical brightness from 10 parsecs.

For example, Sirius is rated at −1.43 apparent magnitude. However, it is 2.7 parsecs from the sun. On the absolute magnitude scale, it becomes +1.5. By the same method the apparent magnitude of −27 for the sun becomes an absolute magnitude of nearly +5.

Luminosity

The system of absolute magnitude allows a true measurement of luminosity, or the rate at which energy is being emitted. The luminosity of the sun is given an arbitrary rating of one. Stars are then rated to show how much more-or-less luminous they are

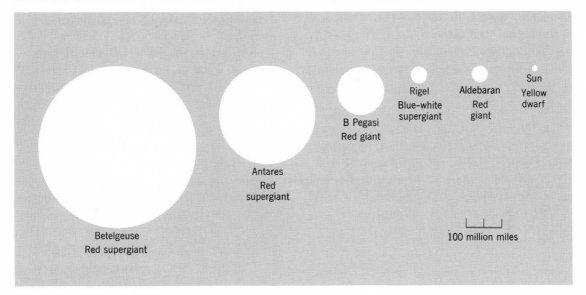

FIGURE 20.3   The range of star sizes.

in relation to the sun (Figure 20.3). They range from 100,000 times more luminous to 1,000,000 times less luminous than our own sun.

The Hertzsprung-Russell Diagram

The Hertzsprung-Russell diagram, named after the men who devised it, shows all the known stars in each group of the Draper classification in relation to four important characteristics. These characteristics are Spectral Class, Temperature, Absolute Magnitude, and Luminosity. When the known stars are plotted into the diagram according to these four characteristics, some very interesting results appear (Figure 20.4). A majority of stars fall into a sequence from upper left to lower right. This band is known as the main sequence of stars. Our sun (arrow) falls nearly at the midpoint of this sequence, making it a rather ordinary star.

Directly above the main sequence is a group of stars called giants, with a smaller number of supergiants directly above them. It is important to notice that these stars are relatively cool stars. However, their immense sizes make them very luminous.

Below the main sequence of stars are groups of subdwarfs and white dwarf stars. Some of these stars are smaller than the Earth. None are very luminous.

The H-R diagram, as it is called, is a valuable tool for interpretation of stellar information. The position of a star on the H-R diagram gives its luminosity. From luminosity, one can calculate the amount of radiated energy and, thus, derive the size, mass, and density of the star. The determination of mass is a difficult procedure, but it is somewhat easier for double stars. In the latter case, the mass of one star can be calculated from the action of its companion. The actions of these double stars are controlled by Kepler's Laws of Planetary Motion. But the determination of mass is a very complex matter, which we shall not discuss in detail. Essentially, we can say that, as stellar mass decreases, luminosity decreases. In addition, as luminosity decreases, the size of the star decreases. Since luminosity is the rate at which energy is emitted, it follows that the more massive stars are emitting energy at a greater rate than the less massive stars.

Stellar densities vary greatly. The white dwarfs are the densest of all, with densities that may be on the order of 100,000 times that of water. A pint of this material might weigh well more than 50 tons. On the other hand, the densities of the supergiants may be so low as to be almost unmeasurable. A supergiant would be considered a vacuum if it were located on the Earth.

The Life and Death of a Star

The H-R diagram has been used to develop a theory of stellar evolution which, although not definitely established, is worthy of consideration.

A star begins as a loose, nebular mass. Then the material begins to contract and condense under the influence of gravitational attraction. As the protostar develops, the temperature increases, allowing the proton-proton reaction to begin converting hydrogen to helium. This beginning might take as long as 50 million years.

Once the nuclear reaction develops, it counteracts further contractions of the mass, and the star becomes a main sequence star. The location of the star on the main sequence is determined by the mass and temperature initially achieved.

As the hydrogen content of the star is utilized in the proton-proton reaction, the star moves up the main sequence of stars from lower right to upper left. The sun has

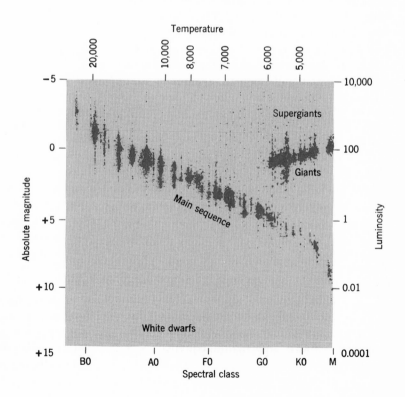

FIGURE 20.4 The Hertzsprung-Russell diagram showing main sequence stars, giants, and dwarfs.

presumably spent five billion years or so in this sequence, and has used about ½ the amount of hydrogen that allows it to remain on the main sequence—7 per cent of the total. In about five billion years, when about 15 per cent of its hydrogen has been converted to helium, it will move off the main sequence. A larger star may last for a shorter period of time on the main sequence. As the hydrogen is depleted, the star expands and cools, moving further up the main sequence.

Once the proton-proton reaction has been set up in massive stars, the carbon cycle also begins, emitting more energy and further depleting fuel sources. About 20 million degrees is required for the carbon cycle to begin; this heat is supplied by the initial proton-proton reaction.

Once the star has used up about 10 to 20 per cent of its hydrogen supply, the outer regions begin rapidly expanding. At the same time, the core becomes highly condensed. It is at this point that the star moves into the red giant class. These stars are marked by large, cooler outer regions of hydrogen with contracted hot interiors of helium.

Finally, by a process that is not completely understood, the core achieves a high, stable temperature, and contraction of the entire star begins. The star then crosses the main sequence and moves down through the H-R diagram to become a white dwarf of low luminosity. At this point, the atoms become tightly compressed and immensely dense.

A star's evolution may include many stages other than the ones discussed in this brief biography. Stars may lose their materials in the processes described earlier in this chapter. New stars may dispose of materials by exploding prior to final contraction. Variable stars may be stars that are moving through the H-R diagram, although no one can be sure.

Star Motions

Stars are not fixed in the sky, but undergo a series of motions. We have discussed the motions of the sun through the galaxy, revolving around the galaxy, and moving with the galaxy. Other stars also move in a similar manner and, as a result, slowly change their positions in respect to the Earth.

The stars move across the celestial sphere each year. Accurate observations of declination and right ascension give an observer the extent of this motion. This movement is known as proper motion.

In addition, stars also have a relative motion toward or away from the Earth that can be determined by the Doppler Shift of the spectrum. Over tens of thousands of years, the relative motions of the stars in the sky will, of course, change the configurations of constellations within our galaxy. Like the sun and stars, the heavens are not fixed, but revolve through many positions.

Star Types

We have been discussing the stars almost as if they were all separate and distinct entities with constant characteristics and properties. But such is not the case.

Variable Stars

There is a large group of variable stars that pulsate between a minimum and maximum brightness. These pulsations vary in length from a single day to several weeks. Depending on the length of pulse, these stars are classified as long- or short-period variables.

The Cepheid variables have a regular variation. In terms of luminosity, there are both Type I and Type II Cepheids. RR Lyrae variables are similar to Cepheids but vary in brightness more rapidly. Irregular

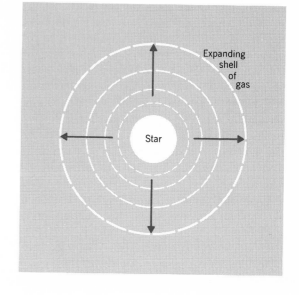

FIGURE 20.5 Novae are produced as expanding gases are thrown outward by unstable stars. A bright absorption spectral band appears as a result of the expansion.

variables exhibit pulses that do not vary with repeated period, but the majority have periods in the range of three months to two and one-half years. Variable star groups usually take their names from the stars first examined.

It is not quite clear why variables behave as they do. Apparently, some imbalance within the star causes a buildup of energy that produces an explosion or expansion. When some maximum size is reached, gravity takes over and contraction occurs. As pressure builds up the cycle repeats itself.

Variable stars have proved to be of great value in the development of the *period-luminosity law*. It has been observed that luminosity varies according to a specific period, especially in Cepheid variables. It is also known that the brightness of variables increases as the period of variation increases. Thus a standard was established in order that great distances could be accurately measured. Since the luminosity varies according to the inverse square law, determination of the variance in the luminosity can give us

the distance of a star. The period tells the relative luminosity, and determining the luminosity of one star establishes a standard for all other variables. This method has led to more accurate estimates of the distances of galaxies beyond our own.

Binary Stars

Binary or double stars are very common. These stars are two companions revolving about some common center of gravity. Some are so close that they are nearer each other than our own sun and the large planets. Others are at great distances but still travel as pairs. Some binaries periodically pass in front of one another, producing an eclipse. Binary stars often distort each other. In addition, materials from one may move to the other (Figures 20.7 and 20.8).

Observation of binary stars has led to information about stellar masses. By observing the effects of one mass on the other, the masses of the binaries are readily calculated. Also, the stellar atmospheres yield much information during eclipses. Beta Lyrae and 61 Cygni (61 Cygni is actually three stars) are two of the more famous binary stars.

Star Clusters

Stars are also found in groups or clusters traveling together. For example, pairs of binaries have been found in relatively close quarters. Coster, a star in the constellation Gemini, is such a double binary.

Stars are also found in large clusters held together by a common gravitational attraction. Star clusters may have 40 members, such as the Big Dipper, or several thousand, and the large globular clusters have well over 50,000 stars (Figure 20.9).

Novae

"Novae" literally means "new stars." Novae are produced when a star suddenly and quite unexpectedly undergoes an increase in magnitude (Figure 20.5). Its brilliance may increase tens of thousands of times.

FIGURE 20.6 This expanding nebula, named Nova Persei, was photographed in 1901. The expulsion of gases can be seen (Mt. Wilson and Palomar Observatories).

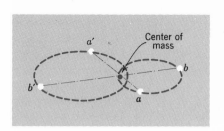

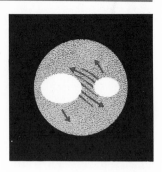

FIGURE 20.7 Binary stars rotate about a common center of gravity. The proximity of these stars may also result in an exchange of gases between the two.

FIGURE 20.8 Analysis of the activity of Barnard's star reveals that it possesses a large planet. Although the planet has not been seen, the nature of the behavior of the star leads to this conclusion (Yerkes Observatory).

It is believed that novae flare up because of an instability in the internal nuclear processes in the star. A sudden and tremendous burst of energy results, during which the star increases in magnitude. After the explosion, the nova decreases rapidly to a faint star.

Some novae are repetitive; that is, they explode several times during their lifetimes. Each explosion usually lasts for a few months. Each time the new minimum is less than the prior magnitude. These recurrent novae, as well as the normal types, are of great value and interest to the astronomer.

On occasion, a supernova with an intensity of millions of suns occurs. These novae expand rapidly, sending out tremendous amounts of energy and material before settling down.

It may be that the masses of dust clouds in space, known as nebulae, are the remains of nova explosions. In fact, it has been definitely established that the famous Crab Nebula is the resulting debris of a supernova explosion that took place more than 900 years ago.

Nebulae are of two types. Bright nebulae are closely associated with stars from within, and they take on a glow from the light given off by their associated stellar members. Dark nebulae obscure starlight behind them, and they produce a dark patch across the

FIGURE 20.9 A globular star cluster (Mt. Wilson and Palomar Observatories).

sky. It is believed that these nebulae become more massive and condensed because of gravitational attraction, which captures dust and other materials. It is thought that this process, carried to its conclusion, results in the formation of new stars (Figures 20.10 and 20.11).

Time and the Calendar

Accurate measurement of the inevitable passage of time has always been one of man's greatest concerns. The desire to predict

seasonal changes, daily changes, and their relationships to weather and climate gave man his initial interest in the heavens. Astronomy has gone far beyond being merely a sophisticated time-measuring study. However, time and seasonal changes are still major concerns to mankind.

Early Attempts at Time Measurement

Some of the features that are as old as history have been retained in calendars by modern man. For example, the ancient

FIGURE 20.10 The Horsehead Nebula. This dust cloud obscures the stars behind, giving it a darkened appearance (Mt. Wilson and Palomar Observatories).

FIGURE 20.11 The Crab Nebula is the remains of a supernova observed in 1054 A.D. It glows with the reflected light of stars in its vicinity (Mt. Wilson and Palomar Observatories).

FIGURE 20.12 An ancient sundial, a forerunner of the sundials with which we are all familiar (New York Public Library, Picture Collection).

Sumerians in about 3000 B.C. kept time in terms of a seven-day week made up of twelve-hour days. Each hour would be equivalent to two of our own "hours." Each was divided into smaller units of thirty.

In Roman times the common man deduced morning, noon, and evening by simple observation of the positions of the sun throughout the day, a situation that did not change significantly until the nineteenth century. To be sure, many schemes and devices were developed for recording the changing day, but for the average person they were of little consequence. The advent of proper calendar-making and the fixing of dates will be discussed later in this chapter.

The Development of Clocks

The earliest "clocks" or timepieces date back to the Egyptian civilization. They were similar to the sundials with which we are familiar. An upright post threw a shadow on divisions of time marked around it in a circle. It did not record actual hours and minutes, but simply marked off the passage of time in a general fashion. Sundials did not become truly accurate much before the sixteenth century, since man could not make precise sundials until he had developed the Earth's latitude and longitude system.

The early Romans and Chinese invented a type of timepiece known as the water clock (Figure 20.13). These devices allowed water to slowly fill a marked container showing the passing hours. Later clocks of this type had a float that moved up with the water. Attached to a gear system, the float moved a dial across a clock face. Because they were independent of the sun, water clocks were reliable and quite accurate.

Much development during the thousand years from Caesar's Rome to the seventeenth century led to several refinements and a greater understanding of clockwork mechanisms and timepieces. Mechanical bell ringers for churches made their appearance, and water clocks underwent sophisticated mechanical variations. The sandglass, or hourglass as it is sometimes erroneously called, also made its appearance during this era. In the sixteenth century, just prior to the Renaissance, mainspring watches were developed. At first, they were large and cumbersome; later portable watches were introduced. They were all inaccurate.

The Pendulum

The most significant discovery in time-keeping during the Renaissance was Galileo's discovery of the law of the pendulum. While in the cathedral at Pisa, Galileo noticed the regular swinging of a lamp and timed its oscillation with his pulse. His observations on pendulums were later refined by Christian Huygens, and they led to the development of the pendulum clock. Although it was by no means the final answer, the pendulum enabled man to construct clocks that were accurate to within a few minutes a day.

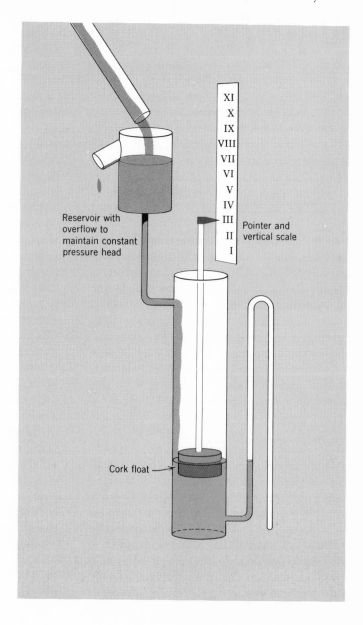

Reservoir with overflow to maintain constant pressure head

Pointer and vertical scale

Cork float

FIGURE 20.13 A water clock built in Alexandria. As water fills the tube (A), the fighter (E) points off passing time on the column (F). The fighter rises atop the float (D). Regulation of the water was accomplished by means of a faucet (C). Devices such as these were even used by early physicians to measure pulse and heartbeat (Bettmann Archive).

To overcome the problems of air pressure and heat affecting the pendulum, clocks were sealed in airtight, temperature-controlled chambers. It was later discovered that gravity also affects the swinging of a pendulum. Gravity, of course, is not uniform over the entire Earth. As a result, the swinging of a pendulum will be affected differently in various parts of the globe.

Using the Stars

The sun can be used to check the accuracy of watches, since it will trace an exact path through the heavens each year.

If a spot of sunlight is allowed to pass through a hole in a wall facing the sun and to fall on the opposite wall, it can be observed that the noon position, when the sun

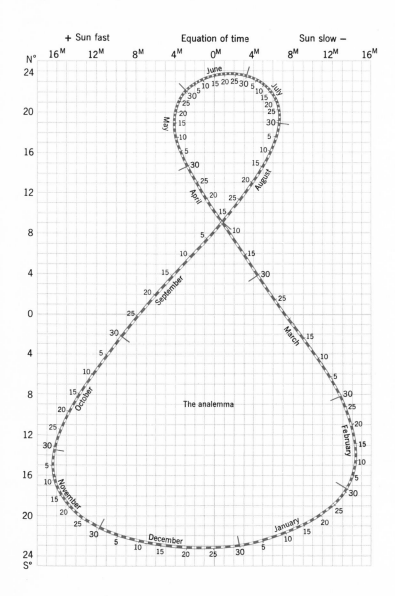

FIGURE 20.14 The analemma, found on most globes and many maps, is used to determine solar time. The right-hand side of the figure represents how many minutes the sun is behind local time.

is at its highest point, changes daily. If the opposition point from the hole has a line drawn to represent the meridian of that position on the Earth, you would find that the 365 noon positions of the sun trace a figure eight around that meridian line. (See Figure 20.14.)

The dates around the figure eight, which is known as an analemma, represent, on the right, how many minutes the sun is behind local time. Those on the left show how many minutes the sun is ahead of local time. This figure may look familiar to you, since it is found on globes and some maps of the world. It is a simple matter to find the date where the analemma crosses the local meridian and determine the accuracy of your clock.

Star positions in relation to the Earth's position around the sun are also used to calculate time. This will be discussed later in this chapter. Suffice it to say, at this point, that our "days" are not of uniform length and that there are several kinds of days, a fact not known to the early timekeepers.

Atomic Clocks

Today, many refinements have been made in timekeeping, such as electric, electronic, and atomic clocks. All of these clocks use electronic and atomic vibrations to maintain accurate time.

The ammonia clock, for example, uses a tube of ammonia gas. Ammonia gas is composed of molecules of three hydrogen atoms bonded to one nitrogen atom. When the molecules are struck by radio waves they oscillate at an exact frequency—23,870 cycles per second. Thus, the "pendulum," the oscillating molecule, vibrates at a set frequency. These clocks have been made with accuracies of one part per billion. This would compare to a mechanical watch with an accuracy of one part in about 87,000—an error of two seconds per day.

A further refinement is the true "atomic clock," which uses cesium atoms. These atoms vibrate at exact frequencies with an error of one in ten billion, a loss of one second in three centuries. These atomic clocks are not true timepieces, of course, but are checks on the accuracy of other devices.

Time Zones

As you can see, timekeeping was for a long time a rather confused subject. But there was another even more practical reason for knowing more about time, and that was navigation. In order for sailors to locate their positions on the Earth, there must be something with which to relate. Man's desire to navigate safely and accurately led to the development of observatories in order that successive passages of the sun and stars might be measured accurately.

Latitude and longitude enable a person to calculate not only location but also time. Longitude tells one where, on any parallel of latitude, a place is located. These lines, called meridians, run from pole to pole, and are known as great circles, since they all bisect the Earth. The meridian passing through the Greenwich Observatory has been designated 0° longitude. All other meridians are either east or west of this meridian. The greatest number of degrees is 180°W or 180°E, depending on the direction of travel. This meridian is at the opposite side of the globe from Greenwich, and is accepted as the international date line.

The longitude meridians are used to establish an arbitrary time zone system. They are usually drawn on the globe at intervals of 15°, producing 24 zones or belts. Each belt traveling east is then one hour later than Greenwich time.

Assume that Greenwich is at high noon. As one travels eastward each zone is an hour later. At the 180° meridian, it is the next day. Thus, crossing the international date line in an eastern direction produces a gain of one day. Crossing it in a westward direction produces a loss of one day.

Although we usually assume each 15° division represents one hour in time and, in fact, use time differences for place location, time calculation is not a simple matter. Daily time changes and annual variations depend on what relative object is used—the sun, the stars, or the Earth's motion, which is not uniform.

Sidereal Versus Solar Time

The system of longitudinal and latitudinal coordinates is also extended into space to form the celestial globe sphere. The parallels and meridians on this sphere give us the declination and hour angles discussed earlier.

In timekeeping, one can use either star lo-cation (sidereal time) or the sun (solar time). A sidereal day is the time between two crossings of the same meridian by a star. In other words, it is the time for the Earth to rotate 360° and to have a star line up with the same meridian, since the motion of stars is really caused by Earth rotation. A sidereal day is 23 hours, 56 minutes, and 45 seconds long.

Everyday existence is based on day and night, not the stars. Day and night of course, depend on the sun's crossing a particular meridian. Since we are near the sun, a single day's duration is affected by the Earth's movement along its orbit around the sun. That is, as we change our position relative to the sun, the Earth must rotate a bit more to bring the sun back to the same meridian as the day before. This is our standard "24-hour day," which varies in length throughout the year. A calendar year is made up of 365 solar days but, actually, a solar year is 365.24 days long.

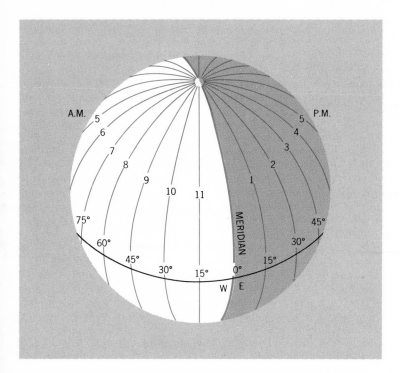

FIGURE 20.15   Lines of longitude run east and west of Greenwich at 0° longitude.

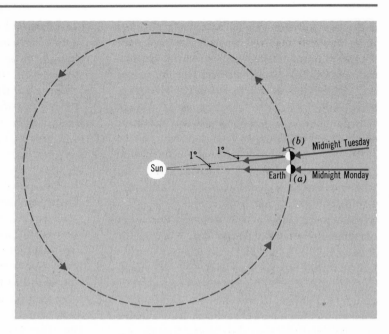

FIGURE 20.16 In order for one complete daily cycle of the Earth to be completed in respect to a star, the Earth must line up twice with the same star. During this 24-hour period, the Earth has also moved in its orbit around the sun. Thus, in order to complete the sidereal cycle, the Earth must actually rotate 361°. A sidereal day is 23 hours, 56 minutes, and 45 seconds long.

Thus, an additional day is added to the calendar every four years making that year a leap year.

To sum up, the 365-day solar year, also called the tropical year, measures one revolution of the Earth around the sun. It is 365.24 days long. The sidereal year measures the time for the Earth to make two passes through the same point in its orbit, relative to a fixed star position, and is 365.25 days long. Although the difference is not more than about a half hour a year, over an extended period of time, problems of synchronization arise.

The Calendar

Time is really a relative quantity. It depends on one's point of view: that is, whether one relates Earth motion to the stars or to the sun. The minor daily differences have little effect on man's daily existence. However, when these minor daily variations add up over several centuries, they can be disconcerting, as the ancients found out.

The Development of the Calendar

The earliest calendars, known as lunar calendars, were related to changes in the phases of the moon. The ancient Hebrew calendar, which is still used today to regulate religious occasions, was based on the moon's cycles. Its twelve-month year has a variable number of days, plus a thirteenth month every four years to calculate the celebration of religious holidays in their proper season of the year. The calculations that are necessary for this calendar are far beyond the average person and are made by scholars for specific religious purposes.

The Egyptians began with a 360-day solar calendar, but soon moved to a 365¼ day lunar calendar. In 2500 B.C., the practice of an extra day for leap year was instituted when it was realized a step was necessary to make up for the ¼ day lost each 365-day year, but this was not universally accepted.

Observation of the sun's shadows thrown by their obelisks soon showed the Egyptians the inaccuracy of their calendar. In part, this resulted from the fact that a lunar month is only 29½ days long, while the calendar used 30-day months.

In 45 B.C., a reformed calendar named after Julius Caesar was introduced. This Julian calendar was calculated by Sosigenes, one of the great mathematicians of the day. The calendar reform was deemed necessary because the original lunar calendar of the Romans, with its periodic additional month, led to abuses when those in power calculated the additional time for their own gains. The Julian calendar based on the Egyptians' solar calendar, used a year of 365.25 days, and retained the leap year every fourth year in the month of February.

The greatest impetus for further calendar reform came from churchmen who needed a faultless calculation of the dates for religious holidays. The calculation for Easter was particularly vexing because it fell progressively earlier in spring.

Finally, in 1582, Pope Gregory instituted the Gregorian calendar. This marked a year of 365.24 days with a leap year every four years. Leap year is skipped every 400 years in century years not divisible by 400 because the continual addition of a day every four years increases the synchronization too rapidly. The year is actually 0.24 days slow and not 0.25 days.

This calendar, which we still follow today, is fairly satisfactory but does present problems. There are a variable number of days each month, so that, from year to year, holidays still fall on different days of the week. Holidays like Easter, which is still calculated from lunar phases, range over several weeks of possible dates.

There are several proposals for calendar reform at present. One calendar that has been suggested is called the World Calendar. It divides the year into four quarters, each of which has three months. The first month in each quarter has 31 days, and the second and third 30 days each. There is a Leap Year Day, the day after July 30; it carries no date and is a World Holiday. There is also a World Holiday on the day after December 30th. All holidays would fall on the same day of the week every year.

Another proposal calls for a 13-month year. In this calendar each month has 28 days, and is exactly four weeks long.

To date, most proposals have not completely satisfied religious holiday requirements and have met much opposition. However, the day may not be far away when a new calendar will be instituted.

Questions

CHAPTER SEVENTEEN

1. Discuss the evolution and the development of the geocentric and heliocentric theories. What are epicycles? Why did they appear in both theories?

2. Why is light described as a photon of energy? In what way does this represent a compromise between two theories?

3. Compare the reflecting and refracting telescopes. What advantages has one over the other? What is a radio telescope?

4. Discuss the evolution of the various theories that attempt to explain the development of the solar system.

5. Describe the general picture of the atom. What is represented by the following: atomic weight, atomic number, isotopes, energy level?

6. Describe the major agreements and disagreements between the evolutionary and the steady-state theories of the universe.

CHAPTER EIGHTEEN

1. Discuss the various motions of the Earth and why each occurs.

2. How might a researcher prove that rotation takes place? Revolution?

3. Explain the factors that produce the seasonal changes on the Earth. What would be the consequences if the Earth were not inclined on its axis?

4. Discuss the system used for star location. How is it related to the Earth's surface?

5. What are Kepler's Laws of Planetary Motion? How are they related to the motions of the Earth?

6. Discuss Newton's Laws of Motion. Give an example of each in action.

CHAPTER NINETEEN

1. List and describe the characteristics of the various types of astronomical bodies which make up our solar system.

2. What is an eclipse? How does each type of eclipse occur?

3. Describe the various portions of the sun. What characteristics are a feature of each area?

4. What clues have we to indicate the nature of sunspots? Describe what they are and how they evolve.

5. What is the origin of the craters on the moon? What evidence have we for this assumption?

6. What is the Electromagnetic Spectrum? Explain the various components and what they have in common.

CHAPTER TWENTY

1. What differences exist between apparent and absolute magnitudes? Which system is, in a sense, an artificial one, and why was it developed? What is the inverse square law?

2. What relationships exist on the Hertzsprung-Russell diagram? In what way do we assume that stars' life cycles evolve according to this relationship?

3. What is a light year? A parsec? Why are the two systems used? How are the two units related?

4. List and describe briefly some of the various star types that have been observed. What important relationship was derived from the period-luminosity law and a specific type of star.

5. Define and explain the difference between sidereal and solar days. Sidereal and solar years.

6. What are novae? What relationship exists between novae and nebulae?

Bibliography

BOOKS

Alter, Densmore, *Pictorial Astronomy,* Wadsworth, Belmont, California, 1966.

Asimov, Isaac, *The Universe—From Flat Earth to Quasar,* Walker, New York, 1966.

Baker, Robert, *Introduction to Astronomy,* 7th Ed., Van Nostrand, New York, 1968.

Bellinger, E. W., *Witness of the Stars,* Kregel, Grand Rapids, Michigan, 1967.

Fanning, Anthony, *Planets, Stars, Galaxies,* Dover Publishers, New York, 1966.

Hesse, Walter H., *Astronomy: A Brief Introduction,* Addison-Wesley, Reading, Massachusetts, 1967.

Huffer, C. M., *Introduction to Astronomy,* Holt, Rinehart and Winston, New York, 1967.

Inglis, Stuart J., *Planets, Stars, Galaxies,* John Wiley, New York, 1967.

Ingrao, H. C., *New Techniques in Astronomy,* Gordon and Breack, New York, 1969.

Mehlin, Theodore G., *Astronomy and the Origin of the Earth,* W. C. Brown, Dubuque, Iowa, 1968.

Mihalas, Dimitri, and Routly, Paul M., *Galactic Astronomy,* Freeman, San Francisco, California, 1968.

Motz, L., and Duveen, A., *Essentials of Astronomy,* Wadsworth, Belmont, California, 1966.

Shklovski, T. S., and Sagan, C., *Intelligent Life in the Universe,* Holden Day, San Francisco, California, 1966.

PERIODICALS

Asimov, Isaac, "A Sizable Universe," *Science Digest,* November, 1969.

Asimov, Isaac, "Over the Edge of the Universe," *Harper,* November, 1967.

Asimov, Isaac, "Quasars and the Birth of the Universe," *Unesco Covier,* December, 1969.

Holcomb, R. W., "Galaxies and Quasars, Puzzling Observations and Bizarre Theories," *Science,* March, 1970.

Mumford, G. S., "Origin of Galaxies," *Sky and Telescope,* November, 1967.

Wheeler, J. A., "Our Universe: The Known and the Unknown," *American Scholar,* September, 1968.

Appendix I

When converting Kelvin degrees to Celsius, subtract 273°, the difference between the two scales. When converting from Celsius to Kelvin, *add* the 273° to the C reading.

Example	*Example*
35°C = 35° + 273° or 308°K	400°K = 400° − 273° or 127°C

When converting Fahrenheit to Centigrade, remember that there is a 32° difference in the freezing points. Therefore, it is necessary to subtract 32 from the Fahrenheit reading. There is also a difference in the size of degree points on each scale. Between 32°F and 212°F there is a difference of 180°. Between 0°C and 100°C there are only 100° of difference. Each Fahrenheit degree is $\frac{5}{9}$ of a Centigrade degree. After subtracting 32 and multiplying by $\frac{5}{9}$, the reading will have been converted to Centigrade.

Example

95°F = ?C $(95° − 32°)\frac{5}{9}$ = 35°C

To convert Centigrade to Fahrenheit we reverse the previous operations. The multiplier is $\frac{9}{5}$ and the 32° is added.

Example

80°C = ?F $80° \times \frac{9}{5} + 32°$ = 144°F

Appendix II

BASIC UNITS OF MEASURE: METRIC AND ENGLISH SYSTEMS

1 foot = 0.305 meter
1 inch = 2.54 centimeters
1 mile = 1.60 kilometers

1 meter   =   100 centimeters = 29.27 inches = 3.28 feet
1 centimeter =   10 millimeters =   0.3937 inch = 0.01 meter
1 kilometer = 1000 meters   =   0.621 mile

1 astronomical unit = 1.496×10^{11} meters
1 parsec   = 3.084×10^{16} meters
1 light year   = 9.460×10^{15} meters

1 fathom   = 6 feet
1 statute mile = 5280 feet = 1760 yards
1 nautical mile = 6080 feet

Index